Mathematics and Computer Science
in Medical Imaging

# NATO ASI Series

## Advanced Science Institutes Series

*A series presenting the results of activities sponsored by the NATO Science Committee, which aims at the dissemination of advanced scientific and technological knowledge, with a view to strengthening links between scientific communities.*

The Series is published by an international board of publishers in conjunction with the NATO Scientific Affairs Division

| | | |
|---|---|---|
| A | Life Sciences | Plenum Publishing Corporation |
| B | Physics | London and New York |
| C | Mathematical and Physical Sciences | D. Reidel Publishing Company Dordrecht, Boston, Lancaster and Tokyo |
| D | Behavioural and Social Sciences | Martinus Nijhoff Publishers Boston, The Hague, Dordrecht and Lancaster |
| E | Applied Sciences | |
| F | Computer and Systems Sciences | Springer-Verlag Berlin Heidelberg New York |
| G | Ecological Sciences | London Paris Tokyo |
| H | Cell Biology | |

Series F: Computer and Systems Sciences Vol. 39

# Mathematics and Computer Science in Medical Imaging

Edited by

## Max A. Viergever

Delft University of Technology
Department of Mathematics and Informatics
P.O. Box 356, 2600 AJ Delft, The Netherlands

## Andrew Todd-Pokropek

University College London
Department of Medical Physics
Gower Street, London WC1E 6BT, UK

Springer-Verlag
Berlin Heidelberg New York London Paris Tokyo
Published in cooperation with NATO Scientific Affairs Division

Proceedings of the NATO Advanced Study Institute on Mathematics and Computer Science in Medical Imaging, held in Il Ciocco, Italy, September 21 – October 4, 1986.

Directors
Max A. Viergever, Delft, The Netherlands
Andrew Todd-Pokropek, London, UK

Scientific Committee
Harrison H. Barrett, Tucson, USA
Gabor T. Herman, Philadelphia, USA
Frank Natterer, Münster, FRG
Robert di Paola, Villejuif, France

Conference Secretary
Marjoleine den Boef

Sponsor
NATO Scientific Affairs Division

Co-sponsors
Delft University of Technology, The Netherlands
National Science Foundation, USA
Siemens Gammasonics, The Netherlands
University College London, UK

ISBN 3-540-18672-7 Springer-Verlag Berlin Heidelberg New York
ISBN 0-387-18672-7 Springer-Verlag New York Berlin Heidelberg

This work is subject to copyright. All rights are reserved, whether the whole or part of the material is concerned, specifically the rights of translation, reprinting, re-use of illustrations, recitation, broadcasting, reproduction on microfilms or in other ways, and storage in data banks. Duplication of this publication or parts thereof is only permitted under the provisions of the German Copyright Law of September 9, 1965, in its version of June 24, 1985, and a copyright fee must always be paid. Violations fall under the prosecution act of the German Copyright Law.

© Springer-Verlag Berlin Heidelberg 1988
Printed in Germany

Printing: Druckhaus Beltz, Hemsbach; binding: J. Schäffer GmbH & Co. KG, Grünstadt
2145/3140-543210

# PREFACE

Medical imaging is an important and rapidly expanding area in medical science. Many of the methods employed are essentially digital, for example computerized tomography, and the subject has become increasingly influenced by developments in both mathematics and computer science. The mathematical problems have been the concern of a relatively small group of scientists, consisting mainly of applied mathematicians and theoretical physicists. Their efforts have led to workable algorithms for most imaging modalities. However, neither the fundamentals, nor the limitations and disadvantages of these algorithms are known to a sufficient degree to the physicists, engineers and physicians trying to implement these methods. It seems both timely and important to try to bridge this gap.

This book summarizes the proceedings of a NATO Advanced Study Institute, on these topics, that was held in the mountains of Tuscany for two weeks in the late summer of 1986. At another (quite different) earlier meeting on medical imaging, the authors noted that each of the speakers had given, there, a long introduction in their general area, stated that they did not have time to discuss the details of the new work, but proceeded to show lots of clinical results, while excluding any mathematics associated with the area. The aim of the meeting reported in this book was, therefore, to do exactly the opposite: to allow as much time as needed to fully develop the fundamental ideas relevant to medical imaging, to include a full discussion of the associated mathematical problems, and to exclude (by and large) clinical results.

The meeting was therefore designed primarily for physicists, engineers, computer scientists, mathematicians, and interested and informed clinicians working in the area of medical imaging. This book is aimed at a similar readership.

The Advanced Study Institute was a great success. The weather was beautiful, the food was excellent, the atmosphere was warm and friendly, and the scientific level was judged to be very high. In order to extend the interest of this meeting to an audience greater than the 63 people fortunate enough to have been present in Italy, an attempt has been made to encapsulate the contents of the meeting in the form of this publication.

At the meeting there were three general types of presentation: tutorial lectures, proffered papers, and workshops in specific areas of interest. All of the tutorial lectures are included as chapters in this book. They should represent a good introduction to various areas of importance in medical imaging. All the proffered papers were refereed after the meeting, and some of them were selected as being of a high enough standard for inclusion in this book. These papers represent short accounts of current research in medical imaging. Unfortunately, it was not possible to retain any record of the workshops where often vigorous and heated discussions resulted.

The layout of this book has been divided into two parts. The first part contains all the introductory and tutorial papers and should function as an overview of the subject matter of the meeting. This first part might serve as a text book. The second part contains papers of a more specialized nature divided into four sections; analytic reconstruction methods, iterative methods, display and evaluation, and a collection of papers grouped together under the heading of applications.

The editors would like to thank all the participants, authors, the scientific committee, and, of course, the sponsors of this meeting, and hope that this book will provide useful introductory and reference material suitable for students and workers in the expanding field of mathematics and computer science applied to medical imaging.


M.A. Viergever
A. Todd-Pokropek
Delft, August 1987

TABLE OF CONTENTS

Part 1: Introduction to and Overview of the Field .................... 1

    Introduction to integral transforms
    A. Rescigno ..................................................... 3

    Introduction to discrete reconstruction methods in medical imaging
    M.A. Viergever ................................................. 43

    Image structure
    J.J. Koenderink ................................................ 67

    Fundamentals of the Radon transform
    H.H. Barrett ................................................... 105

    Regularization techniques in medical imaging
    F. Natterer .................................................... 127

    Statistical methods in pattern recognition
    C.R. Appledorn ................................................. 143

    Image data compression techniques: A survey
    A. Todd-Pokropek ............................................... 167

    From 2D to 3D representation
    G.T. Herman .................................................... 197

    VLSI-intensive graphics systems
    H. Fuchs ....................................................... 221

    Knowledge based interpretation of medical images
    J. Fox, N. Walker .............................................. 241

Part 2: Selected Topics ............................................... 267

    2.1   Analytic Reconstruction Methods ............................ 269

          The attenuated Radon transform
          F. Natterer ............................................. 271

          Inverse imaging with strong multiple scattering
          S. Leeman, V.C. Roberts, P.E. Chandler, L.A. Ferrari ........ 279

    2.2   Iterative Methods .......................................... 291

          Possible criteria for choosing the number of iterations in
          some iterative reconstruction methods
          M. Defrise .............................................. 293

          Initial performance of block-iterative reconstruction
          algorithms
          G.T. Herman, H. Levkowitz ............................... 305

          Maximum likelihood reconstruction in PET and TOFPET
          C.T. Chen, C.E. Metz, X. Hu ............................. 319

          Maximum likelihood reconstruction for SPECT using Monte Carlo
          simulation
          C.E. Floyd, S.H. Manglos, R.J. Jaszczak, R.E. Coleman ....... 331

          X-ray coded source tomosynthesis
          I. Magnin ............................................... 339

Some mathematical aspects of electrical impedance tomography
W.R. Breckon, M.K. Pidcock ................................. 351

2.3 Display and Evaluation ....................................... 363

Hierarchical figure-based shape description for medical imaging
S.M. Pizer, W.R. Oliver, J.M. Gauch, S.H. Bloomberg .......... 365

GIHS: A generalized color model and its use for the representation of multiparameter medical images
H. Levkowitz, G.T. Herman ................................... 389

The evaluation of image processing algorithms for use in medical imaging
M. de Belder ................................................ 401

Focal lesions in medical images: A detection problem
J.M. Thijssen ............................................... 415

2.4 Applications ................................................. 439

Time domain phase: A new tool in medical ultrasound imaging
D.A. Seggie, S. Leeman, G.M. Doherty ........................ 441

Performance of echographic equipment and potentials for tissue characterization
J.M. Thijssen, B.J. Oosterveld .............................. 455

Development of a model to predict the potential accuracy of vessel blood flow measurements from dynamic angiographic recordings
D.J. Hawkes, A.C.F. Colchester, J.N.H. Brunt, D.A.G. Wicks, G.H. du Boulay, A. Wallis .................................. 469

The quantitative imaging potential of the HIDAC positron camera
D.W. Townsend, P.E. Frey, G. Reich, A. Christin, H.J. Tochon-Danguy, G. Schaller, A. Donath, A. Jeavons ................. 479

The use of cluster analysis and constrained optimisation techniques in factor analysis of dynamic structures
A.S. Houston ................................................ 491

Detection of elliptical contours
J.A.K. Blokland, A.M. Vossepoel, A.R. Bakker, E.K.J. Pauwels . 505

Optimal non-linear filters for images with non-Gaussian differential distributions
M. Fuderer ................................................. 517

Participants ....................................................... 527

Subject index ...................................................... 533

# Part 1
# Introduction to and Overview of the Field

INTRODUCTION TO INTEGRAL TRANSFORMS

Aldo Rescigno

*Yale University, New Haven, U.S.A.
and
University of Ancona, Italy*

ABSTRACT

*This note is the summary of a series of lectures presented for a number of years by the author to graduate students interested in Mathematical Modeling in Biology. Its aim is essentially practical; proofs are given when they help understanding the essence of a problem, otherwise the approach is mostly intuitive. The operational calculus is presented from an algebraic point of view; starting from the convolution integral, the extension from functions to operators is defined as analogous to the extension from natural numbers to rational numbers. Laplace and Fourier transforms are presented as special cases. The note concludes showing the connection between Fourier transforms and Fourier series.*

1. CONVOLUTION INTEGRAL

Let $g(x) \cdot dx$ measure the intensity of the signal emitted by a one-dimensional source from the small interval $(x, x+dx)$, $f(u)$ the density of the signal received at the point of coordinate $u$ of a one-dimensional observer, and $h(u,x)$ the transmittance along the straight line connecting the points $u$, $x$. Then

$$f(u) = \int_{-\infty}^{+\infty} g(x) h(u,x) \, dx \tag{1}$$

or, with an appropriate choice of the $u$ and $x$ axes,

$$f(u) = \int_{-\infty}^{+\infty} g(x) h(u-x) \, dx. \tag{2}$$

If $g(x) = 0$ for $x < 0$ and $h(u-x) = 0$ for $x > u$, then

$$f(u) = \int_{0}^{u} g(x) h(u-x) \, dx. \tag{3}$$

This is the well-known *convolution integral*.

If Eq. (3) is true in general, then for two arbitrary functions $g_1$ and $g_2$ we can write

$$f_1(u) = \int_0^u g_1(x) h(u-x) \, dx \qquad (4)$$

$$f_2(u) = \int_0^u g_2(x) h(u-x) \, dx \qquad (5)$$

and by adding them

$$f_1(u) + f_2(u) = \int_0^u (g_1(x) + g_2(x)) h(u-x) \, dx \qquad (6)$$

which shows that the transformation $g(x) \rightarrow f(u)$ described by Eq. (3) is *linear*.

Define the function

$$\begin{aligned} g_1(x) &= 0 & \text{for } 0 \leq x < u_0 \\ &= g(x-u_0) & \text{for } x \geq u_0. \end{aligned} \qquad (7)$$

For this function we have

$$\begin{aligned} \int_0^u g_1(x) h(u-x) \, dx &= \int_{u_0}^u g(x-u_0) h(u-x) \, dx \\ &= \int_0^{u-u_0} g(x) h(u-u_0-x) \, dx \end{aligned} \qquad (8)$$

and using Eq. (3)

$$\begin{aligned} \int_0^u g_1(x) h(u-x) \, dx &= 0 & \text{for } 0 \leq u < u_0 \\ &= f(u-u_0) & \text{for } u \geq u_0. \end{aligned} \qquad (9)$$

In other words, the shifting of $g(x)$ along its axis causes the same shift of $f(u)$ along its axis. This shows that the transformation $g(x) \rightarrow f(u)$ described by Eq. (3) is *invariant*.

These two properties - linearity and invariance - of a system described by Eq. (3) are usually expressed as the *principle of superposition*.

Function h(u) can be computed from the integral in Eq. (3) if g(x) is known and f(u) is measured. Of course f(u) would exactly reproduce h(u) if the signal were emitted only by an infinitesimal interval of g(x); but this requires an infinitely large value of g(x) in that interval.

## 2. FUNCTIONS OF CLASS $\mathbb{C}$

Consider the class $\mathbb{C}$ of continuous functions of the real variable $u \geq 0$; we shall represent by $\{f\}, \{g\}, \ldots$, the functions of class $\mathbb{C}$ whose values at u are $f(u), g(u), \ldots$ .

We can define the sum of functions by

$$\{f\} + \{g\} = \{f + g\} \tag{10}$$

and the product of functions by

$$\{f\} \cdot \{g\} = \left\{ \int_0^u f(x) g(u-x) \, dx \right\}. \tag{11}$$

By simple manipulation we can prove the following properties of the addition:
  a) If $\{f\}$ and $\{g\}$ are of class $\mathbb{C}$, their sum is of class $\mathbb{C}$.
  b) There exists an additive identity, i.e. $\{f\} + \{0\} = \{f\}$.
  c) There exists an additive inverse, i.e. $\{f\} + \{-f\} = \{0\}$.
  d) $\{f\} + \{g\} = \{g\} + \{f\}$.
  e) $(\{f\} + \{g\}) + \{h\} = \{f\} + (\{g\} + \{h\})$.

For the multiplication we can prove that
  f) If $\{f\}$ and $\{g\}$ are of class $\mathbb{C}$, their product is of class $\mathbb{C}$.
  g) $\{f\} \cdot \{g\} = \{g\} \cdot \{f\}$.
  h) $(\{f\} \cdot \{g\}) \cdot \{h\} = \{f\} \cdot (\{g\} \cdot \{h\})$.

Finally we can easily prove that
  i) $\{f\} \cdot (\{g\} + \{h\}) = \{f\} \cdot \{g\} + \{f\} \cdot \{h\}$.

It follows that the functions of class $\mathbb{C}$ with the two operations shown above constitute a *commutative ring*.

The following product is particularly interesting:

$$\{1\} \cdot \{f\} = \left\{ \int_0^u f(x)\,dx \right\}. \tag{12}$$

The power of functions can be defined by

$$\{f\}^2 = \{f\} \cdot \{f\}, \quad \{f\}^{n+1} = \{f\} \cdot \{f\}^n; \quad n = 2,3,\ldots. \tag{13}$$

It follows that

$$\{1\}^2 = \{u\}, \quad \{1\}^3 = \{u^2/2\}, \quad \ldots, \quad \{1\}^{n+1} = \{u^n/n!\}. \tag{14}$$

According to the theorem of Titchmarsh (1926), the product of functions is zero if and only if one of the factors is zero, i.e.

$$\{f\} \cdot \{g\} = \{0\} \iff \{f\} = \{0\} \text{ or } \{g\} = \{0\}. \tag{15}$$

We can define the quotient of two functions by

$$\{f\}/\{g\} = \{h\} \tag{16}$$

if $\{g\} \neq \{0\}$ and if a function $\{h\}$ exists such that

$$\{f\} = \{g\} \cdot \{h\}. \tag{17}$$

The theorem of Titchmarsh guarantees that if the quotient of two functions exists, it is unique. To prove this, suppose that we can find two functions $\{h_1\}$ and $\{h_2\}$ such that

$$\{f\} = \{g\} \cdot \{h_1\}, \quad \{f\} = \{g\} \cdot \{h_2\}, \quad \{g\} \neq \{0\}. \tag{18}$$

Substract one identity from the other:

$$\{0\} = \{g\} \cdot \{h_1\} - \{g\} \cdot \{h_2\}. \tag{19}$$

Using property (i) above, we have

$$\{0\} = \{g\} \cdot (\{h_1\} - \{h_2\}) \tag{20}$$

which, by the theorem of Titchmarsh implies that

$$\{h_1\} = \{h_2\}, \quad q.e.d. \tag{21}$$

## 3. OPERATORS

The uniqueness of the quotient makes it possible to extend the ring of functions to a field; the elements of this field are called *operators*, defined as quotients of two functions, the second of them not $\{0\}$ (Mikusinski, 1959).

We define equality, sum and product of operators by:

$$\frac{\{f\}}{\{g\}} = \frac{\{\phi\}}{\{\psi\}} \quad \text{if and only if} \quad \{f\}\cdot\{\psi\} = \{g\}\cdot\{\phi\} \tag{22}$$

$$\frac{\{f\}}{\{g\}} + \frac{\{\phi\}}{\{\psi\}} = \frac{\{f\}\cdot\{\psi\} + \{g\}\cdot\{\phi\}}{\{g\}\cdot\{\psi\}} \tag{23}$$

$$\frac{\{f\}}{\{g\}} \cdot \frac{\{\phi\}}{\{\psi\}} = \frac{\{f\}\cdot\{\phi\}}{\{g\}\cdot\{\psi\}}. \tag{24}$$

Note that equality of operators is reflexive, symmetric and transitive; that addition and multiplication of operators are commutative and associative; and that multiplication is distributive with respect to addition.

The additive identity of operators is $\{0\}/\{f\}$, with $\{f\} \neq \{0\}$; the multiplicative identity is $\{f\}/\{f\}$, with $\{f\} \neq \{0\}$.

The formal properties of operators are identical with those of rational numbers. For instance,

$$\frac{\{f\}\cdot\{\phi\}}{\{g\}\cdot\{\phi\}} = \frac{\{f\}}{\{g\}}. \tag{25}$$

From the identity

$$\frac{\{f\}\cdot\{\phi\}}{\{\phi\}} = \{f\} \tag{26}$$

we can say that the ring of functions is *embedded* in the field of operators.

The identity

$$\frac{\{a\cdot f\}}{\{f\}} = \frac{\{a\cdot g\}}{\{g\}}, \quad a = \text{constant}, \quad \{f\} \neq \{0\}, \quad \{g\} \neq \{0\} \tag{27}$$

allows us to define the *numerical operator*

$$a = \frac{\{a\cdot f\}}{\{f\}} \tag{28}$$

not depending upon the function $\{f\}$. It is easy to verify that the formal properties of numerical operators are identical with the properties of real numbers.

Two important properties are $\{a\cdot f\} = a\cdot\{f\}$ and $\{0\} = 0$.
The first is a direct consequence of the definition of numerical operators; to prove the second one writes

$$0 = \frac{\{0\}}{\{1\}} = \frac{\{0\}\cdot\{1\}}{\{1\}} = \{0\}. \tag{29}$$

If $\{f\} \neq \{0\}$, $\{g\} \neq \{0\}$, then $\{g\}/\{f\}$ is called the *inverse* of $\{f\}/\{g\}$. The product of an operator by its inverse is the numerical operator 1.

Put

$$s = 1/\{1\} \tag{30}$$

and call it the *differential operator*.

If a function $f(u)$ has a derivative $f'(u)$, then

$$\int_0^u f'(x)\,dx = f(u) - f(0). \tag{31}$$

Thence

$$\left\{\int_0^u f'(x)\,dx\right\} = \{f\} - \{f(0)\} \tag{32}$$

and dividing both sides by $\{1\}$

$$\{f'\} = s\cdot\{f\} - f(0). \tag{33}$$

If f(u) has higher derivatives, we can write

$$\{f''\} = s^2 \cdot \{f\} - s \cdot f(0) - f'(0) \tag{34}$$

or in general

$$\{f^{(n)}\} = s^n \cdot \{f\} - s^{n-1} \cdot f(0) - s^{n-2} \cdot f'(0) - \ldots - f^{(n-1)}(0). \tag{35}$$

If f(u) does not have a derivative, then the expression $s \cdot \{f\} - f(0)$ does not correspond to any function, but it is a well-defined operator with a number of properties formally similar to those of a derivative.

From the identities

$$\frac{d\, e^{-\alpha u}}{du} = -\alpha \cdot e^{-\alpha u}, \quad e^0 = 1 \tag{36}$$

we can write

$$\{-\alpha \cdot e^{-\alpha u}\} = s \cdot \{e^{-\alpha u}\} - 1 \tag{37}$$

and hence

$$\{e^{-\alpha u}\} = \frac{1}{s + \alpha}. \tag{38}$$

From the identities

$$\frac{d\, \sin\beta u}{du} = \beta \cdot \cos\beta u, \quad \sin 0 = 0 \tag{39}$$

$$\frac{d\, \cos\beta u}{du} = -\beta \cdot \sin\beta u, \quad \cos 0 = 1 \tag{40}$$

we can write

$$\{\beta \cdot \cos\beta u\} = s \cdot \{\sin\beta u\} \tag{41}$$

$$\{-\beta \cdot \sin\beta u\} = s \cdot \{\cos\beta u\} - 1 \tag{42}$$

which yields

$$\{\sin\beta u\} = \frac{\beta}{s^2 + \beta^2} \qquad (43)$$

$$\{\cos\beta u\} = \frac{s}{s^2 + \beta^2}. \qquad (44)$$

Other interesting functional correlates are

$$\{1\} = 1/s, \quad \{u\} = 1/s^2, \quad \{u^2\} = 2/s^3, \ldots, \{u^n\} = n!/s^{n+1}. \qquad (45)$$

## 4. OPERATIONAL TRANSFORMATIONS

For a function $f(u)$ of class $\mathbb{C}$ define the operation $T_\alpha$ by

$$T_\alpha\{f\} = \{e^{-\alpha u} f(u)\}. \qquad (46)$$

This operation has the properties

$$T_\alpha(T_\beta\{f\}) = T_{\alpha+\beta}\{f\} \qquad (47a)$$

$$T_\alpha(\{f\} + \{g\}) = T_\alpha\{f\} + T_\alpha\{g\} \qquad (47b)$$

$$T_\alpha(\{f\}\cdot\{g\}) = T_\alpha\{f\}\cdot T_\alpha\{g\} \qquad (47c)$$

$$\{f\} = \{g\}/\{h\} \Rightarrow T_\alpha\{f\} = T_\alpha\{g\}/T_\alpha\{h\}. \qquad (47d)$$

To prove the property (47c) we can write

$$T_\alpha(\{f\}\cdot\{g\}) = T_\alpha\left\{\int_0^u f(x)g(u-x)\,dx\right\} = \left\{e^{-\alpha u}\int_0^u f(x)g(u-x)\,dx\right\}$$

$$= \left\{\int_0^u e^{-\alpha x}f(x)\cdot e^{-\alpha(u-x)}g(u-x)\,dx\right\}$$

$$= \{e^{-\alpha u}f(u)\}\cdot\{e^{-\alpha u}g(u)\}, \quad \text{q.e.d.} \qquad (48)$$

For an operator $p = \{f\}/\{g\}$ define the operation $T_\alpha$ by

$$T_\alpha p = T_\alpha\{f\}/T_\alpha\{g\}. \qquad (49)$$

This operation has the properties

$$T_\alpha(T_\beta p) = T_{\alpha+\beta} p \tag{50a}$$
$$T_\alpha(p+q) = T_\alpha p + T_\alpha q \tag{50b}$$
$$T_\alpha(p \cdot q) = T_\alpha p \cdot T_\alpha q \tag{50c}$$
$$T_\alpha(p/q) = T_\alpha p / T_\alpha q \tag{50d}$$

analogous to the corresponding properties shown for a function.

It is easy to prove that

$$T_\alpha a = a. \tag{51}$$

In fact,

$$T_\alpha a = T_\alpha \frac{\{a\}}{\{1\}} = \frac{T_\alpha\{a\}}{T_\alpha\{1\}} = \frac{\{e^{-\alpha u} \cdot a\}}{\{e^{-\alpha u}\}} = \frac{a \cdot \{e^{-\alpha u}\}}{\{e^{-\alpha u}\}} = a, \quad \text{q.e.d.} \tag{52}$$

We can also prove that

$$T_\alpha s = s + \alpha. \tag{53}$$

For,

$$T_\alpha s = T_\alpha \left(\frac{1}{\{1\}}\right) = \frac{T_\alpha 1}{T_\alpha \{1\}} = 1/\{e^{-\alpha u}\} = s + \alpha, \quad \text{q.e.d.} \tag{54}$$

It is obvious now that

$$T_\alpha s^2 = (s+\alpha)^2, \quad \ldots, \quad T_\alpha s^n = (s+\alpha)^n. \tag{55}$$

If $R(s)$ is a rational expression in $s$, then

$$T_\alpha R(s) = R(s+\alpha). \tag{56}$$

For instance

$$\{e^{-\alpha u} \sin\beta u\} = \frac{\beta}{(s+\alpha)^2 + \beta^2}, \tag{57}$$

$$\{e^{-\alpha u} \cos\beta u\} = \frac{s+\alpha}{(s+\alpha)^2 + \beta^2}. \tag{58}$$

Define the operation D on a function of class $\mathbb{C}$ by

$$D\{f(u)\} = \{-u \cdot f(u)\}. \tag{59}$$

This operation has the properties

$$D(\{f\} + \{g\}) = D\{f\} + D\{g\} \tag{60a}$$
$$D(\{f\} \cdot \{g\}) = D\{f\} \cdot \{g\} + \{f\} \cdot D\{g\} \tag{60b}$$
$$\{f\} = \{g\}/\{h\} \Rightarrow D\{f\} = \frac{D\{f\} \cdot \{g\} - \{f\} \cdot D\{g\}}{\{g\}^2}. \tag{60c}$$

To prove the property (60b) we can write

$$D(\{f\} \cdot \{g\}) = D \left\{ \int_0^u f(x) g(u-x) dx \right\} = \left\{ -u \int_0^u f(x) g(u-x) dx \right\}$$

$$= \left\{ \int_0^u (-x) f(x) \cdot g(u-x) dx \right\} + \left\{ \int_0^u f(x) \cdot (-u+x) g(u-x) dx \right\}$$

$$= \{-u \cdot f(u)\} \cdot \{g(u)\} + \{f(u)\} \cdot \{-u \cdot g(u)\}, \quad \text{q.e.d.} \tag{61}$$

Define the operation D on an operator $p = \{f\}/\{g\}$ by

$$Dp = \frac{D\{f\} \cdot \{g\} - \{f\} \cdot D\{g\}}{\{g\}^2}. \tag{62}$$

This operation has the properties

$$D(p+q) = Dp + Dq \tag{63a}$$
$$D(p \cdot q) = Dp \cdot q + p \cdot Dq \tag{63b}$$
$$D(p/q) = \frac{Dp \cdot q - p \cdot Dq}{q^2} \tag{63c}$$

analogous to the corresponding properties shown for a function.

We can prove that

$$D\alpha = 0. \tag{64}$$

In fact

$$D\alpha = D\left(\frac{\{\alpha\}}{\{1\}}\right) = \frac{D\{\alpha\} \cdot \{1\} - \{\alpha\} \cdot D\{1\}}{\{1\}^2} = \frac{\{-\alpha u\} \cdot \{1\} - \{\alpha\} \cdot \{-u\}}{\{1\}^2} = 0. \tag{65}$$

Similarly we can prove that

$$Ds = 1, \quad D(s^2) = 2 \cdot s, \quad \ldots, \quad D(s^n) = n \cdot s^{n-1}. \tag{66}$$

In general, if $R(s)$ is a rational expression in $s$, then

$$DR(s) = dR(s)/ds \tag{67}$$

where the symbol at the right hand side represents the formal differentiation of $R(s)$ with respect to $s$, as though $s$ were a variable.

For instance

$$\{t \cdot \sin\beta u\} = 2\beta s/(s^2 + \beta^2)^2 \tag{68}$$
$$\{t \cdot \cos\beta u\} = (s^2 - \beta^2)/(s^2 + \beta^2)^2. \tag{69}$$

## 5. TRANSLATION OPERATOR

Call $\mathbb{K}$ the class of functions $f(u)$ of the real variable $u \geq 0$ such that $\int_0^u f(x)\,dx$ is of class $\mathbb{C}$; class $\mathbb{K}$ includes the functions of class $\mathbb{C}$ plus all functions having a finite number of discontinuities in any finite interval of $u$, and such that $\int_0^u |f(x)|\,dx$ is finite for the whole domain of $u$. From the identity

$$\{f\} = \left\{\int_0^u f(x)\,dx\right\} / \{1\} \tag{70}$$

we see that the functions of class $\mathbb{K}$ are embedded in the field of operators defined as quotients of functions of class $\mathbb{C}$.

If two functions $f(u)$ and $f_1(u)$ differ only of a set of values of $u$ of measure zero, then their integrals from 0 to any value of $u$ are identical, therefore they correspond to the same operator; in other terms we can say that $f(u)$ and $f_1(u)$ cannot be distinguished by any physical instrument.

Define the *jump function*

$$H_\lambda(u) = 0 \text{ for } 0 \leq u < \lambda \qquad (71)$$
$$= 1 \text{ for } u \geq \lambda$$

and the *translation operator*

$$h_\lambda = s \cdot \{H_\lambda\}. \qquad (72)$$

For any function $\{f\}$ we have

$$h_\lambda \cdot \{f\} = s \cdot \{H_\lambda\} \cdot \{f\}$$
$$= s \cdot \left\{ \int_0^u H_\lambda(u-x) f(x) dx \right\}. \qquad (73)$$

By the definition of $H_\lambda$ this integral vanishes for $x > u - \lambda$, therefore

$$h_\lambda \cdot \{f\} = s \cdot \left\{ \int_0^{u-\lambda} f(x) dx \right\} \text{ for } u \geq \lambda \qquad (74)$$
$$= 0 \qquad \qquad \text{ for } u < \lambda$$

and finally

$$h_\lambda \cdot \{f\} = 0 \qquad \text{ for } u < \lambda \qquad (75)$$
$$= \{f(t-\lambda)\} \text{ for } u \geq \lambda.$$

The jump function can now be written in operational form

$$\{H_\lambda\} = h_\lambda / s. \qquad (76)$$

The *gate function*

$$G_{a,b}(u) = 0 \text{ for } 0 \leq u < a \text{ or } u > b \qquad (77)$$
$$= 1 \text{ for } a \leq u \leq b$$

can be written

$$\{G_{a,b}\} = (h_a - h_b)/s. \qquad (78)$$

A single sinusoidal pulse

$$S_\omega(u) = \sin\omega t \quad \text{for } 0 \leq u \leq \pi$$
$$= 0 \quad \text{for } u > \pi \tag{79}$$

can be written

$$\{S_\omega\} = \omega \cdot \frac{1 + h_\omega}{s^2 + \omega^2}. \tag{80}$$

Many other discontinuous functions, or functions with a discontinuous derivative, can be written in operational form using the translation operator.

Consider the operator $f(\lambda)$ containing a parameter $\lambda$; if

$$f(\lambda) = p \cdot \{f_1(\lambda, u)\} \tag{81}$$

where $p$ is an operator not containing $\lambda$, and $f_1(\lambda, u)$ is a function with a partial derivative with respect to $\lambda$, continuous for $u \geq 0$ and for a certain domain of $\lambda$, then we can write

$$df(\lambda)/d\lambda = p \cdot \{\partial f_1(\lambda, u)/\partial \lambda\}, \tag{82}$$

called *continuous derivative* of the operator $f(\lambda)$. The continuous derivative of an operator, if it exists, is unique. In fact, suppose that an operator can be decomposed in two different ways:

$$f(\lambda) = p \cdot \{f_1(\lambda, u)\} = q \cdot \{f_2(\lambda, u)\}. \tag{83}$$

We can then choose a function $\phi(u)$ such that

$$p = \{g_1\}/\{\phi\}, \quad q = \{g_2\}/\{\phi\} \tag{84}$$

with $\{g_1\}$ and $\{g_2\}$ functions of $u$ of class $\mathbb{C}$. It follows that

$$\{g_1\} \cdot \{f_1\} = \{g_2\} \cdot \{f_2\} \Rightarrow$$

$$\left\{ \int_0^u g_1(x) f_1(\lambda, u-x)\, dx \right\} = \left\{ \int_0^u g_2(x) f_2(\lambda, u-x)\, dx \right\} \Rightarrow$$

$$\left\{ \int_0^u g_1(x) \frac{\partial}{\partial \lambda} f_1(\lambda, u-x)\, dx \right\} = \left\{ \int_0^u g_2(x) \frac{\partial}{\partial \lambda} f_2(\lambda, u-x)\, dx \right\} \Rightarrow$$

$$\{g_1\} \cdot \left\{\frac{\partial}{\partial \lambda} f_1\right\} = \{g_2\} \cdot \left\{\frac{\partial}{\partial \lambda} f_2\right\} \tag{85}$$

and dividing both sides by $\{\phi\}$

$$p \cdot \left\{\frac{\partial}{\partial \lambda} f_1\right\} = q \cdot \left\{\frac{\partial}{\partial \lambda} f_2\right\}, \quad \text{q.e.d.} \tag{86}$$

To compute $dh_\lambda/d\lambda$ we cannot use definition (72) directly because $H_\lambda(u)$ does not have a continuous partial derivative with respect to $\lambda$; but we can define

$$\begin{aligned} H^*_\lambda(u) &= 0 \quad \text{for } 0 \leq u < \lambda \\ &= u - \lambda \quad \text{for } u \geq \lambda, \end{aligned} \qquad \begin{aligned} H^{**}_\lambda(u) &= 0 \quad \text{for } 0 \leq u < \lambda \\ &= \tfrac{1}{2}(u-\lambda)^2 \quad \text{for } u \geq \lambda. \end{aligned} \tag{87}$$

Then, observing that

$$\begin{aligned} \partial H^{**}_\lambda/\partial \lambda &= 0 \quad \text{for } 0 \leq u < \lambda \\ &= -(u-a) \quad \text{for } u \geq \lambda \end{aligned} \tag{88}$$

and that

$$\{H_\lambda\} = s \cdot \{H^*_\lambda\} = s^2 \cdot \{H^{**}_\lambda\} \tag{89}$$

we can write

$$h_\lambda = s^3 \cdot \{H^{**}_\lambda\} \tag{90}$$

and from definition (82) we compute

$$dh_\lambda/d\lambda = s^3 \cdot \{\partial H^{**}_\lambda/\partial \lambda\} = -s^3 \cdot \{H^*_\lambda\} = -s^2 \cdot \{H_\lambda\} = -s \cdot h_\lambda. \tag{91}$$

Other properties of the translation operator are

$$h_0 = 1, \quad h_\lambda \cdot h_\mu = h_{\lambda+\mu}. \tag{92}$$

The only ordinary function with these three properties is the exponential function; therefore we can define

$$h_\lambda = e^{-\lambda s}. \tag{93}$$

Other properties of this operator justifying the exponential symbol are

$$T_\alpha e^{-\lambda s} = e^{-\lambda(s+\alpha)}; \quad D e^{-\lambda s} = -\lambda \cdot e^{-\lambda s}. \tag{94}$$

## 6. INTEGRAL OF AN OPERATOR

Given the operator $f(\lambda)$ containing the parameter $\lambda$, such that

$$f(\lambda) = p \cdot \{f_1(\lambda, u)\} \tag{95}$$

where $p$ is an operator not containing $\lambda$, we can define

$$\int_\alpha^\beta f(\lambda) \, d\lambda = p \cdot \left\{ \int_\alpha^\beta f_1(\lambda, u) \, d\lambda \right\} \tag{96}$$

if the integral on the right hand side exists. The integral of an operator, if it exists, is unique. To prove this, suppose that an operator can be decomposed in two different ways:

$$f(\lambda) = p \cdot \{f_1(\lambda, u)\} = q \cdot \{f_2(\lambda, u)\}. \tag{97}$$

Again we can choose a function $\phi(u)$ such that

$$p = \{g_1\}/\{\phi\}, \quad q = \{g_2\}/\{\phi\}, \tag{98}$$

with $\{g_1\}$ and $\{g_2\}$ functions of $u$ of class $\mathbb{C}$. It follows that

$$\{g_1\} \cdot \{f_1\} = \{g_2\} \cdot \{f_2\} \Rightarrow$$

$$\left\{ \int_0^u g_1(x) f_1(\lambda, u-x) \, dx \right\} = \left\{ \int_0^u g_2(x) f_2(\lambda, u-x) \, dx \right\}. \tag{99}$$

Integrate both sides with respect to $\lambda$ and change the order of integration:

$$\left\{ \int_0^u g_1(x) \left( \int_\alpha^\beta f_1(\lambda, u-x) \, d\lambda \right) dx \right\} = \left\{ \int_0^u g_2(x) \left( \int_\alpha^\beta f_2(\lambda, u-x) \, d\lambda \right) dx \right\} \Rightarrow$$

$$\{g_1\} \cdot \left\{\int_\alpha^\beta f_1(\lambda,u)\,d\lambda\right\} = \{g_2\} \cdot \left\{\int_\alpha^\beta f_2(\lambda,u)\,d\lambda\right\} \qquad (100)$$

and divide both sides by $\{\phi\}$

$$p \cdot \left\{\int_\alpha^\beta f_1(\lambda,u)\,d\lambda\right\} = q \cdot \left\{\int_\alpha^\beta f_2(\lambda,u)\,d\lambda\right\}, \quad \text{q.e.d.} \qquad (101)$$

For instance consider the ordinary function $g(\lambda)$; by definition

$$e^{-\lambda s} \cdot g(\lambda) = s \cdot \{H_\lambda\} \cdot g(\lambda). \qquad (102)$$

Thence

$$e^{-\lambda s} \cdot g(\lambda) = s \cdot \{g_1(\lambda,u)\} \qquad (103)$$

where

$$\begin{aligned} g_1(\lambda,u) &= 0 \quad \text{for } 0 \leq u < \lambda \\ &= g(\lambda) \quad \text{for } u \geq \lambda. \end{aligned} \qquad (104)$$

By integration

$$\int_0^{\lambda_1} e^{-\lambda s} \cdot g(\lambda)\,d\lambda = s \cdot \left\{\int_0^{\lambda_1} g_1(\lambda,u)\,d\lambda\right\}. \qquad (105)$$

But

$$\begin{aligned} \int_0^{\lambda_1} g_1(\lambda,u)\,d\lambda &= \int_0^u g(\lambda)\,d\lambda \quad \text{for } u \leq \lambda_1 \\ &= \int_0^{\lambda_1} g(\lambda)\,d\lambda \quad \text{for } u > \lambda_1, \end{aligned} \qquad (106)$$

therefore

$$\begin{aligned} \int_0^{\lambda_1} e^{-\lambda s} \cdot g(\lambda)\,d\lambda &= s \cdot \left\{\int_0^u g(\lambda)\,d\lambda\right\} = \{g\} \quad \text{for } u \leq \lambda_1 \\ &= s \cdot \left\{\int_0^{\lambda_1} g(\lambda)\,d\lambda\right\} = 0 \quad \text{for } u > \lambda_1. \end{aligned} \qquad (107)$$

For $\lambda_2 > \lambda_1$,

$$\int_{\lambda_1}^{\lambda_2} e^{-\lambda s} \cdot g(\lambda) d\lambda = 0 \quad \text{for } 0 \leq u < \lambda_1$$
$$= \{g\} \quad \text{for } \lambda_1 \leq u \leq \lambda_2 \qquad (108)$$
$$= 0 \quad \text{for } u > \lambda_2.$$

Using the symbol $\int_0^\infty$ for $\lim_{\substack{\lambda_1 \to 0 \\ \lambda_2 \to \infty}} \int_{\lambda_1}^{\lambda_2}$, we get the formula

$$\int_0^\infty e^{-\lambda s} g(\lambda) d\lambda = \{g\}, \qquad (109)$$

formally similar to the definition of Laplace transform; observe though that here s is the differential operator, while in the Laplace transform s is a complex variable.


## 7. APPROXIMATE FUNCTIONAL CORRELATES

Define the function

$$K_\lambda(u) = 0 \quad \text{for } 0 \leq u \leq \lambda$$
$$= \phi(u) \quad \text{for } \lambda < u < \lambda+\mu \qquad (110)$$
$$= 1 \quad \text{for } u \geq \lambda+\mu$$

where $\mu > 0$ is a "small" number and $\phi(u)$ is a function such that $K_\lambda(u)$ has derivatives up to the order we need to consider; this function is identical to the jump function except in the small interval $\lambda < u < \lambda+\mu$. Then

$$e^{-\lambda s} = s \cdot \{H_\lambda\} \approx s \cdot \{K_\lambda\} = \{dK_\lambda/du\} \qquad (111)$$

where

$$dK_\lambda/du = 0 \quad \text{for } 0 \leq u \leq \lambda$$
$$= d\phi/du \quad \text{for } \lambda < u < \lambda+\mu \qquad (112)$$
$$= 0 \quad \text{for } u \geq \lambda+\mu.$$

Thus the function $\{dK_\lambda/du\}$, approximating the translation operator $e^{-\lambda s}$, is zero everywhere except in the small interval $\lambda < u < \lambda+\mu$, where it assumes unspecified values such that

$$\int_0^\infty \frac{dK_\lambda}{du} \, du = 1. \tag{113}$$

The function $\{dK_\lambda/du\}$ corresponds to the Dirac delta function.

In a similar way we obtain

$$s^n \cdot e^{-\lambda s} \approx s^n \cdot \{dK_\lambda/du\} = \{d^{n+1}K_\lambda/du^{n+1}\} \tag{114}$$

where

$$\begin{aligned} d^{n+1}K_\lambda/du^{n+1} &= 0 & \text{for } 0 \leq u \leq \lambda \\ &= d^{n+1}\phi/du^{n+1} & \text{for } \lambda < u < \lambda+\mu \\ &= 0 & \text{for } u \geq \lambda+\mu \end{aligned} \tag{115}$$

and

$$\int_0^\infty du \int_0^u du \ldots \int_0^u \frac{d^{n+1}K_\lambda}{du^{n+1}} \, du = 1. \tag{116}$$

Making $\lambda = 0$ in the expressions above we obtain the functions approximating the operators $1, s, s^2, \ldots, s^n$.

## 8. RATIONAL OPERATORS

An expression of the form

$$\frac{p(s)}{q(s)} = \frac{p_m s^m + p_{m-1} s^{m-1} + \ldots + p_1 s + p_0}{q_n s^n + q_{n-1} s^{n-1} + \ldots + q_1 s + q_0} \tag{117}$$

where $p_0, p_1, \ldots, q_0, q_1, \ldots$ are real numbers, and $q_n \neq 0$, is called a *rational operator*.

*Lemma.* If

$$a_n s^n + a_{n-1} s^{n-1} + \ldots + a_1 s + a_0 = b_n s^n + b_{n-1} s^{n-1} + \ldots + b_1 s + b_0 \tag{118}$$

then

$$a_0 = b_0, \quad a_1 = b_1, \quad \ldots, \quad a_n = b_n. \tag{119}$$

This statement can be proved by multiplying each side of the identity above by $\{1\}^{n+1}$, thereby transforming it into an identity of polynomials in u.

*Theorem.* If

$$\frac{a_m s^m + a_{m-1} s^{m-1} + \ldots + a_1 s + a_0}{b_n s^n + b_{n-1} s^{n-1} + \ldots + b_1 s + b_0} = \frac{c_p s^p + c_{p-1} s^{p-1} + \ldots + c_1 s + c_0}{d_q s^q + d_{q-1} s^{q-1} + \ldots + d_1 s + d_0} \tag{120}$$

then for any number z (real or complex) such that

$$\begin{aligned} b_n z^n + b_{n-1} z^{n-1} + \ldots + b_1 z + b_0 &\neq 0 \\ d_q z^q + d_{q-1} z^{q-1} + \ldots + d_1 z + d_0 &\neq 0 \end{aligned} \tag{121}$$

it holds that

$$\frac{a_m z^m + a_{m-1} z^{m-1} + \ldots + a_1 z + a_0}{b_n z^n + b_{n-1} z^{n-1} + \ldots + b_1 z + b_0} = \frac{c_p z^p + c_{p-1} z^{p-1} + \ldots + c_1 z + c_0}{d_q z^q + d_{q-1} z^{q-1} + \ldots + d_1 z + d_0}. \tag{122}$$

In fact, from the identity

$$\left(a_m s^m + \ldots + a_1 s + a_0\right) \cdot \left(d_q s^q + \ldots + d_1 s + d_0\right) =$$
$$= \left(b_n s^n + \ldots + b_1 s + b_0\right) \cdot \left(c_p s^p + \ldots + c_1 s + c_0\right) \tag{123}$$

and from the lemma it follows that

$$\begin{aligned} a_0 d_0 &= b_0 c_0 \\ a_1 d_0 + a_0 d_1 &= b_1 c_0 + b_0 c_1 \\ a_2 d_0 + a_1 d_1 + a_0 d_2 &= b_2 c_0 + b_1 c_1 + b_0 c_2 \end{aligned} \tag{124}$$

and so forth, thence the statement of the thesis.

Because of this theorem, any rational operator can be transformed as though the differential operator it contains were a complex variable. In particular, if $m \geq n$, by an algebraic division we get

$$\frac{p(s)}{q(s)} = f(s) + \frac{p^*(s)}{q(s)} \tag{125}$$

where $f(s)$, the quotient, is a polynomial of degree $m-n$, and $p^*(s)$, the remainder, is a polynomial of degree not higher than $n-1$. If $m < n$, we can write the rational operator as a sum of partial fractions of the form

$$\frac{A}{(s-\alpha)^\mu}, \quad \frac{Bs+C}{((s-\beta)^2+\gamma^2)^\nu} \tag{126}$$

where $A$, $B$, $C$, $\alpha$, $\beta$, $\gamma$, are real numbers and $\mu$, $\nu$ are natural numbers.

It is easy to see that each $\alpha$ is a real root of multiplicity $\mu$, and each $\beta \pm i\gamma$ is a pair of complex roots of multiplicity $\nu$ of equation $q(z) = 0$; the coefficients $A$, $B$, $C$ can be computed with the rules of elementary algebra.

For example, let the rational operator $p(s)/q(s)$, where the degree $m$ of $p(s)$ is less than the degree $n$ of $q(s)$, be given. Consider the meromorphic function $p(z)/q(z)$ of the complex variable $z = x+iy$. The polynomial $q(z)$ has $n$ zeroes; suppose for simplicity they are all of multiplicity one. We can write

$$\frac{p(z)}{q(z)} = \frac{R_1}{z-\alpha_1} + \frac{R_2}{z-\alpha_2} + \ldots + \frac{R_n}{z-\alpha_n} \tag{127}$$

where $\alpha_1, \alpha_2, \ldots, \alpha_n$ are the (real or complex) roots of equation $q(z) = 0$, and $R_1, R_2, \ldots, R_n$ the corresponding residues. To compute a residue, say $R_1$, multiply both sides of the above expression by $z-\alpha_1$:

$$\frac{p(z)}{q(z)/(z-\alpha_1)} = R_1 + (z-\alpha_1) \cdot \left( \frac{R_2}{z-\alpha_2} + \ldots + \frac{R_n}{z-\alpha_n} \right) \tag{128}$$

and evaluate (128) at $z = \alpha_1$. Thus in general

$$R_j = \left[ \frac{p(z)}{q(z)/(z-\alpha_j)} \right]_{z=\alpha_j}. \tag{129}$$

For the theorem proved above, we can substitute the differential operator $s$ for the complex variable $z$, therefore

$$\frac{p(s)}{q(s)} = \sum_{j=1}^{n} \left\{ R_j \, e^{\alpha_j u} \right\}. \tag{130}$$

Suppose now that a particular root of $q(z) = 0$, and consequently its corresponding residue, is complex. Call them respectively $\beta + i\gamma$ and $B + iC$; because the coefficients of $p(z)/q(z)$ are real, to each complex term corresponds its conjugate, therefore in the sum on the right hand side of Eq. (127) there will be the two fractions

$$\frac{B + iC}{s - \beta - i\gamma}, \quad \frac{B - iC}{s - \beta + i\gamma}. \tag{131}$$

We can compute

$$\begin{aligned}
\frac{B + iC}{s - \beta - i\gamma} &= \left\{ (B + iC) \cdot e^{(\beta + i\gamma)u} \right\} = \\
&= \left\{ (B + iC) \cdot e^{\beta u} \cdot (\cos\gamma u + i \sin\gamma u) \right\} \\
&= \left\{ e^{\beta u} \cdot [(B \cos\gamma u - C \sin\gamma u) + i(C \cos\gamma u + B \sin\gamma u)] \right\}
\end{aligned} \tag{132}$$

whence

$$\frac{B + iC}{s - \beta - i\gamma} + \frac{B - iC}{s - \beta + i\gamma} = \{ 2 \cdot e^{\beta u} \cdot (B \cos\gamma u - C \sin\gamma u) \}. \tag{133}$$

## 9. MULTIDIMENSIONAL OPERATORS

The operational calculus can easily be extended to more variables (Rescigno, 1984); as an example I shall show here some properties of the operators in two variables.

If $f(u,v)$, $g(u,v)$ are two-variable functions of class $\mathbb{C}$, then define

$$\{f\} + \{g\} = \{f + g\} \tag{134}$$

$$\{f\} \cdot \{g\} = \left\{ \int_0^v \int_0^u f(x,y) g(u-x, v-y) \, dx \, dy \right\}. \tag{135}$$

With these functions we can construct a field with the same properties as the one-variable operators. Define

$$p = \left\{\int_0^u f(x,v)\,dx\right\} / \{f\}, \quad q = \left\{\int_0^v f(u,y)\,dy\right\} / \{f\}. \tag{136}$$

The operators p and q are well-defined because they do not depend on the function f(u,v). In fact,

$$\left\{\int_0^u f(x,v)\,dx\right\} \cdot \{g\} = \left\{\int_0^u g(x,v)\,dx\right\} \cdot \{f\} \tag{137}$$

and

$$\left\{\int_0^v f(u,y)\,dy\right\} \cdot \{g\} = \left\{\int_0^v g(u,y)\,dy\right\} \cdot \{f\} \tag{138}$$

for any {f} and {g}; obviously

$$\{1\} \cdot \{f\} = \left\{\int_0^v \int_0^u f(x,y)\,dxdy\right\} \text{ and } p \cdot q = \{1\}, \tag{139}$$

therefore p, q, {1} can be called *integral operators*.

For any functions f(u) and g(u) of one variable,

$$\frac{\{f(u)\}}{q} + \frac{\{g(u)\}}{q} = \frac{\{f(u)+g(u)\}}{q} \tag{140}$$

$$\frac{\{f(u)\}}{q} \cdot \frac{\{g(u)\}}{q} = \frac{\left\{\int_0^u f(x)g(u-x)\,dx\right\}}{q}, \tag{141}$$

therefore any operator of the form {f(u)}/q follows the rules of the one-variable operational calculus; the same is true for the operators of the form {f(v)}/p. It is convenient to introduce the notation

$$[f(u)] = \{f(u)\}/q, \quad [f(v)] = \{f(v)\}/p. \tag{142}$$

The following identities hold:

$$a \cdot [f(u)] = [a \cdot f(u)], \quad a = \text{constant} \tag{143}$$

$$[f(u)] \cdot [g(u)] = \left[\int_0^u f(x)g(u-x)\,dx\right] \tag{144}$$

$$[f(u)] \cdot [g(v)] = \{f \cdot g\} \tag{145}$$

$$[f(u)] \cdot \{g(u,v)\} = \left\{ \int_0^u f(x) g(u-x,v) \, dx \right\}. \tag{146}$$

Define the inverse operators

$$r = 1/p, \quad s = 1/q. \tag{147}$$

If $f(u,v)$ has all derivatives we are going to consider, then

$$\left\{ \int_0^u \partial f/\partial u \cdot du \right\} = \{f\} - \{f(0,v)\} \tag{148}$$

and dividing by p,

$$\{\partial f/\partial u\} = r \cdot \{f\} - [f(0,v)]. \tag{149}$$

Similarly

$$\{\partial f/\partial v\} = s \cdot \{f\} - [f(u,0)], \tag{150a}$$

$$\{\partial^2 f/\partial u^2\} = r^2 \cdot \{f\} - r \cdot [f(0,v)] - [\partial f/\partial u \big|_{u=0}] \tag{150b}$$

$$\left\{ \frac{\partial f^2}{\partial u \partial v} \right\} = r \cdot s \cdot \{f\} - r \cdot [f(u,0)] - s \cdot [f(0,v)] + f(0,0) \tag{150c}$$

$$\{\partial^2 f/\partial v^2\} = s^2 \cdot \{f\} - s \cdot [f(u,0)] - [\partial f/\partial v \big|_{v=0}]. \tag{150d}$$

It is easy to prove the following identities:

$$\{1\} = \frac{1}{r \cdot s}, \quad \{u\} = \frac{1}{r^2 s}, \quad \{v\} = \frac{1}{rs^2} \tag{151}$$

$$\{u^i v^j\} = \frac{\Gamma(i+1) \cdot \Gamma(j+1)}{r^{i+1} \cdot s^{j+1}} \quad \text{for } i,j > 0. \tag{152}$$

$$e^{au+bv} = \frac{1}{(r-a) \cdot (s-b)} \tag{153}$$

$$\{\sin(au+bv)\} = \frac{rs - ab}{(r^2+a^2) \cdot (s^2+b^2)}. \tag{154}$$

If

$$\begin{aligned} H_{\lambda,\mu}(u,v) &= 0 \quad \text{for } u < \lambda \text{ or } v < \mu \\ &= 1 \quad \text{for } u \geq \lambda \text{ and } v \geq \mu \end{aligned} \tag{155}$$

then

$$\{H_{\lambda,\mu}\} = h_{\lambda,\mu}/(rs) \tag{156}$$

where $h_{\lambda,\mu}$ is the *translation operator* such that

$$\begin{aligned} h_{\lambda,\mu}\{f\} &= \{f(u-\lambda,v-\mu)\} \text{ for } u \geq \lambda \text{ and } v \geq \mu \\ &= 0 \quad \text{everywhere else.} \end{aligned} \tag{157}$$

Define the operation $R_a$ on a function by

$$R_a\{f\} = \{e^{-au} \cdot f(u,v)\} \tag{158}$$

and on an operator by

$$R_a(\{f\}/\{g\}) = R_a\{f\}/R_a\{g\}. \tag{159}$$

We can prove that if $\Phi(r,s)$ is a rational operator in $r$ and $s$, then

$$R_a \Phi(r,s) = \Phi(r+a,s). \tag{160}$$

Similarly if $S_b\{f\} = \{e^{-bv} \cdot f(u,v)\}$ and $S_b(\{f\}/\{g\}) = S_b\{f\}/S_b\{g\}$, we can prove that

$$S_b \Phi(r,s) = \Phi(r,s+b). \tag{161}$$

If we define the two operations

$$A\{f(u,v)\} = \{f(u/a,v)\}, \quad B\{f(u,v)\} = \{f(u,v/b)\} \tag{162}$$

$$A(\{f\}/\{g\}) = A\{f\}/A\{g\}, \quad B(\{f\}/\{g\}) = B\{f\}/B\{g\} \tag{163}$$

with $a$ and $b$ constant, then we can prove that, if

$$\frac{\{f(u,v)\}}{\{g(u,v)\}} = \Phi(r,s) \tag{164}$$

then

$$\frac{\{f(u/a,v/b)\}}{\{g(u/a,v/b)\}} = \Phi(a \cdot r, b \cdot s). \tag{165}$$

## 10. SPECIAL OPERATORS

The operators described in section 9 are analogous to the corresponding operators of the one-variable operational calculus; but there are some very important multidimensional operators that have no one-dimensional counterpart. This of course is a consequence of the fact that the partial differential equations have some properties that are not a trivial extension of analogous properties of ordinary differential equations. We shall examine here some operators that have no counterparts in the operational calculus in one variable.

Define the operation

$$I\{f\} = \left\{ \int_0^{\min(u,v)} f(u-x, v-x)\, dx \right\} \tag{166}$$

that transforms a two-variable function into another two-variable function. Then

$$r \cdot I\{f\} = \left\{ \begin{array}{l} \displaystyle\int_0^v f_u(u-x, v-x)\, dx \\[2ex] \displaystyle\int_0^u f_u(u-x, v-x)\, dx + f(0, v-u) \end{array} \right. \tag{167}$$

$$s \cdot I\{f\} = \left\{ \begin{array}{l} \displaystyle\int_0^v f_v(u-x, v-x)\, dx + f(u-v, 0) \\[2ex] \displaystyle\int_0^u f_v(u-x, v-x)\, dx \end{array} \right. \tag{168}$$

where $f_u$ and $f_v$ represent $\partial f/\partial u$ and $\partial f/\partial v$ respectively, and on the right hand sides the first line inside the brackets corresponds to the domain $u \geqslant v$ and the second line to the domain $u < v$. It follows that

$$(r+s) \cdot I\{f\} = \left\{ \begin{array}{l} f(u-v, 0) + \displaystyle\int_0^v (f_u + f_v)\, dx \\[2ex] f(0, v-u) + \displaystyle\int_0^u (f_u + f_v)\, dx \end{array} \right. , \tag{169}$$

therefore

$$I\{f\} = \{f\}/(r+s). \tag{170}$$

Define the operation

$$L[f] = \left\{ \begin{matrix} f(u-v) \\ 0 \end{matrix} \right\} \tag{171}$$

that transforms a one-variable function into a two-variable function; then

$$p \cdot L[f] = \left\{ \begin{matrix} \int_0^u f(x-v)\,dx \\ 0 \end{matrix} \right\} = \left\{ \begin{matrix} \int_0^u f(y)\,dy \\ 0 \end{matrix} \right\} \tag{172}$$

$$q \cdot L[f] = \left\{ \begin{matrix} \int_0^v f(u-x)\,dx \\ \int_0^u f(u-x)\,dx \end{matrix} \right\} = \left\{ \begin{matrix} \int_{u-v}^u f(y)\,dy \\ \int_0^u f(y)\,dy \end{matrix} \right\} \tag{173}$$

$$(p+q) \cdot L[f] = \left\{ \int_0^u f(y)\,dy \right\} \tag{174}$$

$$L[f] = [f]/(r+s). \tag{175}$$

We have found that the same operator $1/(r+s)$ transforms $\{f\}$ into $I\{f\}$ and $[f]$ into $L[f]$.

Consider now the operation

$$M\{f\} = \left\{ \int_0^u f(u-x, v+x)\,dx \right\} \tag{176}$$

and compute

$$r \cdot M\{f\} = \left\{ \int_0^u f_u(u-x, v+x)\,dx + f(0, u+v) \right\} \tag{177}$$

$$s \cdot M\{f\} = \left\{ \int_0^u f_v(u-x, v+x)\,dx \right\} \tag{178}$$

$$(r-s) \cdot M\{f\} = \left\{ f(0, u-v) + \int_0^u (f_u - f_v)\,dx \right\} = \{f\}. \tag{179}$$

Equations (176) and (179) jointly yield

$$\left\{\int_0^u f(u-x,v+x)\,dx\right\} = \frac{\{f\}}{(r-s)}. \tag{180}$$

Define the operation

$$N[f(u)] = \{f(u+v)\} \tag{181}$$

and compute

$$p \cdot N[f(u)] = \left\{\int_0^u f(x+v)\,dx\right\} = \left\{\int_v^{u+v} f(x)\,dx\right\} \tag{182}$$

$$q \cdot N[f(u)] = \left\{\int_0^v f(u+x)\,dx\right\} = \left\{\int_u^{u+v} f(x)\,dx\right\} \tag{183}$$

$$(q-p) \cdot N[f(u)] = \left\{\int_u^v f(x)\,dx\right\} = \left\{\int_0^v f(x)\,dx\right\} - \left\{\int_0^u f(x)\,dx\right\} \tag{184}$$

which shows that

$$\{f(u+v)\} = \frac{[f(v)] - [f(u)]}{r - s}. \tag{185}$$

The operator $1/(r-s)$ thus transforms $\{f\}$ into $M\{f\}$ and $[f(v)] - [f(u)]$ into $N[f(u)]$.


## 11. THE FOURIER TRANSFORM

We have seen in section 6 that

$$\{f(u)\} = \int_0^\infty \exp(-us) \cdot f(u)\,du. \tag{186}$$

From the results of section 8 we expect $\{f\}$ to be a rational operator with the algebraic properties of a meromorphic function of the complex variable z. To the integral above we can substitute the integral

$$\int_0^\infty \exp(-zu) \cdot f(u)\,du \tag{187}$$

or, if $f(u) = 0$ for $u < 0$,

$$\int_{-\infty}^{+\infty} \exp(-zu) \cdot f(u)\, du \tag{188}$$

where z is a complex variable. But such a function is completely determined in a domain Q where it is analytic if its values are known on an infinite set having a limiting point within Q. As such a set we choose the imaginary axis (provided $\{f\}$ does not have a pole on this axis). We put $z = x + iy$, $y = 2\pi\omega$, and define

$$F(\omega) = \int_{-\infty}^{+\infty} \exp(-2\pi i\omega u) \cdot f(u)\, du. \tag{189}$$

All operational properties of $\{f\}$ are preserved in this complex function; it is also easy to prove that

$$f(u) = \int_{-\infty}^{+\infty} \exp(+2\pi i\omega u) \cdot F(\omega)\, d\omega. \tag{190}$$

The transformations $f(u) \to F(\omega)$ and $F(\omega) \to f(u)$ defined by Eqs. (189) and (190) are called respectively *the Fourier transform* and *the inverse Fourier transform*. Their computation requires that the given function $f(u)$ or $F(\omega)$ be known analytically through all its domain.


## 12. THE DISCRETE FOURIER TRANSFORM

Often the function to be transformed is known only at a finite number of points, and an approximate numerical integration is the only possibility. The transformations (189) and (190) can then be substituted by

$$F(j) = \frac{1}{n} \sum_{k=0}^{n-1} \exp(-2\pi i k j/n) \cdot f(k) \qquad j = 0, 1, \ldots, n-1 \tag{191}$$

$$f(k) = \sum_{j=0}^{n-1} \exp(+2\pi i k j/n) \cdot F(j) \qquad k = 0, 1, \ldots, n-1. \tag{192}$$

For simplicity we can put

$$W_n = \exp(2\pi i/n), \tag{193}$$

where $W_n$ is the *principal n-th complex root of one*; the transformations above become

$$F(j) = \frac{1}{n}\sum_{k=0}^{n-1} f(k) \cdot W_n^{-kj}, \qquad j = 0,1,\ldots,n-1 \tag{194}$$

$$f(k) = \sum_{j=0}^{n-1} F(j) \cdot W_n^{+kj}, \qquad k = 0,1,\ldots,n-1. \tag{195}$$

These two transformations, called the *discrete Fourier transform* and the *inverse discrete Fourier transform* respectively, map the sequence f(0), f(1),...,f(n-1) into the sequence F(0),F(1),...,F(n-1), and vice-versa.

For example, given the sequence

$$f(k) = 0,1,2,3 \tag{196}$$

we compute

$$F(j) = 1/4 \cdot \left[ f(0) \cdot W_4^{-0 \cdot j} + f(1) \cdot W_4^{-1 \cdot j} + f(2) \cdot W_4^{-2 \cdot j} + f(3) \cdot W_4^{-3 \cdot j} \right] \tag{197}$$
$$j = 0,1,2,3$$

where $W_4 = \sqrt[4]{1} = i$; therefore

$$F(j) = 3/2,\ -1/2 + i/2,\ -1/2,\ -1/2 - i/2. \tag{198}$$

For the inverse transform we compute

$$f(k) = F(0) \cdot W_4^{k \cdot 0} + F(1) \cdot W_4^{k \cdot 1} + F(2) \cdot W_4^{k \cdot 2} + F(3) \cdot W_4^{k \cdot 3} \tag{199}$$
$$k = 0,1,2,3$$

by which we get back the original values of f(k).

We can show here a few of the important properties of the transformations (194) and (195). We shall use the symbol $\rightarrow$ to represent the mapping (194) of a sequence into its Fourier transform, and the symbol $\leftarrow$ to represent the mapping (195) of a sequence into its inverse Fourier transform.

*Theorem of reciprocity*: If $f(k) \rightarrow F(j)$, then $F(j) \leftarrow f(k)$. To prove this statement, first observe that

$$\sum_{x=0}^{n-1} w_n^{ax} = 0 \quad \text{if } a \neq 0$$
$$= n \quad \text{if } a = 0. \tag{200}$$

Using Eq. (194), we get

$$\sum_{j=0}^{n-1} F(j) \cdot w_n^{+kj} = \sum_{j=0}^{n-1} \left( 1/n \cdot \sum_{l=0}^{n-1} f(l) \cdot w_n^{-lj} \right) w_n^{+kj}$$

$$= 1/n \cdot \sum_{j=0}^{n-1} \sum_{l=0}^{n-1} f(l) \cdot w_n^{(k-l)j} = f(k), \tag{201}$$

which is identical with Eq. (195).

*Theorem of periodicity*: $F(j+n) = F(j)$, $f(k+n) = f(k)$. This property is easy to prove by substitution. It is useful for certain numerical computations, nevertheless it is important to remember that the so-called periodicity of $f(k)$ and $F(j)$ is purely formal; in fact $f(k)$ and $F(j)$ are sets of n terms, defined only for k and j integers from 0 to n-1. In the same context we can also write $F(-j) = F(n-j)$, $f(n-k) = f(-k)$.

*Theorem of linearity*: If $f \to F$ and $g \to G$, then $a \cdot f + b \cdot g \to a \cdot F + b \cdot G$.
*First theorem of shifting*: If $f(k) \to F(j)$, then $f(k-a) \to F(j) \cdot w_n^{-ja}$.
*Second theorem of shifting*: If $f(k) \to F(j)$, then $f(k) \cdot w_n^{ak} \to F(j-a)$.
*Transform of a transform*: If $f(k) \to F(j)$, then $F(j) \to 1/n \cdot f(n-k)$.


## 13. THE FAST FOURIER TRANSFORM

Numerical computation of the discrete Fourier transform of a sequence of n terms using formula (194) requires $n^2$ complex operations (additions and multiplications). The Fast Fourier Transform algorithm introduced by Cooley and Tukey (1965) can considerably reduce the computation time in many circumstances.

First suppose that n is an even number, and put $m = n/2$; then split the sequence $f(k)$ into the two sequences

$$f_1(k) = f(2k), \quad f_2(k) = f(2k+1), \quad k = 0, 1, \ldots, m-1. \tag{202}$$

For instance, from the sequence (196) we get

$$f_1 = [0,2], \quad f_2 = [1,3]. \tag{203}$$

The discrete Fourier transforms of the sequences $f_1$ and $f_2$ are

$$F_1(j) = 1/m \cdot \sum_{k=0}^{m-1} f_1(k) W_m^{-kj}, \quad F_2(j) = 1/m \cdot \sum_{k=0}^{m-1} f_2(k) W_m^{-kj}, \tag{204}$$
$$j = 0,1,\ldots,m-1.$$

On the other hand,

$$F(j) = 1/n \cdot \left[ \sum_{k=0}^{m-1} f(2k) W_n^{-2kj} + \sum_{k=0}^{m-1} f(2k+1) W_n^{-(2k+1)j} \right]. \tag{205}$$

Since

$$W_n^{-2kj} = \exp(-2\pi i (2k) j/n) = W_m^{-kj} \tag{206}$$

$$W_n^{-(2k+1)j} = \exp(-2\pi i (2k+1) j/n) = W_m^{-kj} W_n^{-j} \tag{207}$$

it follows that

$$F(j) = 1/n \cdot \left[ \sum_{k=0}^{m-1} f_1(k) W_m^{-kj} + W_n^{-j} \cdot \sum_{k=0}^{m-1} f_2(k) W_m^{-kj} \right], \quad j = 0,1,\ldots,n-1 \tag{208}$$

which yields

$$F(j) = 1/2 \cdot \left( F_1(j) + W_n^{-j} F_2(j) \right), \quad j = 0,1,\ldots,n-1. \tag{209}$$

This requires $(n/2)^2$ operations each for $F_1$ and for $F_2$, plus $n$ additions, for a total of $2(n/2)^2 + n = n^2/2 + n$ complex operations, versus $n^2$ if formula (194) is used; the computation time is reduced by a factor $1/2 + 1/n$.

Now suppose that $n$ is divisible by 4; in this case we can apply the same algorithm to the sequences $f_1$ and $f_2$. More generally suppose that $n$ is a power of 2; this algorithm is then applied repeatedly until only sequences of two terms are left, and no more savings are possible. Call $\phi(x)$ the number of complex operations required for the computation with this algorithm of the Fourier transform of a sequence of $x$ terms; then

$$\phi(x) = 2 \cdot \phi(x/2) + x \tag{210}$$

i.e. we have to perform $\phi(x/2)$ operations to compute each of the two sequences of $x/2$ terms each, plus $x$ additions; the solution of this equation is

$$\phi(x) = x \cdot \log_2 x + x. \tag{211}$$

The saving in computation time for a sequence of $n = 2^p$ terms is

$$\frac{n \cdot \log_2 n + n}{n^2} = \frac{p + 1}{2^p}. \tag{212}$$

Although much sequences do not have exactly $2^p$ terms, we can always complete a sequence with zeroes to reach such a convenient number of terms; the extra length of the sequence is more than compensated by the enormous savings afforded by the use of this algorithm.


## 14. VECTOR SPACES

A vector is a directed line segment; if A is its initial point and B its terminal point, the vector can be denoted by $\overrightarrow{AB}$, or simply by $\alpha$. Two vectors are considered equal if they have the same length and the same direction; in other words, parallel displacements do not change a vector.

The sum $\alpha + \beta$ of two vectors $\alpha = \overrightarrow{OA}$ and $\beta = \overrightarrow{OB}$ is the diagonal $\overrightarrow{OC}$ of the parallelogram with sides $\overrightarrow{OA}$ and $\overrightarrow{OB}$. Since $\overrightarrow{OB} = \overrightarrow{AC}$, the sum $\alpha + \beta$ can be written (triangle law)

$$\overrightarrow{OA} + \overrightarrow{AC} = \overrightarrow{OC}. \tag{213}$$

The addition of vectors is commutative and associative; there exists a vector $0 = \overrightarrow{AA}$ such that $0 + \alpha = \alpha$ for any $\alpha$; for any vector $\alpha = \overrightarrow{AB}$ there is a vector $-\alpha = \overrightarrow{BA}$ such that $\alpha + (-\alpha) = 0$.

The product $k \cdot \alpha$ of a vector $\alpha$ by a real number $k$ is a vector with the same direction of $\alpha$ and a length $k$ times that of $\alpha$. This definition is compatible with that of addition of vectors, in the sense that, if k is an integer,

$$k \cdot \alpha = \underbrace{\alpha + \alpha + \ldots + \alpha}_{k \text{ times}}. \tag{214}$$

The following properties of the product can be verified:

$$(k + 1) \cdot \alpha = k \cdot \alpha + 1 \cdot \alpha \tag{215a}$$
$$(kl) \cdot \alpha = k \cdot (l \cdot \alpha) \tag{215b}$$
$$k \cdot (\alpha + \beta) = k \cdot \alpha + k \cdot \beta. \tag{215c}$$

In a Euclidean n-dimensional space a vector $\alpha$ can be represented by an ordered set of n numbers called the *components* of the vector, thus

$$\alpha = (a_1, a_2, \ldots, a_n). \tag{216}$$

The vector

$$\beta = (b_1, b_2, \ldots, b_n) \tag{217}$$

is equal to $\alpha$ if

$$a_i = b_i, \quad i = 1, 2, \ldots, n. \tag{218}$$

The sum of vectors is thus given by

$$\alpha + \beta = (a_1 + b_1, a_2 + b_2, \ldots, a_n + b_n). \tag{219}$$

The zero vector is

$$0 = (0, 0, \ldots, 0). \tag{220}$$

Similarly,

$$\alpha - \beta = (a_1 - b_1, a_2 - b_2, \ldots, a_n - b_n). \tag{221}$$

The product of a vector $\alpha$ by a scalar k is the vector

$$k \cdot \alpha = (k \cdot a_1, k \cdot a_2, \ldots, k \cdot a_n). \tag{222}$$

The collection of all n-dimensional vectors with real components together with the two operations of addition and multiplication by a scalar is called an *n-dimensional vector space*.

A system of vectors $\alpha_1, \alpha_2, \ldots, \alpha_r$, $(r > 1)$, is *linearly dependent* if there exists a set of numbers $k_1, k_2, \ldots, k_r$, not all zero, such that

$$k_1 \alpha_1 + k_2 \alpha_2 + \ldots + k_r \alpha_r = 0. \tag{223}$$

As an example consider the n vectors

$$\begin{aligned} \varepsilon_1 &= (1,0,0,\ldots,0) \\ \varepsilon_2 &= (0,1,0,\ldots,0) \\ &\cdots \\ \varepsilon_n &= (0,0,0,\ldots,1) \end{aligned} \tag{224}$$

called *unit vectors*. They are linearly independent; in fact, the sum

$$k_1 \varepsilon_1 + k_2 \varepsilon_2 + \ldots + k_n \varepsilon_n = (k_1, k_2, \ldots, k_n) \tag{225}$$

is zero only if

$$k_1 = k_2 = \ldots = k_n = 0. \tag{226}$$

On the other hand it is easy to prove that, for $p > n$, any p vectors of an n-dimensional vector space are linearly dependent.

It follows that any n-dimensional vector can be represented as a linear combination of an arbitrary set of n linearly independent vectors; this set is called a *basis* of the vector space. We assume here as a basis the set of unitary vectors (224). Some definitions follow.

Length of vector $\alpha = (a_1, a_2, \ldots, a_n)$:

$$\|\alpha\| = \sqrt{\sum_{i=1}^{n} a_i^2}. \tag{227}$$

Thence, length of vector $\alpha - \beta$:

$$\|\alpha - \beta\| = \sqrt{\sum_{i=1}^{n} (a_i - b_i)^2}. \tag{228}$$

Angle $\phi$ between vectors $\alpha$ and $\beta$:

$$\cos\phi = \frac{\sum_{i=1}^{n} a_i b_i}{\sqrt{\sum_{i=1}^{n} a_i^2} \cdot \sqrt{\sum_{i=1}^{n} b_i^2}}. \tag{229}$$

Scalar product of vectors $\alpha$ and $\beta$:

$$(\alpha, \beta) = \sum_{i=1}^{n} a_i b_i. \tag{230}$$

The scalar product is thus the product of the lengths of the two vectors and the cosinus of their angle, i.e. the length of one vector times the projection on it of the other. It follows that

$$(\alpha, \beta) = 0 \tag{231}$$

is the condition of orthogonality of two vectors.

From the definition of scalar product the following properties follow:

$$(\alpha, \beta) = (\beta, \alpha), \tag{232a}$$
$$(k \cdot \alpha, \beta) = k \cdot (\alpha, \beta), \tag{232b}$$
$$(\alpha, \beta_1 + \beta_2) = (\alpha, \beta_1) + (\alpha, \beta_2), \tag{232c}$$
$$(\alpha, \alpha) \geq 0, \text{ with } (\alpha, \alpha) = 0 \text{ only if } \alpha = 0, \text{ i.e. } a_1 = a_2 = \ldots = a_n = 0. \tag{232d}$$

The scalar product of a vector by itself is equal to the square of its length,

$$(\alpha, \alpha) = a_1^2 + a_2^2 + \ldots + a_n^2. \tag{233}$$

If $\alpha$ and $\beta$ are two orthogonal vectors, then their sum is the vector $\alpha + \beta$; in fact, using the properties of the scalar product,

$$(\alpha+\beta, \alpha+\beta) = (\alpha,\alpha) + (\alpha,\beta) + (\beta,\alpha) + (\beta,\beta) \tag{234}$$

and from the orthogonality of the vectors it follows that

$$(\alpha+\beta, \alpha+\beta) = (\alpha,\alpha) + (\beta,\beta). \tag{235}$$

This is Pythagoras' theorem valid for two vectors lying in a plane; but the above proof is valid in any number of dimensions and for any number of orthogonal vectors; in fact for k pairwise orthogonal vectors $\alpha_1, \alpha_2, \ldots, \alpha_k$ in an n-dimensional space (with $n \geqslant k$) we can write

$$(\alpha_1+\alpha_2+\ldots+\alpha_k, \alpha_1+\alpha_2+\ldots+\alpha_k) = (\alpha_1,\alpha_1) + (\alpha_2,\alpha_2) + \ldots + (\alpha_k,\alpha_k). \tag{236}$$

## 15. HILBERT SPACE

A function $y = f(x)$ defined over the interval a, b may be described with some approximation by specifying the values $y_1, y_2, \ldots, y_n$ it assumes over a set of values $x_1, x_2, \ldots, x_n$ of its domain. The set $y_1, y_2, \ldots, y_n$ may be regarded as a vector of an n-dimensional space. The whole function f(x) may be regarded, in an analogous way, as a vector in an infinite-dimensional space.

In analogy with the n-dimensional vector space, we make the following definitions for any function specified as a vector in an infinite-dimensional space over the domain a, b.

Length of a function:

$$\sqrt{\int_a^b f^2(x)\,dx}. \tag{237}$$

Distance between two functions:

$$\sqrt{\int_a^b (f(x)-g(x))^2\,dx}. \tag{238}$$

Angle between two functions:

$$\cos\phi = \frac{\int_a^b f(x)g(x)\,dx}{\sqrt{\int_a^b f^2(x)\,dx} \cdot \sqrt{\int_a^b g^2(x)\,dx}}. \tag{239}$$

Scalar product of two functions:

$$(f,g) = \int_a^b f(x)g(x)\,dx. \tag{240}$$

If the scalar product of two functions is zero, then the angle between them is $\pi/2$, i.e. $\cos\phi = 0$; therefore two functions such that

$$(f,g) = \int_a^b f(x)g(x)\,dx = 0 \tag{241}$$

are called *orthogonal*.

If $f_1(x)$, $f_2(x)$, ..., $f_n(x)$ are n pairwise orthogonal functions, and

$$f(x) = f_1(x) + f_2(x) + \ldots + f_n(x), \tag{242}$$

then Pythagoras' theorem in Hilbert space is

$$\int_a^b f^2(x)\,dx = \int_a^b f_1^2(x)\,dx + \int_a^b f_2^2(x)\,dx + \ldots + \int_a^b f_n^2(x)\,dx. \tag{243}$$

For a function $f(x)$ to be considered as a vector in a Hilbert space it is necessary that $\int_a^b f^2(x)\,dx$ always exists.

## 16. ORTHOGONAL FUNCTIONS

In a Hilbert space consider a system of functions $f_1(x)$, $f_2(x)$, ..., $f_n(x)$ such that

$$\int_a^b f_i(x) f_j(x)\,dx = 0 \quad \text{for } i \neq j. \tag{244}$$

This system is called *orthogonal*. If in addition

$$\int_a^b f_i^2(x)\,dx = 1 \quad \text{for any } i \tag{245}$$

i.e. all functions have unit length, the system is called *orthonormal*.

For instance the functions

$$1,\ \cos x,\ \sin x,\ \cos 2x,\ \sin 2x,\ \ldots,\ \cos nx,\ \sin nx,\ \ldots \tag{246}$$

are an orthogonal system on the interval $(-\pi, +\pi)$, as is easy to verify, while the functions

$$\frac{1}{\sqrt{2\pi}},\ \frac{\cos x}{\sqrt{\pi}},\ \frac{\sin x}{\sqrt{\pi}},\ \frac{\cos 2x}{\sqrt{\pi}},\ \frac{\sin 2x}{\sqrt{\pi}},\ \ldots,\ \frac{\cos nx}{\sqrt{\pi}},\ \frac{\sin nx}{\sqrt{\pi}},\ \ldots \tag{247}$$

are an orthonormal system over the same interval.

An orthogonal system of functions is called *complete* if it is impossible to find another function orthogonal to all other functions of the system and not identically zero.

In an n-dimensional space any vector $\phi$ can be represented in the form

$$\phi = \sum_{k=1}^{n} a_k \varepsilon_k, \tag{248}$$

where the $a_k$ are n scalars and the $\varepsilon_k$ are n pairwise orthogonal unitary vectors. A similar proposition can be stated in a Hilbert space, thus: If a complete orthogonal system $f_1(x)$, $f_2(x)$, ... is given in a Hilbert space, then every function $f(x)$ in that space can be represented in the form

$$f(x) = a_1 f_1(x) + a_2 f_2(x) + \ldots = \sum_{i=1}^{\infty} a_i f_i(x) \tag{249}$$

where the series on the right hand side converges in the sense that the distance in the Hilbert space between the two functions $f(x)$ and

$$S_n(x) = \sum_{i=1}^{n} a_i f_i(x) \tag{250}$$

approaches zero as $n \to \infty$, i.e.

$$\lim_{n\to\infty} \int_a^b (f(x) - S_n(x))^2 dx = 0. \tag{251}$$

The coefficients $a_1, a_2, \ldots$ are the projections of the vector $f(x)$ on the elements of the orthogonal system, i.e.

$$a_i = (f, f_i) = \int_a^b f(x) f_i(x) dx. \tag{252}$$

Let $f_1(x), f_2(x), \ldots$ be a system of orthonormal functions and put

$$f(x) = a_1 f_1(x) + a_2 f_2(x) + \ldots + a_n f_n(x) + r_n(x), \tag{253}$$

where the coefficients $a_1, a_2, \ldots$ are given by (252); compute the scalar product of all terms with $f_i(x)$, $i = 1, 2, \ldots, n$:

$$(f, f_i) = a_1(f_1, f_i) + a_2(f_2, f_i) + \ldots + a_n(f_n, f_i) + (r_n, f_i). \tag{254}$$

Since the system $f_1(x), f_2(x), \ldots$ is orthonormal, all products on the right hand side are zero except $(f_i, f_i) = 1$ and $(r_n, f_i)$, while the product on the left hand side is $a_i$ by definition; therefore

$$(r_n, f_i) = 0, \quad i = 1, 2, \ldots, n \tag{255}$$

i.e. $r_n(x)$ is orthogonal to all functions $f_1(x), f_2(x), \ldots, f_n(x)$. Using Pythagoras' theorem for Eq. (253)

$$\|f\|^2 = \sum_{i=1}^n a_i^2 + \|r_n\|^2. \tag{256}$$

All terms in this equation are positive, therefore the sum in it converges and

$$\sum_{i=1}^\infty a_i^2 \leq \|f\|^2. \tag{257}$$

Furthermore, if the sign of equality holds, the orthonormal system is complete. Otherwise

$$\lim_{n\to\infty} \|r_n\|^2 = \lim_{n\to\infty} \int_a^b r_n^2(x) dx \neq 0 \tag{258}$$

meaning that there exists another function not identically zero but orthogonal to $f_1(x)$, $f_2(x)$, ...

If we choose as a basis the system (247), the coefficients of the series (253) are called *Fourier coefficients*. They can be expressed in the form

$$A_0 = \int_{-\pi}^{+\pi} f(x)\,dx/\sqrt{2\pi} \qquad (259a)$$

$$A_n = \int_{-\pi}^{+\pi} \cos nx \cdot f(x)\,dx/\sqrt{\pi} \qquad (259b)$$

$$B_n = \int_{-\pi}^{+\pi} \sin nx \cdot f(x)\,dx/\sqrt{\pi} \qquad (259c)$$

which gives

$$f(x) = \frac{A_0}{\sqrt{2\pi}} + \sum_{n=1}^{\infty} A_n \frac{\cos nx}{\sqrt{\pi}} + \sum_{n=1}^{\infty} B_n \frac{\sin nx}{\sqrt{\pi}}. \qquad (260)$$

This is the Fourier series expansion of function $f(x)$. Observe that $f(x)$ has been defined in a Hilbert space over the domain $-\pi$, $+\pi$; nevertheless, in a purely formal way, the expression on the right hand side of (260) can be evaluated for x outside that domain, and in this sense $f(x)$ can be viewed as a periodic function with period $2\pi$ (see section 12, theorem of periodicity).


REFERENCES

Cooley, J.W. and Tukey, J.W. (1965). An algorithm for the machine calculation of complex Fourier series, *Math. Comput.* 19, pp. 297-308.
Mikusinski, J. (1959). *Operational calculus*, Panstwowe Wydawnictwo Naukove, Warszawa.
Rescigno, A. (1984). Multidimensional operators for the construction of mathematical models. In: *Modeling and analysis in biomedicine*, C. Nicolini (ed.), World Scientific, Singapore, pp. 31-62.
Titchmarsh, E.C. (1926). The zeros of certain integral functions, *Proc. London Math. Soc.* 25, pp. 283-302.

# INTRODUCTION TO DISCRETE RECONSTRUCTION METHODS IN MEDICAL IMAGING

Max A. Viergever

*Delft University of Technology
The Netherlands*

ABSTRACT

*In this paper a brief introduction is given to algebraic reconstruction of medical images from projections.*

## 1. INTRODUCTION

Reconstruction problems in medical imaging are generally formulated in terms of an integral equation, the integral operator being the Radon transform or a closely related transform (X-ray, divergent beam, attenuated Radon; see, e.g., Lewitt, 1983; Louis and Natterer, 1983). The techniques used to solve the image reconstruction problem (i.e., to invert the integral transform), are analytical by nature; discrete versions of the inverse transform are introduced only to evaluate the resulting formulas on a computer.

In some applications, however, analytical inversion of the integral equation is not possible, or leads to unacceptably large errors. For instance, in ultrasound tomography and in microwave tomography the effects of refraction and diffraction prevent energy propagation to be modelled along straight lines, which causes loss of resolution in the images reconstructed by integral transform concepts. In electrical impedance tomography, the curvature of the equipotential lines even more strongly prohibits analytical inversion of the model equation. In both single photon and positron emission tomography, the effects of attenuation and scatter are so large that inversion of the (attenuated) Radon transform does not yield acceptable reconstructions. Severely incomplete data problems (which may occur in any image modality) neither admit the use of analytical reconstruction methods. In all of these cases, it has proven successful to consider the reconstruction problem as discrete from the very beginning (Andersen and Kak, 1984; Schomberg, 1981; Gullberg, 1979; Shepp and Vardi, 1982; Oppenheim, 1977).

The purpose of the present work is to give a brief introduction to *discrete* (or: *algebraic*) *reconstruction methods* in the context of medical imaging. The organisation of the paper is as follows. In Sections 2 and 3, it will be shown that the discrete reconstruction problem usually takes the form of a system of linear algebraic equations. Section 4 describes the properties of the system matrix in medical imaging applications, and other relevant limitations imposed by the problem environment. Section 5 discusses the two main concepts underlying approximate solutions to the system of equations. In Section 6 the concepts are converted into solution methods. It is shown that only iterative methods are suitable to solve the discrete reconstruction problem in medical imaging.

I would like to emphasize that these notes are *an introduction* to, *not a survey* of algebraic methods of (medical) image reconstruction. The interested reader is encouraged to consult the various excellent reviews on (part of) this subject, notably Herman and Lent (1976a,b), Artzy, Elfving, and Herman (1979), Herman (1980; chapters 6, 11, 12), and Censor (1981, 1983). Most of the material presented here has been derived from these sources.


2. THE DISCRETE RECONSTRUCTION PROBLEM

Assume that we are given a set of linear functionals[1] $R_i, i = 1(1)M$, which map any two-dimensional picture (i.e., a real-valued function of two variables) $\phi(x,y)$ into real numbers $R_i\phi$. Only two-dimensional (2D) pictures will be considered in these notes; the generalisation to 3D pictures is obvious, however. The functionals $R_i$ are characteristic of some *projection* process. For instance, $R_i\phi$ may be the integral of $\phi$ over a - not necessarily straight - thin strip (a *ray*). Let $g_i$ be the measured value of $R_i\phi$; in practice, $g_i$ will differ slightly from $R_i\phi$ owing to experimental error. The image reconstruction problem is to find the picture $\phi$, given measurement data $g_i$ and the functionals $R_i$.

The formulation of the reconstruction problem can be rendered discrete by introducing a fixed set of *basis pictures* $b_j(x,y)$, $j = 1(1)N$, whose linear combinations provide us with acceptable approximations $\tilde{\phi}$ to any picture $\phi$ we

---

[1] *A functional produces a real number when acting on a function.*

might wish to consider:

$$\hat{\phi}(x,y) = \sum_{j=1}^{N} f_j b_j(x,y). \tag{1}$$

The coefficients $f_j$ are real numbers which describe the picture relative to the chosen basis set (Examples of basis sets will be discussed in the next section). The linearity of the functionals $R_i$ ensures that

$$\begin{aligned}
R_i \hat{\phi}(x,y) &= R_i \sum_{j=1}^{N} f_j b_j(x,y) \\
&= \sum_{j=1}^{N} f_j R_i b_j(x,y).
\end{aligned} \tag{2}$$

In practical implementations, the numbers $R_i b_j$ are often approximated, either since the functionals $R_i$ are not known exactly (as e.g. in ultrasound imaging, because of the nonlinearity of the ray paths), or for computational reasons, an example of which is given in Section 3. Denoting the approximations by $a_{ij}$, we have[2]

$$g_i \approx R_i \phi \approx R_i \hat{\phi} = \sum_{j=1}^{N} f_j R_i b_j \approx \sum_{j=1}^{N} f_j a_{ij}. \tag{3}$$

Apparently, the discrete formulation of the imaging process may be afflicted with errors of various origin: measurement inaccuracy, discretisation of the picture, and approximation of the forward projection process.

The difference between the measured value $g_i$ of $R_i \phi$ and its counterpart $\sum_{j=1}^{N} f_j a_{ij}$ in the discrete model is called the residual $r_i$ of the projection. Equation (3) thus can be written as

$$g_i = \sum_{j=1}^{N} f_j a_{ij} + r_i \tag{4}$$

or in **matrix form**

$$g = Af + r \tag{5}$$

where g is an M-dimensional data vector, r an M-dimensional residual vector, f an N-dimensional object vector (or: image vector), and A an M×N projection matrix.

---

[2] *It is assumed that $\phi \approx \hat{\phi} \Rightarrow R_i \phi \approx R_i \hat{\phi}$. This property is sometimes called continuity of the functionals.*

The *discrete reconstruction problem* now reads: *given A and g, estimate f*. If an estimate f* of f has been found, an estimate $\phi^*$ of the discrete picture $\hat{\phi}$ (and hence of the continuous picture $\phi$) follows from

$$\phi^*(x,y) = \sum_{j=1}^{N} f^*_j b_j(x,y). \tag{6}$$

## 3. SELECTION OF BASIS FUNCTIONS

In the field of medical image reconstruction, the so-called *pixel basis* is almost invariably selected to serve as the set of basis pictures. To obtain this set we suppose that the supports of all relevant pictures lie inside a square, say $|x| \leq 1$, $|y| \leq 1$ if x and y are Cartesian coordinates. This square is divided into $N(=n^2)$ identical subsquares referred to as picture elements or *pixels* (the 3D analogue is called *voxels*), identified by an index j, $j = 1(1)N$. The pixel basis then consists of basis functions $b_j$ of the form

$$b_j(x,y) = \begin{cases} 1 & \text{inside pixel } j, \\ 0 & \text{elsewhere,} \end{cases} \quad j = 1(1)N \tag{7}$$

while the coefficients $f_j$ in Eq. (1) are defined as the average values of $\phi(x,y)$ in the corresponding pixel. The thus defined **approximation is what is commonly understood by an n×n digitization of the picture** $\phi$, although other choices of the $b_j$ likewise yield digitized versions of $\phi$.

The pixel basis is the simplest possible set of basis pictures. It has the advantage of being computationally very cheap. Firstly, the vast majority of the values of $R_i b_j$ are equal to zero in most applications, notably when $R_i$ denotes integration over a ray; and secondly, the non-zero values of $R_i b_j$ are readily evaluated since they represent the area of intersection of the i-th ray with the j-th pixel. A further reduction in computational cost can be accomplished by replacing $R_i b_j$ with either 1 or 0, depending on whether or not the i-th ray contains the centre of the j-th pixel. The projection matrix A in Eq. (5) thus becomes binary-valued, which not only simplifies the computations but also significantly reduces the storage requirements of the reconstruction algorithm.

The major disadvantage of the pixel basis is that it leads to a *discontinuous* picture representation. This might make it necessary to use a rather fine digitization so as to adequately approximate a smoothly varying image, which will at least partly undo the computational efficiency of the pixel basis. Therefore, other selections of basis pictures have been suggested. The simplest choice yielding a continuous representation is the *bilinear basis*, originally proposed by Courant (1943) and now the customary set of basis functions in finite element discretisations. The bilinear basis functions are pyramid-shaped and have a support extending over a square region the size of four pixels (the sampling grid will be the same as before). The apices of the pyramids are most conveniently chosen as either the vertices or the centres of the pixels. While a recent article reports favourably on the use of bilinear elements (Andersen and Kak, 1984), the question of whether the bilinear basis is preferable to the pixel basis is still open. More sophisticated choices of basis pictures as provided, e.g., by Fourier sampling or Karhunen-Loève sampling (Rosenfeld and Kak, 1982) are inappropriate for our purpose, because they would render evaluation of the forward projection process (that is, finding the matrix A) too time-consuming a task.

## 4. PROBLEM ENVIRONMENT

There is a large variety of methods to solve systems of linear algebraic equations. The environment of medical imaging greatly reduces the number of options, however, owing partly to the properties of the projection matrix and partly to the demands set by clinical practice.

The system matrix A of Eq. (5) has the following features:
- *large*. The number of entries of A is not seldom of the order of $10^{10}$. In most applications, M is larger than N (overdetermined system).
- *sparse*. The number of non-zero entries of A is small, typically less than 1%.
- *singular*. The rank of A will generally be less than the number of columns N.
- *unstructured*. The distribution of the non-zero entries throughout A usually has no detectable or usable structure.

Routine clinical application of the reconstruction software requires that the computation time be restricted. The time needed to acquire the projection data may be considered an upper limit. Furthermore, the computer set-up of many hospitals is still such that the algorithms must be run on a machine with low memory specifications.

The suitability of methods to solve the discrete reconstruction problem in medical imaging will be discussed in Section 6. First, however, the ideas on which solution methods are based need be outlined. This is the subject of the next section.

## 5. SOLUTION CONCEPTS

In aiming to solve Eq. (5) we have to face the fact that the residual vector r is unknown. In some applications the noise may be modelled on account of a priori knowledge, for example by specifying a stochastic distribution for r. Quite often, however, such knowledge is not at our disposal. Of course, one could simple neglect r (or, better formulated, put r equal to zero), but the ensuing inconsistency of the system of equations might prove detrimental to any solution algorithm. A more sophisticated concept of minimizing the residue in some sense is mandatory. Two approaches which pursue this idea are now discussed.

### a. *Optimization criteria*

Suppose we are interested in finding a least squares (LS) solution of Eq. (5), i.e. an f* which minimizes[3]

$$||r||^2 = ||g-Af||^2 = (g-Af)^T(g-Af) = \sum_{i=1}^{M} (g_i - \sum_{j=1}^{N} a_{ij} f_j)^2. \qquad (8)$$

A geometric illustration of this minimization is given in Fig. 1.

The LS criterion will yield a unique solution only if A has full column rank. Since A is generally singular, an additional criterion is needed to

---

[3] *Here, and in the sequel*, $||.||$ *denotes the Euclidean norm in* $R^M$ *or* $R^N$; *i.e.*
$||r||^2 = r^T r = \sum_{i=1}^{M} r_i^2$, *and* $||f||^2 = f^T f = \sum_{j=1}^{N} f_j^2$.

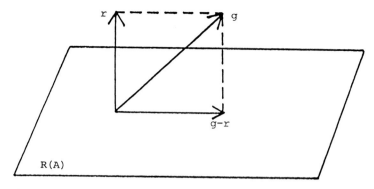

*Fig. 1. Geometric illustration of the least squares solution of an inconsistent system of equations. The space R(A), the range (or column space) of the system matrix A, is the set of all possible vectors Af. Since g does not lie in R(A), the system g = Af has no exact solution. Any residual vector r which satisfies g-r ∈ R(A) makes the system consistent and thus defines an approximate solution (or a class of approximate solutions) of g = Af. LS solutions are defined by ||r|| being minimum, which obviously is accomplished by letting r be orthogonal to R(A); accordingly, g-r is the orthogonal projection of g on R(A).*

choose between the infinite number of solutions of Eq. (8). A customary choice is to take the solution with the smallest norm ||f||, which will give us the so-called least squares minimum norm (LSMN) solution of Eq. (5).

The procedure outlined in this example is characteristic of the *optimization approach*. Some function of f is minimized (or maximized); if this yields more than one solution, a selection amongst these is made by a second minimization (or maximization) criterion. The aim is, as stated before, to minimize the residual r = g-Af in some sense.

The above presented estimation procedure (LSMN) is *deterministic*. The optimization function is generally *stochastic*, however, which means that it depends on the probability density function(s) of g and/or f (often through r).

The absence of stochastic criteria in LS estimation does not imply that the

LS criterion has nothing to do with the stochastic properties of g and f. In fact, LS estimation has a well-established probabilistic interpretation, notably in terms of maximum likelihood (ML) estimation. In order to show this, I introduce the likelihood function p(g/f), that is the conditional probability of the projection data g given an object vector f. A reasonable estimate of f is that vector for which p(g/f) is maximum at the measured projection data. This estimate is called the ML estimate of f given g. Now suppose that the residual vector r has a multivariate zero-mean normal distribution. Then the likelihood function is

$$p(g/f) = \frac{1}{(2\pi L)^{1/2}} \exp(-\frac{1}{2} r^T L^{-1} r) \qquad (9)$$

where L is the (known) covariance matrix of r. Obviously, maximizing the likelihood is equivalent to minimizing $r^T L^{-1} r$, which is weighted LS estimation. If L is the identity matrix, or differs from the identity matrix by an arbitrary multiplicative constant, we have the LS criterion (8). This proves that the deterministic LS estimation procedure depends on the stochastical properties of r, and hence that one should be careful with interpreting the results of LS estimation when the distribution of r is not known to be white Gaussian. While in such a case the LS (or LSMN) solution may still be quite useful, it no longer can be expected to have the maximum likelihood background.

The probabilistic interpolation of the LS criterion can even be extended to include prior knowledge about the image vector f, but a thorough discussion of this would address a fundamental controversy between statistician schools and thus clearly falls outside the scope of this introduction [4].

Least square minimization of the residual vector r is a special case of quadratic optimization. The general form of the quadratic minimization function for Eq. (5) is

---

[4] *If p(f) may be regarded as the probability density ascribed to f before the projections were measured (the controversy alluded to in the main text is whether it is allowed to postulate prior densities, see e.g. Bard, 1974), then the conditional density $p(f/g) = \frac{p(g/f)p(f)}{\int p(g/f)p(f)df}$ which combines this prior knowledge with the information contained in the data by means of the likelihood function, is the density we must ascribe to f after having recorded the data. Maximizing this so-called posterior density is the most frequently used type of Bayesian estimation. If p(f) is uniform, then p(f/g) is proportional to p(g/f) which provides a further probabilistic in-interpretation of LS estimation.*

$$(r-r_0)^T W_1 (r-r_0) + c(f-f_0)^T W_2 (f-f_0) \tag{10}$$

where $W_1$ and $W_2$ are positive definite[5] symmetric matrices of dimension M×M and N×N respectively, c is a nonnegative constant, and $r_0$ and $f_0$ are M- and N-dimensional vectors.

The weighting matrices $W_1$ and $W_2$, the reference vectors $r_0$ and $f_0$, and the regularizing parameter c all allow for incorporating a priori knowledge about the data and the solution in the optimization process. For instance, if one knows the probability density functions of r and f, one may take $r_0$ and $f_0$ to be the mean vectors, and $W_1$ and $W_2$ the inverses of the covariance matrices of r and f. Often, the residual is assumed to have zero mean, so that $r_0 = 0$. The parameter c determines the relative importance of the terms in (10). For low values of c the first term, representing weighted LS minimization of the residue (for $r_0 = 0$), is largest; this implies that the measurement data are considered to be reliable. For high values of c the opposite is true: the solution will only slightly depend on the data. The special case of $W_1 = W_2 = I$, $r_0 = f_0 = 0$ is known as regularized least squares estimation, or also as Tikhonov-Phillips regularization.

Minimization of (10) results in a unique solution if $c \neq 0$. In the case that $c = 0$, there will be infinitely many solutions if A is singular (which is supposedly true, see Section 4). To decide amongst these, a secondary criterion is needed. This can be the MN criterion, as before, or more generally minimizing $||Df||$ where D is a positive symmetric N×N matrix. An alternative secondary criterion is maximization of the entropy (ME)

$$- \sum_{j=1}^{N} (f_j/N\bar{f}) \ln(f_j/N\bar{f}) \tag{11}$$

which, from the minimizers of (10) with $c = 0$, will select the one having the smallest information content in order to avoid spurious features in the reconstruction. The ME criterion requires knowledge of the average picture density $\bar{f}$, which is not unrealistic since $\bar{f}$ may be estimated from the projection data in various applications, including notably CT. If $\bar{f}$ is known, another secondary criterion might be to minimize the variance $||f-\bar{f}||^2$.

---

[5] *A matrix W is positive definite on $R^N$ if $h^T W h > 0$ for every non-zero vector $h \in R^N$.*

However, since

$$||f-\bar{f}||^2 = \sum_{j=1}^{N} (f_j - \bar{f})^2 = \sum_{j=1}^{N} f_j^2 - 2\bar{f} \sum_{j=1}^{N} f_j + N\bar{f}^2 =$$
$$= ||f||^2 - N\bar{f}^2, \qquad (12)$$

minimizing the variance is equivalent with the MN criterion. Note that the equivalence requires that $\bar{f}$ be fixed, so in general minimization of the norm does not imply minimization of the variance as well.

A priori knowledge about the picture to be reconstructed can also be included in the optimization procedure in a quite different manner, viz. by means of inequality constraints that will restrict the feasibility region of the solution. An often employed constraint is to prescribe that an object vector is acceptable only if it does not contain negative elements:

$$f_j \geq 0, \quad j = 1(1)N. \qquad (13)$$

Other useful constraints may be an upper bound for the $f_j$, or an upper bound for the elements $r_i$ of the residual vector.
The latter constraint can, alternatively, be used as primary criterion. This constitutes an approach which differs fundamentally from the methods discussed, because it is not based on optimization concepts; it will be dealt with in subsection b.

Two remarks conclude this subsection. First, the secondary criteria mentioned above cannot be used as primary criteria unless additional constraints (usually inequalities) are imposed. Without such constraints the minimum norm criterion would yield the solution $f = 0$, the maximum entropy criterion the solution $f = \bar{f}$. Second, while the quadratic form (10) is quite general (e.g., under the reasonable assumption of multivariate Gaussian distributions for r and f it may be shown to include maximum posterior density Bayesian estimation, see Herman and Lent, 1976a,b), it does not comprise all optimization functions proposed in the literature. For example, maximum entropy principles cannot be brought within the framework of (10).

*b. Tolerance limits*

The second approach to arrive at a solution of Eq. (5) is to prescribe *tolerances* for all elements of the residual vector r. Customarily (in analogy with setting $r_0 = 0$ in Eq. (10)), the bounds are symmetric with respect to zero error in the projection data, which gives

$$|r_i| \leq \rho_i, \quad \text{or} \quad |g_i - \sum_{j=1}^{N} a_{ij} f_j| \leq \rho_i. \tag{14}$$

The problem here is to find suitable values for the elements of the tolerance vector $\rho$. If one of the elements $\rho_i$ is too small, (14) will allow no solution at all. However, if the $\rho_i$ are generally too large, the inequalities do not provide us with information about the solution to be obtained. A good choice of $\rho$ will yield a restricted set of solutions (the *feasibility region*), from which a unique one may be determined using one of the secondary criteria mentioned in subsection a. Obviously, additional inequality constraints, as for instance (13), can be included without difficulty in order to further restrict the feasibility region.

## 6. SOLUTION METHODS

Methods of solving the system of equations (5) can be either direct or iterative.

*Direct methods* give the solution of a minimization problem for the residual vector in explicit form (The feasibility approach of subsection 5b does not admit of straightforward solution methods). An example of how direct methods may be derived is presented in subsection 6a.

*Iterative methods* are indirect since it is not the solution that is formulated, but only a recursive relation which determines a new iterate of the solution from previous ones. The convergence of an iterative process to a specific solution as defined by either optimization criteria or tolerance limits for the residue (and possibly by additional constraints) is generally difficult to prove, if at all. Examples of iterative solution methods are given in subsection 6b.

## a. *Direct methods*

The derivation of direct solution methods for Eq. (5) is well illustrated by solving the LS minimization problem for the residue r as formulated in Eq. (8). Taking partial derivatives of $||g-Af||^2$ with respect to each of the $f_k$ of f, and equating the results to zero, gives

$$\sum_{i=1}^{M} (g_i - \sum_{j=1}^{N} a_{ij} f_j) a_{ik} = 0, \quad k = 1(1)n \tag{15}$$

which, in matrix notation, takes the form

$$A^T A f = A^T g. \tag{16}$$

So every LS solution of g = Af + r must satisfy (16). The inverse also holds: every solution of (16) is an LS solution. To see this, consider the expression

$$||g-Ah||^2 - ||g-Af||^2 = (g-Ah)^T(g-Ah) - (g-Af)^T(g-Af)$$
$$= g^T g - 2h^T A^T g + h^T A^T A h - g^T g + 2f^T A^T g - f^T A^T A f$$
$$= (h-f)^T A^T A (h-f) + 2(h-f)^T A^T (Af-g). \tag{17}$$

If f satisfies Eq. (16), we have

$$||g-Ah||^2 - ||g-Af||^2 = ||A(h-f)||^2 \geq 0 \tag{18}$$

which proves the statement. The equality sign is valid when h also is an LS solution. Consequently, the problem of finding a least squares solution of Eq. (5) is equivalent to solving Eq. (16), which is the *least squares normal equation* of g = Af + r.

If rank(A) = N, $A^T A$ is invertible and the solution of (16) is

$$f = (A^T A)^{-1} A^T g. \tag{19}$$

Generally, however, rank(A) will be less than N (see Section 4) which makes the number of solutions infinite. Following the example set in subsection 5a, we choose the minimum norm criterion to define a unique solution. It can be shown that this LSMN solution has the form

$$f = A^{\dagger} g \tag{20}$$

with $A^{\dagger}$ the Moore-Penrose inverse[6] of A.

The derivation of Eq. (20) is outside the scope of this introduction. For our purpose it suffices to know that $A^{\dagger}$ can be constructed using the singular value decomposition of the matrix A (e.g., Golub and Van Loan, 1983). Accordingly, formulas (19) and (20) both give direct estimates of the object vector f, and hence through Eq. (6) of the picture to be reconstructed.

The theoretical attractiveness of these direct reconstruction methods is more than undone by the computational costs of implementing the formulas. For example, inverting the matrix $A^T A$ in (19) by a standard technique as Gaussian elimination requires approximately $\frac{1}{3} N^3$ multiplications, which leads to exorbitant CPU times even on a large computer for customary values of N of the order of $10^4$. The implementation of (20) is even more time-consuming.

Other direct methods than the ones discussed in this example either have the same objection or give unacceptable reconstructions (as, for instance, unfiltered discrete backprojection by $f = A^T g$; see Herman, 1980, ch. 7).

*b. Iterative methods*

An iterative method to solve Eq. (5) produces a sequence of vectors $f^{(0)}$, $f^{(1)}$, $f^{(2)}$, ... with the aim to arrive at a meaningful solution; that is, a solution satisfying a - possibly constrained - optimization problem and/or feasibility problem.

While direct methods are derived quite straightforwardly, iterative methods are designed in a more heuristic fashion. The usual procedure is that an iteration scheme is devised which minimizes the residue in some sense. This scheme is tested using certain benchmark pictures and, if approved of,

---

[6] *The Moore-Penrose inverse of an M×N matrix A is a generalized inverse of dimension N×M satisfying $AA^{\dagger}A = A$, $A^{\dagger}AA^{\dagger} = A^{\dagger}$, $(AA^{\dagger})^T = AA^{\dagger}$, and $(A^{\dagger}A)^T = A^{\dagger}A$. See Ben-Israel and Greville (1974) for further details.*

applied to realistic reconstruction problems. Convergence properties are oftentimes found only much later.

I shall present one example of an iterative reconstruction method, viz. the Algebraic Reconstruction Technique (ART) for consistent systems of equations. The method was introduced into the field of medical imaging by Gordon, Bender, and Herman (1970). It is wellknown that ART is a generalization of the method proposed by Kaczmarz (1937) to solve a system of equations with a square, invertible system matrix. Less known is that ART is a form of Gauss-Seidel iteration, one of the basic methods of numerical linear algebra.

Let us consider the consistent system of equations

$$y = Bx \qquad (21)$$

where x is a vector of unknowns (dimension L), y is a data vector (dimension K), and B is a K×L matrix.

Suppose we have an initial estimate $x^{(0)}$ for the vector x. We check whether $x^{(0)}$ satisfies the first equation of the system (21). This is so if the residue

$$z_1^{(0)} = y_1 - B_1 x^{(0)} \qquad (22)$$

is equal to zero; $B_k$ is the k-th row of the matrix B, $y_k$ the k-th element of the vector y. It is assumed that $||B_k|| > 0$ for all k, which implies that measurements to which no object elements contribute are left out of account.

If (22) is satisfied, we set $x^{(1)} = x^{(0)}$. Generally, however, $z_1^{(0)}$ will differ from zero. In this case, we want to add to $x^{(0)}$ an amount $\Delta x^{(0)}$ satisfying

$$y_1 = B_1 (x^{(0)} + \Delta x^{(0)}). \qquad (23)$$

Upon combining (22) and (23) we find

$$B_1 \Delta x^{(0)} = z_1^{(0)}. \qquad (24)$$

This is one equation with L unknowns (namely the elements of $\Delta x^{(0)}$). The solution space thus is an L-1 dimensional hyperplane in $R^L$. A suitable solution is the minimum norm solution

$$x^{(0)} = B_1^\dagger z_1^{(0)} . \qquad (25)$$

The Moore-Penrose inverse of a row matrix $B_k$ is the column matrix

$$B_k^\dagger = \frac{B_k^T}{||B_k||^2} . \qquad (26)$$

So we have as new estimate of the object vector

$$x^{(1)} = x^{(0)} + \frac{(y_1 - B_1 x^{(0)}) B_1^T}{||B_1||^2} . \qquad (27)$$

The geometrical interpretation of (27) is that $x^{(1)}$ is the orthogonal projection on the hyperplane represented by the first equation of (21).

Often a slightly modified version of (27) is used. The second term on the right hand side is multiplied by a relaxation parameter $\lambda^{(0)}$ with the aim to increase the speed of convergence. The vector $x^{(1)}$ remains on the line orthogonal to the hyperplane, but only reaches the hyperplane for $\lambda^{(0)} = 1$; see Fig. 2 for an illustration.

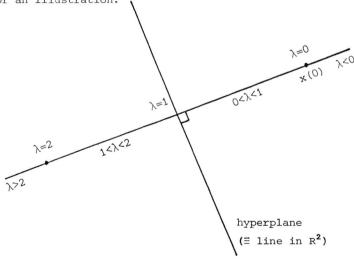

*Fig. 2. Geometric illustration in $R^2$ of the influence of the relaxation parameter.*

After thus having removed (for $\lambda^{(0)} = 1$) or decreased (for $0 < \lambda^{(0)} < 2$) the residue of the first equation of the system (21), we go to the second equation and repeat the procedure. After all equations have been addressed once, one iteration cycle has been completed. Usually it takes more than one cycle to arrive at a satisfactory solution. The general formula of the iterative scheme can be written as

$$x^{(i+1)} = x^{(i)} + \lambda^{(i)} \frac{(y_k - B_k x^{(i)}) B_k^T}{||B_k||^2} , \quad k = i \bmod (K) + 1. \tag{28}$$

It has been shown (Herman, Lent, and Lutz, 1978) that the process (28) converges to the minimum norm solution of (21), provided
- the system is consistent,
- $x^{(0)}$ lies in range$(A^T)$,
- $0 < \underline{\lim} \lambda^{(i)} \leq \overline{\lim} \lambda^{(i)} < 2$.

An illustration of the convergence is given in Fig. 3.

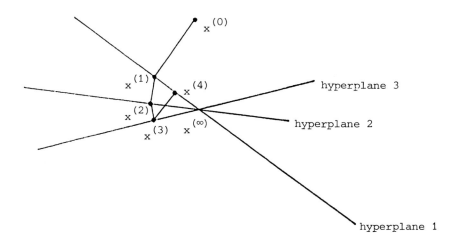

Fig. 3. Geometric illustration in $R^2$ of the way ART for equations converges (without relaxation). The system is consistent - and the solution is unique - since the lines representing the equations of the system all go through one point. Note that changing the order of the equations of the system would give a completely different sequence of estimates; although the solution to which the process converges is the same, changing the order may significantly influence the rate of convergence.

The condition that the process converges only for consistent systems implies that we cannot simply make the conversion x→f, y→g, B→A (and K,L→M,N), since the system g = Af is almost certainly inconsistent. One way to get around this difficulty is to include the components of the residual vector r among the unknowns to be estimated. In order that the thus obtained problem be meaningful, r must be constrained. To stay in line with the example used in subsections 5a and 6a, we impose the so-called least square error constraint. I.e., we require that r be orthogonal to the column space of A (cf. Fig. 1), or, by a wellknown theorem from linear algebra, that r is in the null space of $A^T$:

$$A^T r = 0. \qquad (29)$$

Another interpretation of this equation is that the backprojection of r is constrained to be zero.

The system upon which the iterative procedure is to be performed thus becomes

$$\begin{cases} g = Af + r \\ A^T r = 0 \end{cases} \Rightarrow \begin{pmatrix} A & I \\ O & A^T \end{pmatrix} \begin{pmatrix} f \\ r \end{pmatrix} = \begin{pmatrix} g \\ 0 \end{pmatrix} \qquad (30)$$

which corresponds to the following conversion of variables:

$$x \to \begin{pmatrix} f \\ r \end{pmatrix}, \quad y \to \begin{pmatrix} g \\ 0 \end{pmatrix}, \quad B \to \begin{pmatrix} A & I \\ O & A^T \end{pmatrix}; \quad K, L \to M+N, \; M+N. \qquad (31)$$

So the number of unknowns is M+N rather than N, which makes both the storage demand and the computation time larger than might be expected at first glance (It is recalled that M is usually larger than N, cf. Section 4).

The solution obtained by applying the iterative process (28) to the system (30) is the LSMN solution of Eq. (5), as is immediately obvious from the equivalence of (30) and the LS normal equation (16). Solving the latter system iteratively is much less efficient because the matrix $A^T A$ is generally not sparse. The fact that the number of unknowns is less (N) cannot nearly compensate for this.

One important issue has not been discussed yet, namely up to what point the iteration should be continued.

The data $g_i$ are only an approximation of the true projection values $R_i\phi$ (see Section 2). This measurement error induces uncertainty limits for the solution ($f^{(\infty)}$, $r^{(\infty)}$). Once an estimate ($f^{(i)}$, $r^{(i)}$) is within these limits, the process may be halted. However, finding a reasonable estimate for the uncertainty limits is far from trivial.

Furthermore, for ill-posed problems of the type encountered in image reconstruction, any non-regularized solution (as e.g. the LSMN solution) is unstable in the sense that small changes in the data g may yield large variations in the solution f (Louis and Natterer, 1983; Natterer, 1987). This behaviour occurs when the eventual solution is approached, it is not yet present in the initial iterative estimates of f. So for this type of algorithm the iterative process should be ended at an early stage not just for reasons of efficiency, but in order to avoid entering the region of instability (This, in fact, regularizes the method). But once more, it is quite difficult to determine when the instability will make itself felt. It may be concluded that the question of when the iterative process should be stopped does not yet have a satisfactory answer. While progress is being made in deriving criteria on theoretical grounds (Defrise, 1987), iterative algorithms are at present still terminated on an ad hoc basis.

The idea of removing the inconsistency of the system g = Af by considering both f and the residual vector r = g-Af as unknowns to be estimated, and simultaneously constraining r, is closely related to the optimization concept of subsection 5a. The constraint imposed on r determines to the solution of which optimization problem the iterative algorithm converges. Indeed, there are variants of ART converging e.g. to the regularized least squares solution, to the maximum likelihood solution, and to the maximum posterior density Bayesian solution of Eq. (5).

Alternatively, the inconsistency of g = Af may be dealt with according to the feasibility concept described in subsection 5b. In its most general form, the system of inequalities to be considered reads

$$Bx \leq y \qquad (32)$$

with again x an L-dimensional vector of unknowns, y a K-dimensional data vector, and B a K×L matrix. The system (32) may, for example, consist of tolerance inequalities of the form (14) and additional constraints as (13). As before, we exclude inequalities for which $||B_i|| = 0$. Because the procedure is halted as soon as an x has been found which satisfies (32), the algorithm reads:

$$x^{(i+1)} = \begin{cases} x^{(i)} & \text{if } B_k x^{(i)} \leq y_k, \quad k = i \bmod (K)+1 \\ x^{(i)} + \lambda^{(i)} \dfrac{(y_k - B_k x^{(i)}) B_k^T}{||B_k||^2} & \text{if } B_k x^{(i)} > y_k \end{cases} \quad (33)$$

with $x^{(0)}$ arbitrary.

ART for inequalities converges to a solution of (32) provided (Herman et al., 1978)

- there exists an x satisfying (32),
- $0 < \underline{\lim} \lambda^{(i)} \leq \overline{\lim} \lambda^{(i)} < 2$.

While these convergence criteria are more lenient than those of ART for equations, the convergence itself has the weakness of being unspecified; it is not necessarily optimal according to criteria discussed in subsection 5a. Figure 4 shows that the solution is not even unique, but depends on the order in which the inequalities are addressed. This non-unicity can be removed by an additional optimization criterion distinguishing between the solutions in the feasibility region. Another possibility of restricting the range of solutions is to reduce the tolerance limits during the iterative procedure, e.g. after each iteration cycle. This approach has proven very efficient in combination with a special version of (33), called ART3 (Herman and Lent, 1976a,b; Van Giessen, Viergever, and De Graaf, 1985). In fact, solutions quite similar to those obtained by (28) are produced at significantly lower computational cost.

ART is a so-called *row-action method*, because in a single iterative step access is required to only one row of the matrix; furthermore, in a single step the only iterate needed to calculate $f^{(i+1)}$ is its immediate predecessor $f^{(i)}$. Since no changes are made to the original matrix, the sparseness of this matrix is preserved which makes the calculation of the second term on the right hand side of Eq. (28) cheap. All properties together make the method pre-eminently suited for implementation on computers with low memory specifications. This is especially true for ART for inequalities because of the lesser number of unknowns (N vs. M+N in ART for equations).

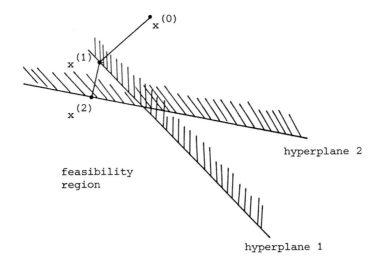

*Fig. 4. Geometric illustration in $R^2$ of ART for inequalities (unrelaxed). The algorithm stops after two iterations because then a solution inside the feasibility region has been obtained. The order in which the inequalities are addressed is even more important here than in ART for equations (see Fig. 3), because not only the rate of convergence of the process but also the eventual solution depends on this order (If $x^{(0)}$ is projected orthogonally on hyperplane 2, a different solution is obtained, in just one step).*

Instead of considering just one equation of the system per iteration step, various classes of iterative methods obtain a new estimate of the object vector by considering all equations simultaneously. I mention three examples of methods of this type: SIRT-type methods, projection methods, and the EM method.

SIRT-type methods are named after the first of such methods proposed in the field of medical imaging, the Simultaneous Iterative Reconstruction Technique (Gilbert, 1972). Similarly to ART, SIRT is a generalization of a method developed several decades before (Cimmino, 1938). In numerical linear algebra, SIRT is known as Richardson iteration.

In spite of their frequent use, SIRT-type iterative methods are suboptimal in various aspects, including notably convergence behaviour. Projection methods (as for instance the conjugate gradients (CG) method) provide a much faster convergence at only slightly increased computational cost and storage demands (e.g., Kawata and Nalcioglu, 1985; Van der Sluis and Van der Vorst, 1987).

SIRT and projection methods converge to the LSMN solution of Eq. (5). A method converging to the more generally valid ML solution is the EM (Expectation-Maximization) algorithm of Dempster et al. (1977). ML estimation is superior to LS estimation particularly in emission tomography, since the assumption of the noise being white and Gaussian breaks down when the counting statistics are poor. The EM algorithm, first applied to reconstruction of emission distributions by Shepp and Vardi (1982), can incorporate physical factors as e.g. the Poisson nature of emission, attenuation, scatter, and variation in spatial resolution directly into the model (Lange and Carson, 1984). The superior convergence properties and its flexibility have made EM a quite popular method of image reconstruction in emission tomography, in spite of the considerable computation time required as compared with non-statistical iterative methods.

As a concluding remark, it should be noted that iterative methods of image reconstruction are computationally much faster than direct algebraic methods, but much slower than methods based on analytic inversion formulas. So, in cases where the latter may be accurately used (which include **especially all** standard CT reconstructions), algebraic methods fall short. However, analytic reconstruction methods cannot be applied to a variety of problems, examples of which have been given in the introduction (Section 1). Furthermore, advanced linear algebra techniques might close or, at least, significantly reduce the gap between algebraic and analytic methods as concerns computational efficiency. In fact, remarkable results have already been obtained by reducing the tolerance limits of the inequalities (14) during the iterative procedure (as discussed earlier in this section), by optimizing relaxation (e.g., ART3 as compared with classical ART for inequalities; CG as compared with SIRT), and by partioning the system of equations (5) into blocks (Eggermont, Herman, and Lent, 1981; Herman and Levkowitz, 1987).


REFERENCES

Andersen, A.H. and Kak, A.C. (1984). Simultaneous algebraic reconstruction technique (SART): a superior implementation of the ART algorithm, *Ultrasonic Im*. 6, pp. 81-94.

Artzy, E., Elfving, T., and Herman, G.T. (1979). Quadratic optimization for image reconstruction II, *Comp. Graph. Im. Proc.* 11, pp. 242-261.

Bard, Y. (1974). *Nonlinear parameter estimation*, Academic Press, New York.

Ben-Israel, A. and Greville, T.N.A. (1974). *Generalized inverses: Theory and applications*, Wiley, New York.

Censor, Y. (1981). Row-action methods for huge and sparse systems and their applications, *Siam Rev.* 23, pp. 444-466.

Censor, Y. (1983). Finite series-expansion reconstruction methods, *Proc. IEEE* 71, pp. 409-419.

Cimmino, G. (1938). Calcolo approsimato per le soluzioni dei sistemi di equazioni lineari, *La Ricerca Scientificia (Roma)* XVI, Ser. II, Anno IX 1, pp. 326-333.

Courant, R. (1943). Variational methods for the solution of problems of equilibrium and vibrations, *Bull. Amer. Math. Soc.* 49, pp. 1-23.

Defrise, M. (1987). Possible criteria for choosing the number of iterations in some iterative reconstruction methods. *These proceedings*.

Dempster, A.P., Laird, N.M., and Rubin, D.B. (1977). Maximum likelihood from incomplete data via the EM algorithm, *J. Royal Stat. Soc.* B 39, pp. 1-37.

Eggermont, P.P.B., Herman, G.T., and Lent, A. (1981). Iterative algorithms for large partitioned linear systems, with applications to image reconstruction, *Lin. Alg. and Appl.* 40, pp. 37-67.

Giessen, J.W. van, Viergever, M.A., and Graaf, C.N. de (1985). Improved tomographic reconstruction in seven pinhole imaging, *IEEE Trans. Med. Im.* 4, pp. 91-103.

Gilbert, P. (1972). Iterative methods for the three-dimensional reconstruction of an object from projections, *J. Theor. Biol.* 36, pp. 105-117.

Gordon, R., Bender, R., and Herman, G.T. (1970). Algebraic reconstruction techniques (ART) for three-dimensional electron microscopy and X-ray photography, *J. Theor. Biol.* 29, pp. 471-481.

Golub, G.H. and Van Loan, C.F. (1983). *Matrix computations*, North Oxford Academic, Oxford.

Gullberg, G.T. (1979). *The attenuated Radon transform: Theory and application in medicine and biology*, Ph.D. thesis, Univ. of California, Berkeley.

Herman, G.T. (1980). *Image reconstruction from projections. The fundamentals of computerized tomography*, Academic Press, New York.

Herman, G.T. and Lent, A. (1976a). Quadratic optimization for image reconstruction I, *Comp. Graph. Im. Proc.* 5, pp. 319-332.

Herman, G.T. and Lent, A. (1976b). Iterative reconstruction algorithms, *Comput. Biol. Med.* 6, pp. 273-294.

Herman, G.T., Lent, A., and Lutz, P.H. (1978). Relaxation methods for image reconstruction, *Comm. ACM.* 21, pp. 152-158.

Herman, G.T. and Levkowitz, H. (1987). Initial performance of block-iterative reconstruction algorithms. *These proceedings*.

Kaczmarz, S. (1937). Angenäherte Auflösung von Systemen linearer Gleichungen, *Bull. Int. Acad. Pol. Sci. Lett. A*, pp. 355-357.

Kawata, S. and Nalcioglu, O. (1985). Constrained iterative reconstruction by the conjugate gradient method, *IEEE Trans. Med. Im.* 4, pp. 65-71.

Lange, K. and Carson, R. (1984). EM reconstruction algorithms for emission and transmission tomography, *J. Comp. Ass. Tom.* 8, pp. 306-316.

Lewitt, R.M. (1983). Reconstruction algorithms: Transform methods, *Proc. IEEE* 71, pp. 390-408.

Louis, A.K. and Natterer, F. (1983). Mathematical problems of computerized tomography, *Proc. IEEE* 71, pp. 379-389.

Natterer, F. (1987). Regularization techniques in medical imaging. *These proceedings*.

Oppenheim, B.E. (1977). Reconstruction tomography from incomplete projections. In: *Reconstruction tomography in diagnostic radiology and nuclear medicine*, M.M. ter Pogossian et al. (eds.), Univ. Park Press, Baltimore, pp. 155-183.

Rosenfeld, A. and Kak, A.C. (1982). *Digital picture processing*, Vol. 1, Academic Press, Orlando.

Schomberg, H. (1981). Nonlinear image reconstruction from projections of ultrasonic travel times and electric current densities. In: *Mathematical aspects of computerized tomography*, G.T. Herman and F. Natterer (eds.), Lecture notes in medical informatics 8, Springer-Verlag, Berlin, pp. 270-291.

Shepp, L.A. and Vardi, Y. (1982). Maximum likelihood reconstruction for emission tomography, *IEEE Trans. Med. Im.* 1, pp. 113-122.

Sluis, A. van der and Vorst, H.A. van der (1987). Numerical solutions of large, sparse linear algebraic systems arising from tomographic problems. In: *Seismic tomography*, G. Nolet (ed.), Reidel, Dordrecht, pp. 53-87.

IMAGE STRUCTURE

Jan J. Koenderink
*Utrecht University*
*The Netherlands*

ABSTRACT

*This paper presents a theoretical introduction into image structure. The topological structure of scalar and vector images is described. Scale-space is treated in some depth, including the problem of sampling and that of canonical projection. Objects are defined by way of a theory of (physical) measurements. Their properties (shape of the boundary, skeleton, natural subparts) are defined and their topological changes over varying ranges of resolution explored. A short section on the theory of local and global operators in scale-space is provided.*

1. THE SUPERFICIAL STRUCTURE OF IMAGE

a. *The definition of an image*

The physical (not psychological!) notion of an "image" presupposes a "substrate" or "image space" as the carrier of some "field". The image space may be *discrete* (e.g. 256 × 256 pixels) or *continuous* (e.g. the unit square of $R^2$). The field is often an illuminance (i.e. a scalar), but it may equally well be a *property list* or a *vector*. (A list of 2 scalar properties is an ordered pair of reals, i.e. an element of $R \times R$, whereas a two-dimensional vector is an element from $R^2$). It is essential that the image space has a *topological structure*, otherwise one has no image but merely an *indexed set* (Koenderink, 1984a, b; van Doorn et al., 1984). If the image space is some subset of the Euclidean plane we use the usual topology, for discrete spaces one has to specify the connectivity (e.g. the neighbourhood of a pixel). Then an "image" is a map from the image space to the space of values of the field (e.g. for the illuminance the positive reals $R^+$; for a digitized image a subset of the cardinals). We will often use a "scalar image"

$$L : R^n \to R^+ \quad (n = 2, 3) \tag{1}$$

or a "vector image"

$$V : R^2 \rightarrow R^2 \tag{2}$$

as important special cases.

These constructs are still too general. One would like an explicit notion of *scale* as in the discrete, digitized image whose entropy (in the informationtheoretical sense) is well defined. Thus we need the *size* of the image, the image *resolution* (size and resolution together specify the number of *degrees of freedom* of the image), and the *accuracy* (the tolerance on the value of the field) as well as the *range* of values the field may assume (accuracy and range define the entropy per degree of freedom).

*b. Local versus global properties*

We sharply discriminate between *point properties* (or *local* properties) and *global* or *multilocal properties*. The illuminance at a pixel is a local property; the histogram of illuminances is a global entity. In many cases we will augment the image with e.g. its partial derivatives. For instance, one may store not only the illuminance but also the components of the illuminance gradient at a pixel (in a "record"): in such a case a "zero crossing" (zero of the Laplacean) is a global property, whereas a stationary point (zero gradient) is a local (or point) property. This dichotomy is crucial because point properties can be the input for "point processors" having no topological expertises whatsoever, whereas global properties need "array processors" (e.g. neighbourhood operations or worse).

Consider an image $L(x,y)$. All functions $H(x,y)$ which coincide with $L$ in all partial derivatives up to the nth order are in the equivalence class known as the "n-jet of L at $(x,y)$" or $j^n_{(x,y)}$. The "n-jet extension of L" is an image $L^{(n)} : R^2 \rightarrow J_n$ where $J_n$ is the "jet space" of n-jets of functions from $R^2$ to $R^+$. All properties that can be expressed in terms of partial derivatives up to the nth order are point properties of the augmented image; e.g. "boundary curvature" is a point property of $L^{(2)}$, not of $L^{(1)}$. Note that $J_n$ has dimension $(n+1)(n+2)/2$.

Similar considerations apply to vector images (e.g. integral curves are global entities, singular points are local entities). In this case the difference is even more pronounced since it may be impossible to compare vectors at different points! (One needs a "connection" or constant field to be able to do that). An example is a vector image with pixels representing samples of points on the unit sphere $S^2$ at $\pi/2$ apart as the discrete image space: there appears to be no way to define a connection! (One cannot comb the hair on a sphere!).

*c. The notion of infinitesimal stability or genericity*

Consider an image $L(x,y;p,q,r,...)$ that depends not only on $(x,y)$ (in "image" or "configuration" space) but also on $(p,q,r,...)$, a set of *control parameters* (a point of "control space"). Suppose we have some way of deciding whether two images $L(x,y)$, $L'(x,y)$ are "equivalent" (or qualitatively identical). In practice we use topological classifications which map $L(x,y)$ on a discrete set (e.g. the number and configuration of stationary points). Then L is *generic* if we have infinitesimal stability: the classification $L(x,y;p+\delta p, q+\delta q, r+\delta r,...)$ does not depend on $(\delta p, \delta q, \delta r,...)$ but changes only for finte perturbation. For instance, $L(x;p) = x^4/24 + px^2/2$ is generic for $p \neq 0$. For $p = 0$ the example is singular: infinitesimal perturbations yield a pair of minima at $x_{1,2} = \pm\sqrt{-6p}$ and a maximum at $x = 0$, or a single minimum at $x = 0$.

Note that the notion of genericity depends on:

- what changes we are prepared to tolerate;
- which perturbations we want to consider.

Next suppose we want to study a *one parameter family* of images (e.g. a time series). Then we will meet singular images whereas the *family as a whole* may be reasonable. For instance, $L(x;p) = x^4/24 + px^2/2$ is reasonable as a family; all members are generic except that for $p = 0$. Such families are called *versal*. In our applications it is sufficient to check whether singular cases for one parameter families occur at isolated points in the (one dimensional) control space: such families are versal.

Closely connected with the notion of genericity is that of *transversality*. We will need it a few times. Two submanifolds (e.g. points and curves in the plane; points, curves and surfaces in $R^3$) are transversal if one of the tangent spaces is not contained in the other. Thus two curves are transversal if their tangents do not coincide at points of intersection, etc.

Precise definitions take much care and may be found in the literature. The present introduction will do here. The reader may want to consult: Bruce and Giblin (1984), Thom (1972), Milnor (1963), Guillemin and Pollack (1974), Golubitski and Guillemin (1973), Eells (1967) and (especially) Poston and Stewart (1978).

*d. Morse functions*

Consider an image $L(\vec{r})$. At almost any point $\vec{r}$ all partial derivatives will be finite and dL will be a nice linear approximation. At some points the gradient dL may vanish, however. Such a point is called a *critical point* and there the image assumes a *critical value*. A Morse function is a function with the following properties:
- all critical points are isolated and there are finitely many of them;
- all critical values are distinct.

The class of Morse functions is very large, in fact an infintesimal transformation will take any image to a Morse function. At a critical point a Morse function can be approximated with the matrix of second derivatives, the *Hessian*. For instance, in 2-D:

$$L(x+\delta x, y+\delta y) = L(x,y) + L_x \delta x + L_y \delta y \quad \text{at regular points} \tag{3}$$

$$L(x+\delta x, y+\delta y) = L(x,y) + \frac{1}{2} \begin{pmatrix} \delta x \\ \delta y \end{pmatrix} \cdot \begin{pmatrix} L_{xx} & L_{xy} \\ L_{xy} & L_{yy} \end{pmatrix} \begin{pmatrix} \delta x \\ \delta y \end{pmatrix} \quad \text{at critical points} \tag{4}$$

The invariants of the Hessian (in 2-D the trace $L_{xx} + L_{yy}$ and the determinant $L_{xx}L_{yy} - L_{xy}^2$) can be used to *classify* the critical point. For instance, in 2-D:

$$L_{xx}L_{yy} - L_{xy}^2 < 0 \quad \text{a saddle ("hyperbolic" point)} \tag{5a}$$

$$L_{xx}L_{yy} - L_{xy}^2 = 0 \quad \text{a "parabolic" point} \tag{5b}$$

$$L_{xx}L_{yy} - L_{xy}^2 > 0 \qquad \text{an extremum} \begin{bmatrix} L_{xx} + L_{yy} > 0 & \text{a minimum} \\ L_{xx} + L_{yy} < 0 & \text{a maximum} \end{bmatrix}. \qquad (5c)$$
$$\text{("elliptic" point)}$$

The parabolic points lie (generally) on curves that are either closed or run into the boundary and do not end or intersect.

*e. Hills and dales*

Consider the image $L(x,y)$ and the *level set* $L_a = \{\vec{x} \in R^2 | L(\vec{x}) \leq a\}$. The level set $L_a$ is a smooth "bounded manifold" with boundary $L^{-1}(a)$. Let $(a,b)$ be an interval of regular levels. If $d,e \in (a,b)$, then $L_d$ en $L_e$ can be deformed continuously into each other (are homologous). The boundary of $L_a$ is an *"isoluminance curve"*. The orthogonal trajectories of the isoluminance curves are the integral curves of the gradient $(L_x, L_y)$; the curves of *steepest ascent*. These curves issue forth from the *minima* and end up in the *maxima* of L. Consider a minimum and the paths of steepest ascent that leave it. One will have a number of *sectors* at the minimum such that each path in a certain sector will end up at a certain maximum belonging to that sector. The sectors are separated by singular paths, the *separatrices*. A separatrix runs into a *saddle point* and *bifurcates* there into two paths that run into two different maxima. Each saddle point receives paths from two minima. A path (from minimum to maximum) that is *not* a separatrix may be called a *regular path*. All regular paths connecting a certain minimum with a certain maximum subtend a *cell*. All cells with the separatrices that end at a maximum subtend the *hill* corresponding to that maximum and the hills subdivide the image. The boundaries of the hills are "water courses". All cells with the separatrices that issue forth from a minimum constitue a *dale* and (like the hills) the dales subdivide the image. The boundaries of the dales are "water sheds". Thus we obtain two dual partitions of the image. These partitions may be represented by adjacency graphs. In one partition the minima are the vertices of the graph, in the other the maxima. In both cases the saddles (or passes) are the edges. Note that one may have "loose edges" in this graph, or mazes bounded by a single edge. This canonical tesselation was proposed by Maxwell (1890) as a decomposition of the landscape into "natural districts" by means of water sheds and water courses.

At a regular point the level set is infinitesimally stable.

At a critical point the level set of the critical value is *not stable*, however. At an extremum the level set near the critical point is either empty or an area including the point. At a saddle the connectivity of the level set changes at the critical level. If the threshold of the level set is continuously raised one has a *merge* at the saddle for the critical value, when it is lowered a *split*. The level curves are an alternative way (relative to the "districts") of partitioning the image. (They provide a "foliation", or partition of the image into arbitrarily thin strips). Note that there exist two globally distinct saddles, one type surrounding two hills (or two dales), the other surrounding a hill and a dale (Koenderink and van Doorn, 1979).

The partition into natural ditricts is not equivalent to the foliation into isoluminance curves. Except from the districts one needs (partial) information about the *order* of the critical values in order to sketch the foliation.

In practice one may want both representations, e.g. a level set may be used to *define an object*. Then a water shed may be used to define a *ridge* which is similar to a "**medial** axis" or "skeleton" of that object, and a water course may define *"natural part boundaries"*. Since these entities are very different in a computational sense they are best treated as **complementary**. Of course one wants to link the two descriptions for a "higher order" shape description.

The foliation easily lends itself to a *partial ordering* (or inclusion) of hills and dales. A saddle may be labelled "false extremum" and be treated as such. For instance, two summits may form a mountain together and can be labelled by their common saddle, etc. Of course this is merely a *partial* order: quite large gray scale deformations will leave the inclusion order invariant.

*f. Change of the topology of the level sets under variation of threshold*

Consider a point $(x,y)$ with value $L(x,y) = L^*$. We consider the level sets at levels $L^* + \delta L$ in a neighbourhood $(x+\delta x, y+\delta y)$ of the point. If the point is *regular* we have as equation for the level sets:

$$L_x \delta x + L_y \delta y - \delta L \geq 0 \tag{6}$$

which describes a *plane* in ($\delta x$, $\delta y$, $\delta L$) space, or a *moving line* in image space ($\delta L$ the "time"). If the point is *critical*, we have:

$$1/2 (L_{xx} \delta x^2 + 2 L_{xy} \delta x \delta y + L_{yy} \delta y^2) - \delta L \geq 0 \tag{7}$$

which is a *quadric* in ($\delta x$, $\delta y$, $\delta L$) space. In suitably rotated and scaled coordinates we have:

$$\tfrac{1}{2} \sigma_1 \delta u^2 + \tfrac{1}{2} \sigma_2 \delta v^2 - \delta L \geq 0 \quad \text{with } \sigma_{1,2} = \pm 1. \tag{8}$$

If $\sigma_1 \sigma_2 = 1$ we have either an expanding *blob* or a shrinking *bubble* in $(u,v)$-space (depending on the sign of $\sigma_{1,2}$). If $\sigma_1 \sigma_2 = -1$ we have a *neck*.

In 3-D images the possibilities are greater because we now have (in suitable coordinates):

$$\tfrac{1}{2} \sigma_1 \delta u^2 + \tfrac{1}{2} \sigma_2 \delta v^2 + \tfrac{1}{2} \sigma_3 \delta w^2 - \delta L \geq 0. \tag{9}$$

We now can have a *"tunnel"* event. (A worm hole forming or collapsing).

g. *Vector images; general*

Consider a vector image $(u(x,y), v(x,y))$. The topological structure is best revealed by study of the Jacobian

$$J = \begin{pmatrix} \frac{\delta u}{\delta x} & \frac{\delta u}{\delta y} \\ \frac{\delta v}{\delta x} & \frac{\delta v}{\delta y} \end{pmatrix}. \tag{10}$$

The condition for degeneracy is $\frac{\delta u}{\delta x} \frac{\delta v}{\delta y} - \frac{\delta u}{\delta y} \frac{\delta v}{\delta x} = 0$ (determiniant of the Jacobian vanishes). This is known as a *fold*. Reason is that the image in "velocity space" folds over there. Example: the image $(x^2/2, y)$. The Jacobian is $\begin{pmatrix} x & 0 \\ 0 & 1 \end{pmatrix}$ with determiant $x$. The determinant vanishes for $x = 0$. Note that the image space is mapped on the half space $x \geq 0$ in velocity space. Worse things happen if the value of the determinant is stationary along the fold. In that case one obtains a *cusp*, e.g. $(xy - x^3/6, y)$. The fold is $y = \frac{x^2}{2}$ with a cusp at the origin.

Whitney (1955) has shown that fold and cusp are the *only* generic singularities in this case. They are *stable* against small deformations and thus very robust and conspicuous features of the vector image.

## h. *Vector images; potential fields*

Vector images possessing a *potential* are a special case. Let $(u(x,y), v(x,y))$ be a vector image with potential $\Phi(x,y)$. (Thus $u = \Phi_x$, $v = \Phi_y$). Then the condition for the *fold* is $u_x v_y - u_y v_x = 0$ or $\Phi_{xx}\Phi_{yy} - \Phi_{xy}^2 = 0$. Thus the folds of the vector image are the parabolic curves of the potential.

If we form the *augmented potential* $\Psi(x,y;u,v) = \Phi(x,y) - ux - vy$, then the condition $d\Psi = 0$ yields $u = \Phi_x$, $v = \Phi_y$. Thus the critical points of $\Psi$, solved for u, v yield the vector image. Similarly the vanishing of the Hessian of $\Psi$ yields the folds. Thus the conditions

$$\Psi_x = 0 \tag{11}$$

$$\Psi_y = 0 \tag{12}$$

$$\Psi_{xx}\Psi_{yy} - \Psi_{xy}^2 = 0 \tag{13}$$

solved for u,v yield the fold locus in (u,v)-space. This representation is useful if we want to apply *catastrophy theory*: we can find the generic singularities by looking up the case of 2 state and 2 control variables in Thom's (1972) table. This is especially important if we consider a one parameter family of vector images (parameter t): $\Psi(x,y;u,v,t)$. Then we have 2 state and 3 control variables. The generic singularities are: fold, cusp, swallowtail and hyperbolic and elliptic umbillic. This classification allows us to foresee the possible changes of folds and cusps when we vary the control parameter t.

Interesting applications have been made concerning the drift of ice in the northern polar seas (from satellite data) by Thorndyke et al. (1978) and by Nye and Thorndyke (1980). These authors provide additional useful theory.

An application of potential vector images is to *solid (3D) shape* (Koenderink and van Doorn, 1986). Consider a 3-D surface $z(x,y)$. The surface normals are given by $(-z_x, -z_y, 1)$. They intersect the plane $z = -1$ in the points

($z_x$, $z_y$). Thus the height function z is the potential for the space of normals (the *Gaussian image*). Folds correspond to inflections or parabolic curves of the surface. Cusps are special points such that in an infinitesimal neighbourhood of such a point one has three parallel normals. The previous classification allows us to categorize the possible events when we deform the shape through a one parameter process (e.g. blurring).

## 2. THE REPRESENTATION OF IMAGES

### a. *Inner and outer scale*

Consider the difference between a typical *tele* and a typical *wide angle* photograph. The difference is twofold:
- the tele view shows *less* than the wide angle view, at least if the pictures were taken with a single camera with different lenses, its *scope* is smaller. We may take scissors and *trim* the wide angle view to the same scope as the tele view. Then their *outer scales* are made equal. Given the pictures we can only trim, not add;
- the tele view shows *more* detail than the wide angle view, at least if the pictures were taken with a single camera with different lenses, its *resolution* is better. We may *blur* the tele view to give it the same resolution as the wide angle view has. Then their *inner scales* are made equal. Given the pictures we can only *blur*, not sharpen.

Thus *inner and outer scale* are two very practical parameters of an image (Koenderink, 1984). Together they define the number of degrees of freedom of an image (constant for a given camera if the resolution is determined by the photographic material rather than by the lenses). We consider three natural operations that change the degrees of freedom:
- *trim* an image, i.e. leave out parts and keep the *region of interest*;
- *blur* an image, e.g. smooth out "noise" and keep the *relevant detail*;
- *zoom in* on the image, i.e. decrease the region of interest but at the same time increase the resolution so as to keep the number of degrees of freedom fixed.

Note that the latter procedure works only if the image is really obtained from a much larger one (e.g. a $512^2$ blurred image taken from a GEOS weather satellite image with $15000^2$ pixels). This notion of "zooming" is different from the one current in computer graphics, which does *not* increase the resolution while decreasing the region of interest.

In practice the situation is often complicated through gray scale limitations (e.g. a newspaper picture with its binary image that looks like a gray scale one). Such cases are not difficult to handle but we will not consider them here.

In order to study an object we must set the inner and outer scale suitably. For instance, when studying people we want to see them completely (outer scale 2m), but we are not interested in their cellular structure (inner scale 1mm, say); whereas when studying chromosomes we want a much smaller inner scale (fraction of a micron) but also a much smaller outer scale (just a few microns). If we do not get our scales right we do not solve the forest-and-the-tree-problem.

b. *Graded inner scale*

There is no reason why an image should be restricted to a single inner or outer scale. In fact, we are all acquainted with *atlasses* with maps of the world, continents, countries, cities, etc. They are simple to use: one starts at a wide scope with low resolution and "zooms in" (just by flipping pages) until one has the finest map that contains our region of interest. Although such a system does not allow us to e.g. measure the distance between the Eiffel Tower and Buckingham Palace we may measure the distance between Paris and London (or Buckingham Palace and Trafalgar Square). Roughly speaking an atlas yields a *fixed relative accuracy*. You will have seen "powers of ten" which shows the universe in a few pages: it does not take much if you do not insist on a depiction of everything small. Practical people have a world atlas and acquire detailed maps of only the places they intend to visit. The human eye uses a similar mechanism: it codes everything on a coarse inner scale and only codes fine detail when you look at a thing. Roughly the eye takes a *series of images* with different inner and outer scale but the same number of degrees of freedom, all centered around the line of sight. A change of visual direction completely changes the high resolution images but hardly changes the wide angle view, thus providing an invariant context (Koenderink and van Doorn, 1978).

This *zoom-lens principle* is also very practical in image processing, in fact it is the only way to process really large images such as the GEOS weather satellite data or blood smear mounts.

Note that such a system is the complete opposite of the so-called *pyramid representations* nowadays so popular in image processing (Rosenfeld, 1984). In the pyramid representation the outer scale is left invariant and we consider a family of images graded with respect to inner scale.

*c. The building blocks*

In *pointillistic* images the building blocks are *records* (point descriptions) at *points* in image space. In many cases this is a very useful way to represent the image. However, nothing prevents us from taking other bases. Any invertible linear transformation of the image would do as well. For instance, one might consider the Fourier transform. Then the building blocks are global rather than local. If you change one Fourier component you change the image all over the place. Nevertheless such bases have their uses, e.g. you may leave out high frequency components without much noticeable effect, whereas setting as many pixels to zero would be immediately apparent. One may also take a middle way and e.g. take "local" (patch wise) Fourier transforms. (It seems that the visual system might do just that). What method is best?

Basically what one wants is:
- a *logical decomposition*, i.e. the building blocks (or multiples of them) should be *natural* in some sense. In anatomy one does not *slice* a cadaver (a neat, regular decomposition if there ever was one!) but instead decomposes it into *organs*. Comparative anatomy shows the sense of this (Koenderink and van Doorn, 1981);
- a *graceful deterioration under truncation*, i.e. a compression of the image (by just dropping some of the building blocks) should not produce garbage. Pointillistic images are very bad here. Fourier transforms score much better for many natural images;
- *robust behaviour with respect to perturbation*, i.e. if I slightly change the geometry of image space or the gray scale, the description should not change very much. (For instance, all home TV-sets are very different and e.g. their Fourier transforms no doubt vary all over the place, yet everyone seems to be happy). Anatomical dissection is completely insensitive to even large perturbation: one easily maps an elephant on a giraffe, organ by organ!

Note that both pointillistic and Fourier techniques score pretty badly on most counts! The only representations that get in the ballpark set by anatomical description are the *topological* ones; for instance, the partition of the image into natural districts, the skeletons and natural part boundaries, the folds and cusps of the vector images. Such descriptions are completely insensitive to even drastic changes in gray scale or geometric deformation. In practice we may want to have the best of everything, which is possible if we tolerate a *redundant representation*. An example based on the human visual system (as guesstimated from electrophysiology and psychophysics) is:

- *a pointillistic basis of records*. This is the most transparent base for geometrical problems and takes advantage of the fact that the world is globally a mere conglomerate of disjunct entities;
- records containing local *Fourier like* coefficients. This makes it easy to exploit translation invariances, do fast searches or correlations, simple neighbourhood operations, etc. This takes advantage of the fact that the world is locally rather homogeneous: entities have extension;
- higher order *topological descriptions* (so-called Gestalt principles, Metzger, 1975). This takes advantage of the fact that entities (because of "common fate") are rather homogeneous and tend to stop "catastrophically" (in a topological sense) where other regimes take over.

3. SCALE SPACE

a. *The problem of correspondence*

The advantages of having images at different inner scales are clear enough (think of an atlas). A problem is how to use them together, as an ensemble. For instance, one wants to be able to identify features on different images. Clearly this is a *one way process*. Anything on a coarse scale image must have a *cause* to be found in a fine scale image but the converse need not be true at all.

How can we define the correspondence for e.g. a scalar image $L(x,y; t)$? We require *metrical identity*, i.e. $L(x',y'; t') = L(x,y; t)$ and *structural proximity*, i.e. $[(x' - x)^2 + (y' - y)^2]$ must be a local minimum. One easily solves the least squares problem

$$\partial\{(x' - x)^2 + (y' - y)^2 + \lambda(L(x', y'; t') - L(x,y; t))\} = 0$$

with Lagrange multiplier $\lambda$:

$$x' = x - \frac{L_x L_t}{L_x^2 + L_y^2}(t' - t)$$

$$y' = y - \frac{L_y L_t}{L_x^2 + L_y^2}(t' - t).$$
(14)

Thus the integral curves of the vector field

$$\vec{s} = (-L_t L_x, -L_t L_y, L_x^2 + L_y^2)$$
(15)

in $(x,y,t)$ space define the correspondences or "projections". Note that the result is not defined if and only if the gradient vanishes ($L_x^2 + L_y^2 = 0$). At such places the vector field may be expected to *bifurcate*. This is where the causality condition comes in: *bifurcations only make sense if the projection is one-too-many for increasing resolution for almost all paths.*

This condition is a strong one. Note that the projections are constrained to lie on surfaces of constant L in $(x,y,t)$-space. In fact, they are curves of steepest ascent on these surfaces. If the surface has a horizontal tangent plane (t = cst) and lies on one side of that plane at some point, then we have a bifurcation; a whole "umbrella" of paths meets at the critical point. This means that such a "dome" must turn its concave side to the direction of decreasing t (greater resolution). The condition is simple: $L(x,y; t) = L_0$ is a surface with principle curvatures

$$k_i = \frac{\lambda_i}{\sqrt{L_x^2 + L_y^2 + L_t^2}} \quad (i = 1, 2)$$
(16)

where the $\lambda_i$ are the roots of the (quadratic!) equation

$$\det \begin{vmatrix} L_{xx} - \lambda & L_{xy} & L_{xt} & L_x \\ L_{xy} & L_{yy} - \lambda & L_{yt} & L_y \\ L_{xt} & L_{yt} & L_{tt} - \lambda & L_t \\ L_x & L_y & L_t & 0 \end{vmatrix} = 0 \qquad (17)$$

or (because we consider the case $L_x = L_y = 0$, $L_t \neq 0$):

$$\lambda^2 - \lambda(L_{xx} + L_{yy}) + (L_{xx}L_{yy} - L_{xy}^2) = 0. \qquad (18)$$

Now the determinant of the Hessian $(L_{xx}L_{yy} - L_{xy}^2)$ is positive by hypothesis (we considered an extremum), thus both roots have equal sign. The sign is given by the trace $(L_{xx} + L_{yy})$ relative to the sign of the surface normal $(L_t)$. Thus the constraint boils down to: $L_t(L_{xx} + L_{yy}) > 0$ whenever $L_x^2 + L_y^2 = 0$ and $L_{xx}L_{yy} - L_{xy}^2 > 0$.

This will certainly be true if

$$L_t = L_{xx} + L_{yy} \quad \text{or} \quad \Delta L = L_t \qquad (19)$$

which is the *diffusion equation* or heat conduction equation.
We have shown that for any solution $L(x,y; t)$ of the diffusion equation the projections bifurcate only in the direction of increasing resolution. Thus these solutions really capture our intuitive notion of a set of mutually blurred images. Moreover, the diffusion equation is the only linear, rotation and shift invariant partial differential equation that is of the first order in t and satisfies the requirement. (And we require all these conditions for obvious reasons, e.g. because we require causality).

Consequently, if I have a master image $L(x,y; 0)$ I can *embed* it in one parameter family $L(x,y; t)$ by using it as the *boundary condition* for a solution of the diffusion equation. The canonical *projections* are the integral curves of the vector field

$$\vec{s} = (-\Delta L \cdot L_x, -\Delta L \cdot L_y, L_x^2 + L_y^2). \qquad (20)$$

## b. *The diffusion equation*

I can write the diffusion equation as

$$\vec{g} = \text{grad } L, \tag{21}$$

$$\text{div } \vec{g} = \frac{\partial L}{\partial t}. \tag{22}$$

From this one sees that (using Stokes theorem) for a volume V with boundary $\partial V$ the flux of $\vec{g}$ appears as the increase of the integrated illuminance:

$$\oint_{\partial V} \vec{g} = \frac{\partial}{\partial t} \int_V L. \tag{23}$$

Thus L is "conserved" and $\vec{g}$ is its "current". On the other hand the current is just the gradient of L, thus L is the "potential" for the current. Thus *maxima* tend to *decrease* whereas *minima increase* because L is transported from the one to the other (at infinite speed!). It follows that *zero loci are strictly removed by blurring*: for adding a zero locus (a loop in x-y space) one has to create an extremum whereas these can only be attenuated. Similarly, one cannot create new extrema through blurring.

An elementary solution of the diffusion equation is:

$$K(x,y; x_0, y_0, t) = \frac{e^{-\frac{(x-x_0)^2 + (y-y_0)^2}{4t}}}{(\sqrt{4\pi t})^2} \tag{24}$$

(written in a form easily generalized to other dimensions). One may form the one parameter family by convolution with this kernel:

$$L(x,y; t) = K * L(x,y; 0) \tag{25}$$

and because of the *concatenation theorem*

$$K(\ldots, t) * K(\ldots, s) = K(\ldots, t+s) \tag{26}$$

one may even do the process in *stages*.

Note that the kernel $K(x,y;t)$ *averages* over an area with diameter $d = 4\sqrt{t}$, that is over a number of pixels proportional to $t$. Thus we may expect the width of the illuminance histogram to diminish proportional to $t^{-\frac{1}{2}}$: the image becomes less articulated under blurring.

This kernel ("a Gaussian") is *automatically* obtained if one iterates convolution with an *arbitrary symmetric everywhere positive kernel*, which may be seen easily in the frequency domain. The arbitrary kernel of width $\Delta$ will have a spectrum $S(\omega) = 1 - \alpha\Delta^2\omega^2 + \ldots$ (if normalized). If I take the N-fold convolution of the $N^{\frac{1}{2}}$ times shrunken kernel I obtain the spectrum $S(\omega; N) = (1 - \alpha\Delta^2\omega^2/N + \ldots)^N$, which in the limit for N to infinity approaches $\exp(-\alpha\Delta^2\omega^2)$, i.e. the Fourier transform of the kernel

$$\frac{e^{-\frac{x^2}{4\alpha\Delta^2}}}{2\Delta\sqrt{\pi\alpha}}.$$

Thus the popular "pyramids" approach the solution to the diffusion equation in a very rough way.

### c. *The bifurcations*

Let us investigate the **bifurcations** in somewhat greater detail. The canonical projection is

$$\vec{s} = (-L_t L_x, -L_t L_y, L_x^2 + L_y^2) \qquad (27)$$

which vanishes whenever $L_x^2 + L_y^2$ vanishes. If the gradient vanishes and the Hessian is negative ($L_{xx}L_{yy} - L_{xy}^2 < 0$) the result depends on the magnitude of the trace of the Hessian ($\Delta L$ or $L_t$). If the trace does not vanish the situation is quite simple: almost all the paths are normal except those on two surfaces made up of orbits through the critical point locus in $(x,y,t)$-space ($L_x = L_y = 0$). An example is $L(x,y;t) = x^2 - \frac{1}{2}y^2 + t$. The locus of critical points is $x = 0$, $y = 0$, i.e. the t-axis. The singular paths are the integral curves of $\vec{s} = (-2x, y, 4x^2 + y^2)$ through the t-axis and are parabolas that lie in the planes $x = 0$ and $y = 0$.

If $L_t$ vanishes the situation is more complicated. Generically this happens only at isolated points in $(x,y,t)$-space. An example is

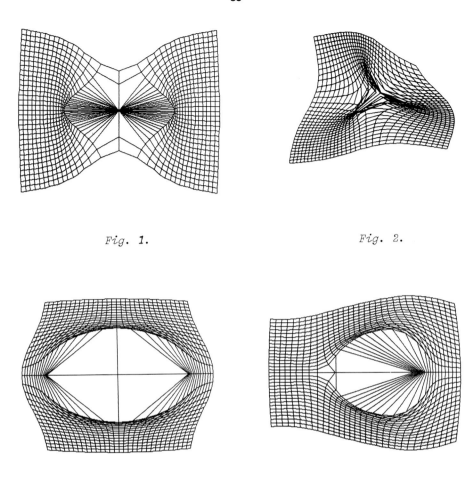

Fig. 1. Projection from a coarse to a fine image plane of a regular grid for a patch that contains a regular saddle.

Fig. 2. Projection from a coarse to a fine image plane of a regular grid. At an intermediate image plane a saddle at which the Laplacian vanishes exists. (The case of Eq. (28)).

Fig. 3. Projection from a course to a fine image plane of a regular grid. The patch contains an extremum. Note that the projection cannot reach the central region.

Fig. 4. Projection from a coarse to a fine image plane of a regular grid. This figure illustrates the model of Eq. (29).

$$L(x,y; t) = 1 + xy + \frac{1}{6}(a_1 x^3 + 3a_2 x^2 y + 3a_3 xy^2 + a_4 y^3) +$$
$$+ \ldots + ((a_1 + a_3)x + (a_2 + a_4)y)t + \ldots \tag{28}$$

Keeping only second order terms we have $L(x,y; t) = 1 + xy + Axt + Byt$. The foliation of scale through the surfaces $L$ = constant is a family of quadrics. For $L = 0$ one has a real cone, for $ABL > 0$ a family of hyperboloids of one sheet, for $ABL < 0$ a family of hyperboloids of two sheets. In fact, we have just two families of the previous type tied together at the origin. (There the "fork" of the bifurcation becomes infinitely thin).

One obtains a good impression of the correspondence if one maps a plane $t$ = cst on another one. The mapping can be appreciated well if we let it deform a regular grid. For instance, the map $\vec{s} = (-2x, y, 4x^2 + y^2)$ maps a certain line piece (inside out!) on another line piece but it is regular outside these horrible areas (Fig. 1, 2).

When the determinant of the Hessian is positive its trace cannot vanish. In such cases we obtain the umbrella like projections (Fig. 3). The mapping maps a point on an area but is regular outside it. This area (in the fine grained image) contains illuminances that have no match in the coarse grained image.

When the determinant of the Hessian vanishes we get the generic singularity according to the catastrophy classification. The local model is

$$L(x,y; t) = L_0 + x^3/6 + y^2/2 + (x+1)t. \tag{29}$$

The surface $L$ = cst is a "shoe-surface", the origin is some kind of average between the saddle and an extremum. The map projects a line piece on a closed closed loop, part of the loop being the image of one endpoint on the line piece. Everything inside the loop finds no match (Fig. 4, 5).

In a computer implementation one may link pixels at a fine level of resolution to pixels at a coarser level with the same or at least as close as possible illuminance within as short a distance as possible. (Here we have lots of room for awkward problems!). The pattern of pointers will be like the canonical projections except that we link right through bifurcation points. In that case the areas that find no match will link to the singular

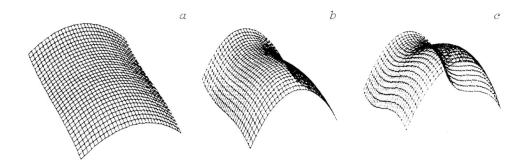

*Fig. 5. An illustration of the "shoe-surface" described by Equation (29). The surfaces L=constant in (x,y,t)-space (with t increasing upwards) are shown as transparent wire meshes. In Fig. 5a the surface is completely regular (so the gradient of L does never vanish). In Fig. 5b the critical case (t=0) is illustrated. There exists one point at which the gradient of L vanishes. Figure 5c illustrates the case in which both a saddle and an extremum exist.*

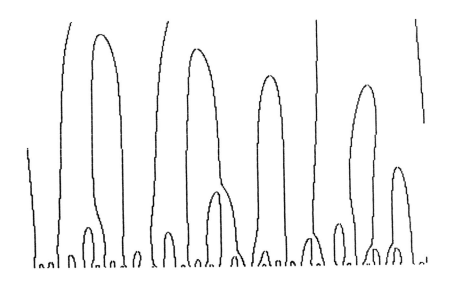

*Fig. 6. Zero crossings in one dimensional scale space (L(x,t)= 0 in (x,t) space; t increases upwards, x horizontal). The characteristic nesting and juxtaposition of "tubes" is evident. Note that all tubes have their openings turned downwards.*

paths through the "shoe-surface" singularities. The umbrellas all link up to the orbits of extrema to form "showers".

The geometry of pointers (the canonical projection) in itself specifies the image and can replace the illuminance! This can be shown as follows. If two images $L_1(x,y; t)$, $L_2(x,y; t)$ have the same canonical projection the surfaces $L_{1,2}$ = cst must coincide. But then they must be equal except for an arbitrary gray scale transformation, say $L_1(x,y; t) = \Phi(L_2(x,y; t))$. Now $L_1$ satisfies the diffusion equation, hence

$$\Phi'' \cdot ||\nabla L_2||^2 + \Phi' \cdot (\Delta L_2 - L_{2_t}) = 0 \tag{30}$$

but $L_2$ also satisfies the diffusion equation, thus

$$\Phi'' \cdot ||\nabla L_2||^2 = 0. \tag{31}$$

Consequently, $\Phi'' = 0$, which means that $L_1$ is just a scaled version of $L_2$ with an additive constant. If we always scale the gray scale to a given interval (say (0,255)), then the pointer structure alone determines the image completely!

d. *Spectral descriptions*

Note that $A \cdot \exp(i\vec{k} \cdot \vec{r} - k^2 t)$ satisfies the diffusion equation. This indicates how fast gratings with wave vector $\vec{k}$ damp out. The spatial frequency at which the grating is attenuated to 50% for a given resolution t is

$$f_{50\%} = 0.11 \ldots/\sqrt{t}$$

Suppose one starts with a Gaussian random signal with a white spectrum (for t = 0). Then the power spectrum at any level of resolution t is $\exp(-2k^2 t)$. From this one simply calculates the density of zero crossings and the density of extrema (for a one dimensional image): there are $0.036 \ldots/t$ zeros per unit length and $0.088 \ldots/t$ extrema. Thus for an outer scale L the image will be featureless if t > 0.088L. The relative number of extrema that can be expected to vanish equals the relative increase of t.

*e. The image deep structure*

In the superficial structure we had a partial order induced by the nesting of isoluminance curves. There was no way of ordering e.g. two summits together forming a "false maximum". Scale space permits the introduction of such an ordering: one of the tops will blur away first (coalesce with the saddle) and thus reveal itself as a subsidiary of the other.

Each extremum (perhaps except for a single one) will eventually coalesce with a saddle point. A certain "tube" of scale space will eventually link up to the singular path. The volume within the tube finds no match at resolutions below the critical one. One might call the tubes the *realms* of the extrema. Many realms coexist, both juxtaposed and nested to arbitrary depth. (Because the surfaces of constant illuminance foliate scale space the boundaries of the tubes can never penetrate each other). We have a veritable *partial order of light and dark blobs* (Figs. 6, 7).

For a certain finite range of resolution the blobs can be *identified* (that is, if t is less than the value at which the extremum meets its saddle point), and in a still more limited range the blob exists in its pure form, unarticulated. For too high a resolution the blob may be difficult to detect because it is articulated with smaller detail (e.g. blurring really helps to find objects in scintigrams), whereas too low a resolution the blobs loose identity (e.g. in a cardio-scintigram the left and right ventricles may merge). Details thus have a limited range of resolution in which they can be said to exist. We can define this range from the top of the realm to the next lower top of any included subrealm.
Some details exist over a long range of resolution, others are more ephemeral and at once desintegrate once you identify them. There is some evidence that "stable features" (those that exist over long ranges) are the visually most conspicuous ones.

Note that you cannot "reconstruct" the original image from a highly blurred one through the device of downprojection: certainly this sharpens or "deblurs" the image, but at the cost of introducing blank spaces (points that find no match). Thus you have to bring in extra information at the levels of resolution where - by downprojection - new realms appear.

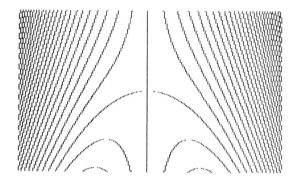

*Fig. 7. Orbits of projection in one dimensional scale space ($L(x, t) = cst$ in $(x, t)$-space) for the simple image $L(x, t) = x^3/6 + xt$.*
*Note the bifurcation (at $x = 0$, $t = 0$): orbits from the upper border can reach only part of the lower border.*

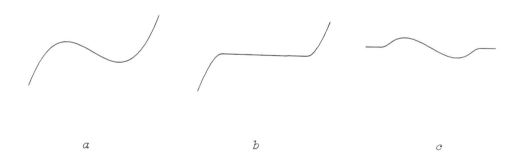

      *a*                 *b*                 *c*

*Fig. 8. Logical filtering is demonstrated for the image the scale space structure of which is illustrated in Fig. 7.*
*Figure 8a shows the image; Fig. 8b the low resolution structure; Fig. 8c the high resolution detail. The sum of the low and high resolution images equals the original image.*

This representation permits *logical filtering in the scale domain*. For every range in the primal image plane you may solve $\Delta L' = 0$ within the range with the boundary value $L' = L$ on the boundary of the range. Then you may "lift off" the detail by defining it as $L - L'$ within the range and zero outisde. In this way the whole primal image can be written as a superposition of light and dark blobs. A subfamily may be defined for each subimage, and because the diffusion equation is linear the original family is just the superposition of the subfamilies. Now you may choose, for insance, to use only summands belonging to features existing in a certain range of scales. This is in effect a logical filtering in the scale domain. You may even compose images in which details in different scale ranges have been blurred differently, etc. (Fig. 8).

## 4. SAMPLING OF THE IMAGE AND OF SCALE SPACE

### a. *General*

An image that is strictly bandlimited at a highest frequency $f_{max}$ can be completely reconstructed from samples spaced at intervals $1/2f_{max}$, the so-called Nyquist theorem. If the image is not strictly bandlimited, but merely of a low-pass character, we can still assign a suitable sample frequency such as to stay below a certain error criterion. For instance, a Gaussian spectrum $\exp(-\omega^2 t)$ has an energy fraction $\exp(-2\Omega^2 t)$ above the frequency $\Omega$. If this fraction is not to exceed some value $e^{-\xi}$, then we may take a sample interval $\pi\sqrt{2t/\xi}$. This indicates how to sample scale space in the planes $t = cst$.

How should one sample scale space in the $t$ dimension? Several approaches are possible. One is to notice that sinewave gratings attenuate at a rate $\exp(-\omega^2 t)$. Thus the fastest rate is $\exp(-\Omega^2 t)$. The spectrum is ($\Phi$ a circular frequency in the t-spectrum domain) $\Omega^2/(\Phi^2 + \Omega^4)$ and has a power fraction $\exp(-\xi)$ above the frequency $\Phi = \frac{2\Omega^2}{\pi} e^{\xi}$. This yields a sampling interval of $\frac{1}{2\Phi} = 2\pi^3 t \ e^{-\xi}/\xi$. Thus the sampling interval is proportional with $t$, which means that one has to sample uniformly on a logarithmically scaled t-axis.

Another "natural" scaling of $(x,y,t)$-space is obtained if we introduce

$\eta = 1/t$. The sampling cells in $(x,y,\eta)$-space have a volume

$$\Psi = \frac{2\pi^3}{t} \frac{e^{-\xi}}{\xi} \cdot (\pi\sqrt{2t/\xi})^2 = 4\pi^5 \frac{e^{-\xi}}{\xi^2} = \text{constant}. \qquad (33)$$

A volume defined by the outer scale L, inner scale $\delta$ (thus an effective t-range of about $\delta^2$ to $L^2$) has an x-y-$\eta$ volume $V = L^2/\delta^2$ and consequently contains $\chi$ degrees of freedom:

$$\chi = V/\Psi = \frac{L^2 \xi^2}{4\pi^5 \delta^2} e^{\xi}. \qquad (34)$$

(This amounts to 6.43 ... $L^2/\delta^2$ for $\xi = 5.55$ ... that is for 8-bit accuracy). Since the original image contains $(L/\delta)^2$ degrees of freedom, scale space contains about 6.43 times as many for a typical accuracy. We see that scale space is not really very much larger than the original image. This is in accord with the "pyramid" constructs, where a $2^N \times 2^N$ pixel image yields a pyramid with $(4^{N+1}-1)/3$ degrees of freedom, i.e., about 33% more than the original image. The pyramids can be seen to undersample scale space severely.

That one has to sample uniformly on a logarithmically scaled t-axis makes sense from the point of view that in the absence of prior information there can be no *preferred scale*. Thus our operations should not depend on any arbitrary unit of length. Then the logarithmic scale is the only solution.

Instead of introducing $\eta = 1/t$ we might have scaled the x,y-axes according to t, i.e., we might have introduced new coordinates

$$u(x,y,t) = \frac{1}{\pi} \sqrt{\frac{\xi}{2t}} \cdot x$$
$$v(x,y,t) = \frac{1}{\pi} \sqrt{\frac{\xi}{2t}} \cdot y \qquad (35)$$
$$w(x,y,t) = \frac{\xi}{2\pi^3} e^{\xi} \cdot \ln t$$

In this space the elementary cells are unit cubes, thus the space can be uniformly sampled. In fact, this is a continuous approximation to the pyramid structures. The "outline" of the space is (for an outer scale L):

$$u, v < \text{cst} \cdot L \cdot e^{-w/2} \qquad (36)$$

Thus the space is exponentially "tapered".

## b. Physical sampling

Instead of thinking of scale space as a *mathematical fiction*, we can use it to define a *physical theory of measurement*. Consider a cloud, which is a somewhat localized region with a high concentration of condensed water droplets in the atmosphere (sizes from ca 3-30 μ for a continental cumulus cloud with a few hundred droplets per cubic centimeter (Mason, 1962)). What is the *density* of condensed water (in mass per volume)? The answer depends critically on the volume of measurement (the "sampling aperture"). Clearly a mathematical definition such as:

$$\text{density} = \lim_{\text{volume} \downarrow 0} \frac{\text{mass inside volume}}{\text{volume}} \qquad (37)$$

leads to *useless* results: the result is anything between zero and the density of liquid water, moreover it varies wildly over space. The correct physical alternative is:

$$\Phi(\vec{r},t) = \frac{\text{mass inside volume of diameter } \sqrt{t}}{\text{volume of diameter } \sqrt{t}} \qquad (38)$$

(no limit!). The density $\Phi$ depends both on position ($\vec{r}$) and on the resolution t. If our sampling volumes are Gaussian windows, then $\Phi$ will satisfy the diffusion equation and the *physical density* will have the structure of scale space. (In practice, the sampling volume will be of different shape, e.g. cubical, which introduces *spurious resolution* but does not affect the main argument).

What then is a cloud? Obviously, mathematics cannot be expected to yield an answer. We have to ask the meteorologist, who has to define a cloud on the basis of some physical measurement. The most reasonable one is to set a limit to the amount of condensed water *and* on the inner scale. (The amount of condensed water will always be large inside an isolated droplet however small it may be, but a single droplet is no cloud!). Thus the definition may be: a cloud is a region of space where the amount of condensed water measured in a resolution interval $(t_{min}, t_{max})$ exceeds $\Phi_0 \text{gm}^{-3}$ (0.4 $\text{gm}^{-3}$ for a small continental cumulus cloud).

Let us introduce the dimensionless function

$$\phi(\vec{r},t) = \Phi(\vec{r},t)/\Phi_0(\vec{r},t) \tag{39}$$

which indicates a cloud (if $\sqrt{t}$ is in the range of $10^{-2}$ to 10m) if its value exceeds unity. Then "the cloud" would be the *level set* given by the characteristic function

$$\begin{aligned}\chi(\vec{r},t) &= 1 \quad \text{if} \quad \phi(\vec{r},t) > 1 \\ \chi(r,t) &= 0 \quad \text{otherwise.}\end{aligned} \tag{40}$$

In some cases we would like to retain some measure of "how cloudy" the cloud really is, e.g. if $\chi = 1$ one could have $\phi = 2$ or $\phi = 10$. To this end we construct a *fuzzy characteristic function* $\mu$:

$$\mu(\vec{r},t) = 1 - \exp(-\phi(\vec{r},t)\ln 2). \tag{41}$$

Note that $\chi = 1$ whenever $\mu > 1/2$ and that $0 \le \mu \le 1$ whereas $\mu$ increases monotonically with $\phi$. The *boundary region* $\{\vec{r}|\tfrac{1}{4} < \mu < \tfrac{3}{4}\}$ defines a measure as how "sharp" the cloud really is (Koenderink and van Doorn, 1986).

### c. *Properties of physical densities*

Note that the normalized density $\phi(\vec{r},t)$ lends itself to set theoretical operations. For example, we might be interested in *treetops* which may be defined in the same manner as clouds (density based on the weighting of biomass inside sampling apertures). Consider how to combine a pear tree with an apple tree. Suitable definitions of intersection and union are:

$$\begin{aligned} P &= \{\vec{r} \mid \phi_p(\vec{r},t) > 1\} \\ A &= \{\vec{r} \mid \phi_a(\vec{r},t) > 1\} \\ P \cap A &= \{\vec{r} \mid \phi_p \cdot \phi_a > 1\} \\ P \cup A &= \{\vec{r} \mid \phi_p + \phi_a > 1\} \end{aligned} \tag{42}$$

Note that $P \cap A = A \cap P$ and $P \cup A = A \cup P$, moreover, $A \cap A = A$ and $P \cap P = P$. But notice that $A \cup A \supset A$! (We could have defined $P \cup A = \{\vec{r}|(\phi_p + \phi_a)/2 > 1\}$ but in that case $A \cup \emptyset \ne A$ which is even worse!). Our construction is quite nice if we are prepared to live with the fact that many set theoretical theorems apply only in attenuated form (e.g. the "law of absorption" $A \cup (A \cap B) = A$ becomes $A \cup (A \cap B) \supseteq A$). An added complication is that we have no concept of *universal set* (that would be a tree with infinite biomass filling the universe, clearly an unphysical concept), thus no notion of *complementation*.

When we study the *shape* of the level sets over scale space we find that their topology is not conserved over variations of the resolution. (Clear enough a cloud will disperse into droplets if you raise the resolution far enough! On the other hand, distinct clouds will coalesce into an unbroken cloud cover if the resolution is lowered by a suitable amount). The possible events can be neatly categorized, however. We have (Fig. 9):
- *emergence processes:* when one raises the resolution a blob or a bubble (hole) emerges out of the blur. This process *never* happens the other way around (no emergence on blurring), the structure of scale space is a guarantee for that;
- *accretion processes:* two blobs or two bubbles meet and coalesce via a "neck". This process runs *both ways*, thus on either blurring or deblurring one has both *merges* and *splits*;
- *versification processes* (only in 3D): A blob or a bubble suffers a "punch through" and obtains a tunnel like hole (like a doughnut), or conversely, a tunnel like hole collapses (a doughnut becomes a mere holeless blob). This process also runs *both ways*.

c. *Ridges and troughs; seams*

Because we have a natural projection in scale space we may follow the singular points of the accretion and versification processes (e.g. the points on the blobs where the "neck" arises) and project them as "scars" or "seams" on the blobs at other levels of resolution. This process is extremely valuable because it induces *natural object boundaries*. For instance, a dumpbell will have a seam encircling the neck dividing two *natural subobjects*.

The seams are closely related to the ridges and troughs of the density. (Remember that the ridges and troughs are the boundary of the natural districts induced by the gradient). The ridges and troughs meet the boundary of the level set in points (2D) or curves (3D) that define "crests" and "indentations" on the surface. When the accretion process occurs the seam coincides with a crest or indentation. The indentations have been implied as natural object boundaries by Hoffman and Richards (1982). *Inside* the

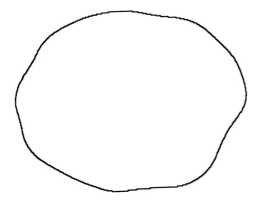

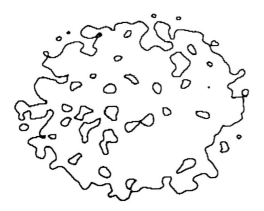

Fig. 9. Several snapshots of a simultated "cloud" or "tree-top" at different levels of resolution. At very low resolution the shape tends to a circular blob whereas at very high resolution it disperses into a swarm of pointlike specks. Emergence processes (both of the blob and the bubble type) and accretion processes (both merges and splits) are evident when you compare successive stages of the process. The figures span a range of t values of about two and a half decades.

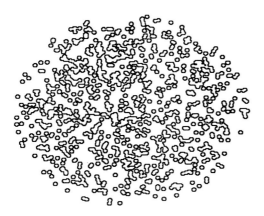

level set the *ridges* are a useful *skeleton* of the object (close to the "medial axis") whereas the troughs are natural boundaries of subobjects.

If we follow the object through scale space we see the pattern of subobjects and the skeleton changing in discrete steps. (For instance, branches in the skeleton will form and break). This induces a *discrete description* of the object useful for recognition purposes.

*d. The shape of the boundary*

The shape of the boundary of a level set can be obtained directly from the density. For instance, consider a blob in 2D given as

$$\phi(x,y) = 1. \tag{43}$$

When we differentiate two times with respect to x we obtain

$$\phi_x + \phi_y \cdot y' = 0 \tag{44}$$

$$\phi_{xx} + 2\phi_{xy} \cdot y' + \phi_{yy}(y')^2 + \phi_y \cdot y'' = 0 \tag{45}$$

from which we have

$$y' = -\phi_x/\phi_y \quad \text{the } direction \text{ of the boundary} \tag{46}$$

$$y'' = \frac{\phi_{xx} \cdot \phi_y^2 - 2\cdot\phi_{xy}\cdot\phi_x\cdot\phi_y + \phi_{yy}\cdot\phi_x^2}{\phi_y^3}. \tag{47}$$

Taking coordinates such that $y' = 0$ (which is always possible) we find $y'' = -\phi_{xx}/\phi_y$, i.e. the *curvature* of the boundary. Note that $K(x,y) = \phi_{xx}/\phi_y$ can be calculated at almost any point (excepting those where the gradient vanishes). Thus we find a "curvature" that makes sense even *outside* the boundary: it is *completely insensitive to faulty boundary detection*.

When we follow the shape throughout scale space we observe changes of the curvature pattern (e.g. inflections are destructed on blurring). For instance, in the 2D case *pairs of inflections* are created or destructed. The stretch of boundary between two inflections that on blurring coalesce may be called the *realm* of the inflection pair. When we follow the shape through scale space we obtain a *realm tree* which starts at a single node (when there is just a convex blob left). When new pairs of inflections are created the realms articulate farther and farther. Such a simple description cannot be used for 3D: there the inflections are *closed loops* (parabolic curves). A realm tree can again be constructed based on the areas enclosed by the loops, but this construction misses the articulation of the loops itself (the *cusps*). Because loops may exchange cusps (by way of the *hyperbolic umbilic*) a simple tree description is not suitable.

## 5. GEOMETRICAL ROUTINES

### a. Partial derivatives

The geometrical description of *shapes* depends in a large part on combinations of partial derivatives of the density at a point. (Example: boundary curvatures as $\phi_{xx}/\phi_y$). Thus the partial derivatives are of paramount importance, a reason why it is so useful to employ *jet extensions* of the image: at each *pixel* we store a *record* containing all partial derivatives up to a certain order. The record in fact is a *local* description of the density: it defines a truncated Taylor series which allows us to study the function in a *neighbourhood* of the pixel.

In scale space we require the partial derivatives for each level of resolution. It is an interesting fact that they can all be obtained from a *single (high resolution) image* by convolution with kernels that may well be called "fuzzy derivatives".

Consider the following set of solutions of the 1D diffusion equation:

$$\phi_n(x; t) = \frac{\partial^n}{\partial x^n} \frac{e^{-\frac{x^2}{4t}}}{\sqrt{4\pi t}} = \left(\frac{-1}{2\sqrt{t}}\right)^n H_n\left(\frac{x}{2\sqrt{t}}\right) \frac{e^{-\frac{x^2}{4t}}}{\sqrt{4\pi t}} \tag{48}$$

where $H_n(x)$ denotes the n-th order Hermite polynomial defined recursively as:

$$H_0(x) = 1 \tag{49}$$

$$H_1(x) = 2x \tag{50}$$

$$H_n(x) = 2x\, H_{n-1}(x) - 2(n-1)\, H_{n-2}(x). \tag{51}$$

Let $*$ denote convolution, then we have for any function $f$:

$$\frac{\partial^n}{\partial x^n}[f(x) * \phi_0(x;t)] = \frac{\partial^n f}{\partial x^n} * \phi_0 = f * \phi_n \tag{52}$$

which can be interpreted as follows: $f * \phi_0$ is the function $f$ blurred to resolution $t$. Thus the *n-th derivative of the blurred function equals the convolution of that function with* $\phi_n$. This allows us to refer to $\phi_n$ as the "n-th order fuzzy derivative". It behaves in all respects like a differentiation, e.g. Leibnitz' rule is

$$(f \cdot g) * \phi_1 = f * \phi_1 \cdot g * \phi_0 + f * \phi_0 \cdot g * \phi_1. \tag{53}$$

Concatenation leads to higher derivatives, but you have to watch the resolution, which also increases. We have a *concatenation theorem:*

$$\phi_n(x;t) * \phi_m(x;s) = \phi_{n+m}(x;t+s). \tag{54}$$

Derivatives of the fuzzy derivatives satisfy simple rules:

$$\frac{\partial}{\partial x}\phi_n(x;t) = \phi_{n+1}(x;t)$$

$$\frac{\partial}{\partial t}\phi_n(x;t) = \phi_{n+2}(x;t) \tag{55}$$

(from which we see that $\frac{\partial^2}{\partial x^2} = \frac{\partial}{\partial t}$, i.e. the $\phi_n$ satisfy the diffusion equation).

Let $a_k(t) = \int_{-\infty}^{+\infty} f(x)\, \phi_k(x;t)\, dx$, then the Taylor expansion of the blurred function is

$$f * \phi_0 = \sum_{k=0}^{\infty} \frac{a_k}{k!} x^k. \tag{56}$$

Less obvious, but easily proven from the orthonormality of the parabolic cylinder functions, is the formula:

$$f(x) \cdot \phi_0(x; t) = \sum_{k=0}^{\infty} \frac{a_k}{k!} 2^{k-1} t^{k-\frac{1}{2}} \phi_k(x; t) \tag{57}$$

which allows you to run the diffusion backwards in a small window set by $\phi_0(x; t)$.

In the spectral domain we have the simple representation:

$$\phi_n(\omega; t) = (i\omega)^n e^{-\omega^2 t} \tag{58}$$

Thus the $\phi_n$ are "bandlimited differentiators". All the previous theorems are really trivial in the spectral domain.

The $\phi_n(x; t)$ are similar to sine (odd n) and cosine (even n) functions of circular frequency $\omega_n \sqrt{\frac{1}{t} [\frac{n+1}{2}]}$, modulated with a Gaussian envelope of half-width $4\sqrt{t}$. The bandwidth of the spectral representation is asymptotically equal to $1/2\sqrt{t}$; thus the spatial width-spectral width product ("uncertainty relation") equals unity: these functions are very similar to "Gabor functions".

In the spectral domain the following notation is particularly useful (Fig. 10):

$$\begin{aligned}
S_m(x; t) &= \omega_m \phi_{2m+1}(x; t) & &\text{("sine")} \\
C_m(x; t) &= \phi_{2m+1}(x; t) & &\text{("cosine")} \\
\text{with } \omega_m &= \sqrt{\frac{m+1}{t}} & &\text{("frequency")} \\
A &= \sqrt{S_m^2 + C_m^2} & &\text{("amplitude")} \\
\eta &= \text{arctg }(S_m/C_m) & &\text{("phase")}
\end{aligned} \tag{59}$$

For a local disturbance $\delta(x)$ the amplitude behaves like $\exp(-x^2/8t)$ whereas the phase varies in the interval $(-m\pi, +m\pi)$, with slope proportional to $\omega_m$ at the origin. Thus *amplitude ratios are independent of position whereas phase angles are proportional with position.*

Two (and higher) dimensional partial derivatives are obtained through multi-

*Fig. 10. A sine-like and cosine-like fuzzy-derivative of the same frequency and amplitude. Figure 10b shows the sum of the squares of the even and odd function, which resembles a Gaussian. (It is asymptotically a Gaussian for very high frequency).*
*These functions behave very much like the trigonometric functions (m cycles of the sine and cosine) modulated with a Gaussian envelope of frequency-independent width.*

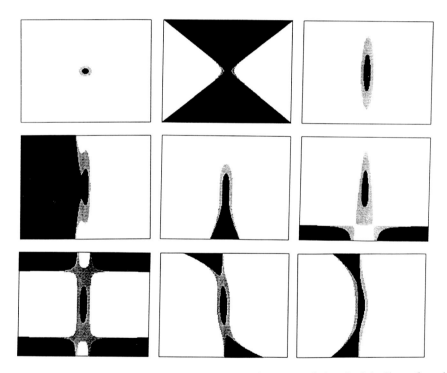

*Fig. 11. The geometrical significance of the second (and third) order jet. Upper row: blobs with β=100% (pure blob), γ=100% (featureless) and α=80%, β=20% (elongated blob). The elongated blob is shown deformed by third order terms: Middle row: edginess, endedness and wedginess; Lower row, final column the curvature term. The first and second columns of the final row illustrate some fourth order perturbation (an "hourglass tendency" and a "curvature trend" or inflection).*

plication and addition. Example:

$$\frac{\partial^2}{\partial x^2} \to \phi_{20}(x,y;\ t) = \phi_2(x;\ t) \cdot \phi_0(y;\ t) = \frac{1}{2t}\left(\frac{x^2}{2t} - 1\right)\frac{e^{-\frac{x^2+y^2}{4t}}}{4\pi t}. \qquad (60)$$

Another example:

$$\Delta = \frac{\partial^2}{\partial x^2} + \frac{\partial^2}{\partial y^2} \to \phi_{20}(x,y;\ t) + \phi_{02}(x,y;\ t) =$$

$$= \frac{1}{\rho}\frac{\partial}{\partial \rho}\rho\frac{\partial}{\partial \rho}\phi_{00}(x,y;\ t). \qquad (61)$$

Thus $\Delta = \frac{1}{\rho}\frac{\partial}{\partial \rho}\rho\frac{\partial}{\partial \rho}$ in polar coordinates.

### b. *Geometrical structure of the local jets*

The fuzzy derivatives yield the Taylor expansion up to the n-th order (for an n-th order jet extension):

$$L(x+\xi, y+\eta;\ t) = \sum_{k=0}^{n}\frac{1}{k!}\left(\xi\frac{\partial}{\partial x} + \eta\frac{\partial}{\partial y}\right)^k L = \sum_{k=0}^{n}\sum_{i=0}^{n}\binom{k}{i} a_{k-i,i}\frac{\xi^{k-i}\eta^i}{k!} =$$

$$= a_{00} + a_{10}\xi + a_{01}\eta + \frac{1}{2}(a_{20}\xi^2 + 2a_{11}\xi\eta + a_{02}\eta^2) + \frac{1}{6}(a_{30}\xi^3 + 3a_{21}\xi^2 +$$

$$3a_{12}\xi\eta^2 + a_{03}\eta^3) + \ldots \qquad (62)$$

where $a_{pq} = L(x,y;\ 0) * \phi_{pq}(x,y;\ t)$.

We usually neglect the zeroth order since it yields no geometric information whatsoever. The first order yields the *gradient* (modulus and direction). If it vanishes, the second order yields a local shape that is best appreciated in a suitably rotated coordinate system:

$$a_{11}x^2 + 2a_{12}xy + a_{22}y^2 = \rho^2\left[H + \sqrt{H^2 - K}\cos\left(2\phi - \text{atg}\frac{2a_{12}}{a_{11} - a_{22}}\right)\right] \qquad (63)$$

with $H = \frac{1}{2}(a_{11} + a_{22})$     one half the trace of the Hessian

and $K = a_{11}a_{22} - a_{12}$     the determinant of the Hessian.

We define the deviation from flatness D:

$$D = \sqrt{a_{11}^2 + a_{22}^2} \tag{64}$$

and the "shape angle" $\mu$:

$$\sin \mu = \frac{a_{11} + a_{22}}{D \cdot \sqrt{2}}$$

$$\cos \mu = \frac{a_{11} - a_{22}}{D \cdot \sqrt{2}} \, . \tag{65}$$

Then small values of D indicate absence of structure, larger values increasing deviation from the first order jet. The interesting range of $\mu$ is limited to $(0, \pi/2)$ because other quadrants can be obtained by interchange of x,y axes and/or contrast reversal. We have:
- $\mu = 0$ "anti umbilical" (like $x^2 - y^2$; equal to its own contrast reversal)
- $0 < \mu < \pi/2$ elongated shapes, especially
  $\mu = \pi/4$ is like $x^2$: a "cylinder" or pure line
- $\mu = \pi/2$ "umbilical", a circular blob like $x^2 + y^2$.

Thus we define fuzzy measures of "elongatedness", "blobness" and "featurelessness" as follows:

$$\begin{aligned}
\alpha \text{ (elongatedness)} &= \sin^2 2\mu \cdot 100\% \\
\beta \text{ (blobness)} &= \cos^2 2\mu \cdot 100\% \text{ if } \cos 2\mu < 0, \text{ zero} \\
&\qquad\qquad\qquad\qquad \text{otherwise} \\
\gamma \text{ (featurelessness)} &= \cos^2 2\mu \cdot 100\% \text{ if } \cos 2\mu > 0, \text{ zero} \\
&\qquad\qquad\qquad\qquad \text{otherwise}
\end{aligned} \tag{66}$$

with $\alpha + \beta + \gamma = 100\%$.

Similar analyses apply to the third order jet, although it is more profitable to regard the third order terms as a *perturbation* on the quadratic ones. If the quadric is *elongated* the cubic form can be interpreted in terms of a *curvature*, a *wedginess*, an *edginess* and an *endedness* with the following meaning:
- *curvature*: measures a curvature of the elongated shape,

- *wedginess:* measures a thickness change along the length of the elongated shape,
- *edginess* : measures how close the elongated shape is to the edge of an extended region,
- *endedness:* measures how close the elongated shape is to a line piece ended on one side (Fig. 11).

*c. Local and multilocal routines*

We differentiate between routines accessing only *single pixels* (the n-th order jet in that record) as *point processors* and routines that access *multiple pixels* as *array processors* (e.g. neighbourhood operators).

*Point processors* can measure local boundary curvature, line curvature, conditions for ridges or troughs, extrema, etc., in short everything that can be expressed in terms of partial derivatives at a point up to the n-th order. *Differential geometry* is the discipline that shows the way to the construction of such routines.

*Array processors* are necessary for inherently multilocal entities as e.g. correlations, gray scale histograms, numerosity (e.g. number of connected components of a shape), etc. They can be ordered with respect to the *geometric expertise* required. For instance, "numerosity" needs no expertise at all (e.g. number of pixels in a shape: we feed all pixels into the routine but the order is immaterial by definition of numerosity). Other routines may use a notion of projective or affine geometry (e.g. line continuation, parallellism of line pieces, etc.). For such routines one needs a *connection* that allows us to notice e.g. whether a field of line elements is *constant in direction*. Still other routines may need a *metric* (e.g. those that measure global distances).

Many calculations may be done both on a *local* and on a *multilocal* basis. An example is the *detection of movement:* locally this may be obtained from phase shifts, globally by correlation.

Multilocal routines may or may not use the full jet (note that the zeroth order jet easily covers everything!). However, it is generally useful to do so. An example is offered by *search routines* (movement, detectors, correlators, ...). One may search on the basis of the low order jet structure and in case of an expected "hit" go to more refined descriptions. Such a divide and conquer strategy is economical.

If one provides an n-th order jet at the Nyquist sample frequency one of course effectively *oversamples*. Because the jet has $(n+1)(n+2)/2$ degrees of freedom the sample distance could be raised by a factor of $\sqrt{(n+1)(n+2)/2}$, or about three times for $n = 3$ (which seems to be a practical value). If this is done right away when one builds scale space the extra amount of work is small whereas one gains because the stored samples have a more explicit descriptive power.

*Acknowledgements*

*The author gratefully acknowledges grants from NATO and from the Dutch Organization for the Advancement of Pure Research (Z.W.O.).*

REFERENCES

Bruce, J.W. and Giblin, P.J. (1984). *Curves and singularities*, Cambridge.
Doorn, A.J. van, Grind, W.A. van de, and Koenderink, J.J. (eds.)(1984). *Limits in Perception*, V.N.U. Science Press, Utrecht.
Eells, J. (1967). *Singularities of smooth maps*, Gordon and Breach, New York.
Golubitski, M. and Guillemin, V. (1973). *Stable mappings and their singularities*, Springer, New York.
Guillemin, V. and Pollack, A. (1974). *Differential topology*, Prentice-Hall, Englewood Cliffs.
Hoffman, D. and Richards, W. (1982). Representing smooth plane curves for visual recognition: implications for figure-ground reversal, *Proc. Am. Ass. Art. Intell.*, pp. 5-8.
Koenderink, J.J. and Doorn, A.J. van (1978). Visual detection of spatial contrast; influence of location in the visual field, target extent and illuminance level, *Biol. Cybern.* 30, pp. 157-167.
Koenderink, J.J. and Doorn, A.J. van (1979). The structure of two-dimensional scalar fields with applications to vision, *Biol. Cybern.* 33, pp. 151-158.
Koenderink, J.J. and Doorn, A.J. van (1981). A description of the structure of visual images in terms of an ordered hierarchy of light and dark blobs. In: *Second International Visual Psychophysics and Medical Imaging Conf.*, IEEE Cat. No. 81 CH 1676-6.
Koenderink, J.J. (1984). Simultaneous order in nervous nets from a functional standpoint, *Biol. Cybern.* 50, pp. 35-41.
Koenderink, J.J. (1984). Geometrical structures determined by the functional order in nervous nets, *Biol. Cybern.* 50, pp. 43-50.
Koenderink, J.J. (1984). The structure of images, *Biol. Cybern.* 50, pp. 363-370.
Koenderink, J.J. and Doorn, A.J. van (1986). Dynamic shape, *Biol. Cybern.* 53, pp. 383-396.
Mason, B.J. (1962). *Rain and rainmaking*, Cambridge U.P.

Maxwell, J.C. (1980). *The scientific papers of James Clerk Maxwell. Vol. II: On Hills and Dales*, W.D. Niven (ed.), Cambridge U.P.

Metzger, W. von (1975). *Gesetze des Sehens*, Waldemar Kramer, Frankfurt a/M.

Milnor, J. (1963). *Morse theory*, Princeton U.P.

Nye, J.F. and Thorndike, A.S. (1980). Events in evolving three-dimensional vector fields, *J. Phys. A: Math. Gen.* 13, pp. 1-14.

Poston, T. and Stewart, I. (1978). *Catastrophe theory and its applications*, Pitman, London.

Rosenfeld, A. (ed.)(1984). *Multiresolution image processing and analysis*, Springer Berlin.

Thom, R. (1972). *Stabilité structurelle et morphogénèse*, Benjamin, Reading.

Thorndike, A.S., Cooley, C.R. and Nye, J.F. (1978). The structure and evolution of flow fields and other vector fields, *J. Phys. A: Math. Gen.* 11, pp. 1455-1490.

Whitney, H. (1955). On singularities of mappings of Euclidean spaces I. Mappings of the plane into the plane, *Ann. Math.* 62, pp. 374-410.

# FUNDAMENTALS OF THE RADON TRANSFORM

Harrison H. Barrett

*University of Arizona*
*Tucson, AZ, USA*

ABSTRACT

*The Radon transform is the mathematical basis of computed tomography and finds application in many other medical imaging modalities as well. In this chapter we present the fundamental mathematics of this transform and its inverse, with emphasis on the central-slice theorem.*


## 1. INTRODUCTION

The Radon transform is a remarkably versatile tool that finds application in many areas of medicine and applied physics. In two dimensions (2D), the Radon transform is the set of all line-integral projections of a 2D function $f(\mathbf{r})$. For example, line integrals of the x-ray attenuation coefficient are obtained in computed tomography by passing a pencil beam of x-rays through a section of the body. In single-photon emission computed tomography (SPECT), a parallel-hole collimator defines a pencil-like region of the body, so again line integrals are obtained, this time of the activity distribution. And in positron-emission tomography (PET), coincident detection of two annihilation gamma rays defines a line through the body, and line integrals of the positron activity are measured.

The Radon transform in 3D refers to integrals over planes rather than lines. Plane integrals occur in some kinds of magnetic resonance imaging and in nuclear medicine if a slat collimator is used.

The Radon transform can also be defined in spaces of higher dimensionality. Although it is difficult to find practical applications, we shall briefly discuss the general mD Radon transform for completeness.

Whatever the dimensionality of the space, the Radon transform is a set of 1D projections of some object. The usual problem of interest is the inverse problem: given a set of projections, find the original object. The inverse Radon transform is the solution to this problem.

Our main goal in these lectures is to present a unified treatment of the Radon transform and the many algorithms that can be used to find an exact or approximate inverse. The unifying theme will be an important theorem called the central-slice theorem, which relates the Fourier transform of a projection to the Fourier transform of the object. It is presumed, therefore, that the reader has a good grasp on multidimensional Fourier theory and some acquaintance with the theory of linear systems.


2. DEFINITIONS AND GEOMETRY

Let us begin by considering the 1D projection of a 2D function $f(x,y)$, with the projection being defined by line integrals along a series of lines parallel to the y axis. The projection along the line $x=p$, denoted $\lambda_0(p)$, is given by

$$\lambda_0(p) = \int_{-\infty}^{\infty} f(p,y)\, dy = \int_{-\infty}^{\infty} dx \int_{-\infty}^{\infty} dy\, f(x,y)\, \delta(p-x) \, . \tag{1}$$

A more general projection is along the line $p = \mathbf{r} \cdot \mathbf{n}$, as shown in Fig. 1. In this case the projection is given by

$$\lambda_\phi(p) = \int_\infty d^2r\, f(\mathbf{r})\, \delta(p - \mathbf{r} \cdot \mathbf{n}) \, , \tag{2}$$

where the integral runs over the infinite plane, and $\phi$ is the angle between the x axis and the unit vector $\mathbf{n}$. Note that the Dirac delta functions in Eqs. (1) and (2) are one-dimensional and can be used to perform only one of the two integrals in each equation.

The projection $\lambda_\phi(p)$ can be regarded either as a 1D function of p, parameterized by $\phi$, or as a function in a 2D space with polar coordinates p and $\phi$. In the latter viewpoint, it is convenient to use a slightly different notation $\lambda(\mathbf{p})$, where $\mathbf{p} = p\mathbf{n}$ is a 2D vector in *Radon space*. The scalar p is thus the magnitude of the vector $\mathbf{p}$, and $\phi$ is its polar angle relative to the x axis. An operator notation is also useful; we may write

$$\lambda_\phi(p) = \mathcal{R}_2\{f(\mathbf{r})\} \tag{3}$$

or simply

$$\lambda = \mathcal{R}_2\{f\}, \tag{4}$$

with $\mathcal{R}_2$ being the integral operator implied by Eq. (2).

The formalism of the Radon transform may be readily extended to three or more dimensions. In 3D, the line integral of Eq. (2) is replaced by an integral over a plane. If $\mathbf{r}$ is a vector in 3D, then $p = \mathbf{r}\cdot\mathbf{n}$ is the equation of a plane with unit normal $\mathbf{n}$ and distance from the origin $p$. The 3D Radon transform for $f(\mathbf{r})$ is defined in terms of integrals over these planes as

$$\lambda_\mathbf{n}(p) = \int_\infty d^3r\, f(\mathbf{r})\, \delta(p-\mathbf{r}\cdot\mathbf{n}), \tag{5}$$

where now the integral runs formally over the entire 3D space, but the 1D delta function causes the integrand to vanish except on the specified plane.

More generally, the m-dimensional Radon transform involves integrals over (m-1)-dimensional hyperplanes, defined by an mD unit vector $\mathbf{n}$ and a scalar distance p from the origin in the mD space.

## 3. THE CENTRAL-SLICE THEOREM

To understand what information about the unknown 2D function $f(\mathbf{r})$ is contained in the projection $\lambda_\phi(p)$, let us take the 1D Fourier transform of the projection with respect to p at fixed $\phi$:

$$\Lambda_\phi(\nu) = \mathcal{F}_1\{\lambda_\phi(p)\} = \int_{-\infty}^{\infty} dp\, \lambda_\phi(p)\, e^{-2\pi i \nu p}$$

$$= \int_\infty d^2r \int_{-\infty}^{\infty} dp\, f(\mathbf{r})\, \delta(p-\mathbf{r}\cdot\mathbf{n})\, e^{-2\pi i \nu p}$$

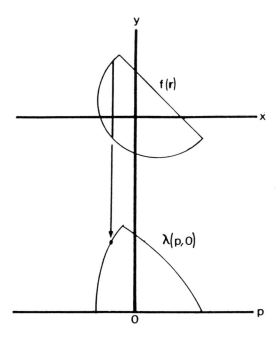

Fig. 1a. *Projection along lines parallel to the y-axis, i.e.* $\phi = 0$.

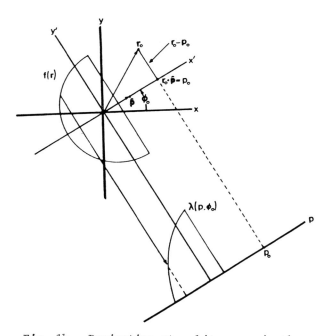

Fig. 1b. *Projection at arbitrary azimuth.*

$$= \int_\infty d^2r \, f(\mathbf{r}) \, e^{-2\pi i \mathbf{r} \cdot \mathbf{n}\nu}$$

$$= F(\mathbf{n}\nu) = \mathscr{F}_2\{f(\mathbf{r})\}_{\boldsymbol{\rho}=\mathbf{n}\nu} \, , \qquad (6)$$

where $\mathscr{F}_2\{f(\mathbf{r})\}$ or $F(\boldsymbol{\rho})$ denotes the 2D Fourier transform of $f(\mathbf{r})$, with $\boldsymbol{\rho}$ being the general 2D spatial frequency vector with Cartesian components $(\xi,\eta)$. Equation (6) is the very important *central-slice theorem*. In words, it says that the 1D Fourier transform of the projection yields one line through the 2D transform of the original object $f(\mathbf{r})$. The line, defined by $\boldsymbol{\rho} = \mathbf{n}\nu$, passes through the origin of the 2D Fourier plane (hence, *central* slice) and makes an angle of $\phi$ to the $\xi$ axis.

If we can obtain $F(\boldsymbol{\rho})$ for all points in the 2D spatial frequency plane, we can reconstruct $f(\mathbf{r})$ by a 2D inverse Fourier transform. The data set is thus said to be complete if all points in the frequency plane are sampled, which will be the case either if the values of p and $\phi$ satisfy

$$-\infty < p < \infty \quad \text{and} \quad 0 < \phi \leq \pi \qquad (7)$$

or if

$$0 \leq p < \infty \quad \text{and} \quad 0 < \phi \leq 2\pi \, . \qquad (8)$$

Note that

$$\lambda_\phi(p) = \lambda_{\phi+\pi}(-p) \, , \qquad (9)$$

since replacing $\phi$ with $\phi+\pi$ is the same as replacing $\mathbf{n}$ with $-\mathbf{n}$, and the vector $\mathbf{p}$ in Radon space can be written either as $p\mathbf{n}$ or $(-p)(-\mathbf{n})$.

The central-slice theorem in 3D, or in general in mD, has the same structure as in 2D. In all cases, the projection is a 1D function of the scalar p, and we may take its 1D Fourier transform just as in Eq. (3):

$$\Lambda_{\mathbf{n}}(\nu) = \mathscr{F}_1\{\lambda_{\mathbf{n}}(p)\} = \int_{-\infty}^{\infty} dp\, \lambda_{\mathbf{n}}(p)\, e^{-2\pi i \nu p}$$

$$= \int_{\infty} d^m r \int_{-\infty}^{\infty} dp\, f(\mathbf{r})\, \delta(p - \mathbf{r}\cdot\mathbf{n})\, e^{-2\pi i \nu p}$$

$$= \int_{\infty} d^m r\, f(\mathbf{r})\, e^{-2\pi i \mathbf{r}\cdot\mathbf{n}\nu}$$

$$= F(\mathbf{n}\nu) = \mathscr{F}_m\{f(\mathbf{r})\}_{\boldsymbol{\rho}=\mathbf{n}\nu}, \tag{10}$$

where $\mathscr{F}_m$ denotes the mD Fourier operator. Thus the general statement of the central-slice theorem is that the 1D Fourier transform of the projection yields one line through the mD Fourier transform of the original function, with the direction of the line given by $\mathbf{n}$. The mD unit vector $\mathbf{n}$ must explore all possible directions in order to obtain a complete data set. Specifically in the case of 3D, $\mathbf{n}$ must range over a hemisphere if $-\infty < p < \infty$ or a full sphere if $0 \leq p < \infty$.

## 4. THE INVERSE RADON TRANSFORM IN TWO DIMENSIONS

The easiest way to find the inverse Radon transform is to take the inverse 2D Fourier transform of $F(\mathbf{n}\nu)$, as expressed by the central-slice theorem. This is most easily accomplished by letting $\boldsymbol{\rho}$ have polar coordinates $(\rho, \theta_\rho)$ and, somewhat unconventionally, taking $-\infty < \rho < \infty$ and $0 < \theta_\rho \leq \pi$. With this choice for the range of the coordinates, the area element is given by $d^2\rho = |\rho| d\rho d\theta_\rho$ and the general 2D inverse Fourier transform is given by

$$f(\mathbf{r}) = \mathscr{F}_2^{-1}\{F(\boldsymbol{\rho})\} = \int_{-\infty}^{\infty} |\rho| d\rho \int_0^{\pi} d\theta_\rho\, F(\boldsymbol{\rho})\, e^{2\pi i \boldsymbol{\rho}\cdot\mathbf{r}}, \tag{11}$$

where $\boldsymbol{\rho}\cdot\mathbf{r} = \xi x + \eta y$.

We now specifically invoke the central-slice theorem, Eq. (6), which requires that we substitute $\boldsymbol{\rho} = \mathbf{n}\nu$ in $F(\boldsymbol{\rho})$ to obtain $\Lambda_\phi(\nu)$. This implies that $\rho = \nu$ and $\theta_\rho = \phi$, so that

$$f(\mathbf{r}) = \int_{-\infty}^{\infty} |\nu| d\nu \int_0^{\pi} d\phi \, \Lambda_\phi(\nu) \, e^{2\pi i \nu \mathbf{r} \cdot \mathbf{n}} . \qquad (12)$$

This is one form of the 2D inverse Radon transform, which may be expressed in a handy operator notation as

$$f(\mathbf{r}) = \mathcal{R}_2^{-1}\{\lambda_\phi(p)\}, \qquad (13)$$

or simply

$$f = \mathcal{R}_2^{-1}\{\lambda\} \qquad (14)$$

A useful alternative form of the inverse Radon transform may be found by regrouping terms to yield

$$f(\mathbf{r}) = \int_0^{\pi} d\phi \left[ \int_{-\infty}^{\infty} d\nu \, K_\phi(\nu) \, e^{2\pi i \nu p} \right]_{p=\mathbf{r} \cdot \mathbf{n}} , \qquad (15)$$

where

$$K_\phi(\nu) = |\nu| \Lambda_\phi(\nu) \qquad (16)$$

represents a filtered projection in the 1D frequency domain, with the filter function given by $|\nu|$. We may now rewrite $K_\phi(\nu)$ as the 1D Fourier transform of a convolution of two terms:

$$f(\mathbf{r}) = \int_0^{\pi} d\phi \, [\lambda_\phi(p) * \mathcal{F}_1^{-1}\{|\nu|\}]_{p=\mathbf{r} \cdot \mathbf{n}}, \qquad (17)$$

where the asterisk denotes 1D convolution.

Equation (17), still another form of the 2D inverse Radon transform, shows that we may reconstruct $f(\mathbf{r})$ in three steps:

1. filter each projection by convolution with $\mathcal{F}_1^{-1}\{|\nu|\}$ (the functional form of which will be discussed later);

2. substitute **r**·**n** for p in the filtered projections;

3. integrate (sum) over all angles.

Let us discuss steps 2 and 3 first. The substitution of **r**·**n** for p is the operation of "back-projection", which creates a 2D function of **r** from the 1D function of p by smearing the 1D function back in the original direction of projection. Step 3 simply adds up all of the back-projected 2D functions for different projection angles $\phi$. In practice, of course, only a finite set of angles is used, and the continuous integral is replaced with a discrete sum.

The remaining problem is to determine the functional form of $\mathscr{F}_1^{-1}\{|\nu|\}$. Note that

$$|\nu| = \nu\, \text{sgn}(\nu) = 2\pi i \nu\, \frac{\text{sgn}(\nu)}{2\pi i}\ , \qquad (18)$$

where

$$\text{sgn}(\nu) \equiv \begin{cases} +1 & \text{if } \nu > 0 \\ 0 & \text{if } \nu = 0 \\ -1 & \text{if } \nu < 0 \end{cases} \qquad (19)$$

From tables of integrals or a simple contour integration, we find that

$$\mathscr{F}_1^{-1}\left[\frac{\text{sgn}(\nu)}{2\pi i}\right] = \left(\frac{1}{2\pi^2}\right) P\{1/p\}\ , \qquad (20)$$

where $P\{1/p\}$ denotes the Cauchy principal value of $1/p$, defined by

$$\int_{-\infty}^{\infty} g(p)\, P\{1/p\}\, dp \equiv P\int_{-\infty}^{\infty} \frac{g(p)\,dp}{p} = \lim_{\varepsilon \to 0}\left[\int_{-\infty}^{-\varepsilon} + \int_{\varepsilon}^{\infty}\right]\frac{g(p)}{p}\, dp\ , \qquad (21)$$

where g(p) is a well-behaved function. In other words, the effect of the principal value is simply to delete an infinitesimal neighborhood around the singularity of 1/p.

The factor $2\pi i\nu$ in Eq. (18) may be handled by the Fourier derivative theorem:

$$\mathscr{F}_1^{-1}\{2\pi i \nu G(\nu)\} = dg(p)/dp ,\tag{22}$$

where $G(\nu)$ is the 1D Fourier transform of $g(p)$. We may also rewrite this theorem by using the fact that the (inverse) Fourier transform of a product is the convolution of the transforms of the individual factors, so that

$$\mathscr{F}_1^{-1}\{2\pi i \nu G(\nu)\} = \mathscr{F}_1^{-1}\{2\pi i \nu\} * g(p).\tag{23}$$

Furthermore, the derivative of a function is obtained by convolving it with the derivative of a delta function:

$$dg(p)/dp = g(p) * \delta'(p).\tag{24}$$

By comparing Eqs. (22) and (23) we see that

$$\mathscr{F}_1^{-1}\{2\pi i \nu\} = \delta'(p).\tag{25}$$

We may now put all the pieces together and write

$$\mathscr{F}_1^{-1}\{|\nu|\} = \frac{1}{2\pi^2}P\left[\frac{1}{p}\right] * \delta'(p) = \frac{1}{2\pi^2}\frac{d}{dp}P\left[\frac{1}{p}\right].\tag{26}$$

This equation involves a number of questionable mathematical operations, including the derivative of a function with an infinite discontinuity. The pure mathematician resolves these problems via the theory of generalized functions (Lighthill, 1962), the applied mathematician approaches them through a limiting process, and the engineer uses a technique called "apodization". We shall briefly consider all three approaches, beginning with generalized functions.

In fact, we have already introduced the concept of a generalized function in Eq. (21) where the function $P(1/p)$ was defined through its action within an integral. Similarly, the function $(d/dp)P(1/p)$ is defined by

$$\int_{-\infty}^{\infty} dp\, g(p)\, \frac{d}{dp}P\left[\frac{1}{p}\right] = -P\int_{-\infty}^{\infty} dp\, \frac{g'(p)}{p} ,\tag{27}$$

where $g(p)$ is a well-behaved test function and $g'(p)$ is its non-singular derivative. The right-hand side of this equation may be formally regarded as resulting from an integration by parts of the left-hand side. With this interpretation in mind, we may adopt the shorthand

$$\frac{d}{dp}P\left[\frac{1}{p}\right] = -\frac{1}{p^2} . \tag{28}$$

This generalized function behaves just as the ordinary function $-1/p^2$ for $p \neq 0$, but there is an extreme positive singularity, which may be loosely described as an infinite-weight delta function, at $p=0$. In terms of this generalized function, we have, finally,

$$\mathcal{F}_1^{-1}\{|\nu|\} = -\frac{1}{2\pi^2 p^2} . \tag{29}$$

With the shorthand of Eq. (28), the inverse Radon transform becomes

$$f(\mathbf{r}) = -\frac{1}{2\pi^2} \int_0^\pi d\phi \left[\lambda_\phi(p) * \frac{1}{p^2}\right]_{p=\mathbf{r}\cdot\mathbf{n}} . \tag{30}$$

Since $\mathbf{r}\cdot\mathbf{n} = r[\cos(\theta-\phi)]$, with $(r,\theta)$ being the polar coordinates of $\mathbf{r}$, we may also write

$$f(\mathbf{r}) = -\frac{1}{2\pi^2} \int_0^\pi d\phi \int_{-\infty}^\infty dp \frac{\lambda_\phi(p)}{[r\cos(\theta-\phi)-p]^2} . \tag{31}$$

This is a very common form for the inverse Radon transform, but care must be exercised in interpreting it because of the implied singularity in the integrand. Without this singularity, the result would be nonsensical. Consider what happens in the common case where $f(\mathbf{r})$ is non-negative. Then its projection in any direction is also non-negative, the entire integrand in Eq. (31) must be non-negative (except for the singularity), and there is an overall minus sign in front of the integral. The right-hand side is therefore negative, yet it must represent the non-negative quantity on the left; the paradox is resolved through the singularity.

Let us now turn to apodization as another (and more practical) way of dealing with the singularities in the inverse Radon transform. The term apodization comes from the radar literature. It literally means a "cutting off of feet", and in radar it refers to the process of weighting a waveform with some smooth function to eliminate sidelobes ("feet") in the response after signal processing. Here we use it in a somewhat more general sense to refer to any beneficial use of smooth weighting functions. For example, most of the mathematical problems discussed above arose because we were trying to take the Fourier transform of a function, $|\nu|$, that is not absolutely integrable and hence does not possess a Fourier transform in the usual sense. Instead of defining a generalized function to represent the Fourier transform of $|\nu|$, we could instead modify the function to be transformed:

$$|\nu| \rightarrow |\nu|A(\nu) , \qquad (32)$$

where $A(\nu)$ is some broad, slowly varying function. For example, a suitable choice for $A(\nu)$ would be

$$A(\nu) = \exp(-\pi\alpha^2\nu^2) . \qquad (33)$$

If $\alpha$ approaches zero, $|\nu|A(\nu)$ approaches $|\nu|$, so we may always recover the original function, but in practice the "roll-off" parameter $\alpha$ is a very useful way to control the response of the system at high spatial frequencies and hence the noise properties of the reconstruction.

We shall now consider the modified form of the inverse Radon transform when the substitution (32) is made. Since we are departing from the mathematically exact formalism at this point, the resulting inverse transform will not represent $f(\mathbf{r})$ but rather some approximation to it which we shall denote by $\hat{f}(\mathbf{r})$. The space-domain counterpart of (32) is

$$-\frac{1}{2\pi^2 p^2} \rightarrow -\frac{1}{2\pi^2 p^2} * a(p) , \qquad (34)$$

where $a(p)$ is the 1D Fourier transform of $A(\nu)$. It is convenient to define

$$h(p) = -\frac{1}{2\pi^2 p^2} * a(p) , \qquad (35)$$

in terms of which the approximate (apodized) form of Eq. (31) becomes

$$\hat{f}(\mathbf{r}) = \int_0^\pi d\phi \int_{-\infty}^\infty dp \ [\lambda_\phi(p) * h(p)]_{p=r\cos(\theta-\phi)} \ . \tag{36}$$

The steps in the reconstruction are now:

1. filter each projection by convolution with $h(p)$;

2. back-project (substitute $\mathbf{r}\cdot\mathbf{n}$ or $r\cos(\theta-\phi)$ for $p$ in the filtered projections);

3. integrate (sum) over all angles.

This procedure is often referred to as the convolution-backprojection algorithm or simply as the convolution algorithm.

Since convolution is associative, the substitution (34) is equivalent to replacing $\lambda_\phi(p)$ with $\lambda_\phi(p) * a(p)$ in Eq. (30). One easy way to see the effect of this substitution is to use the central-slice theorem to write

$$F_1\{\lambda_\phi(p) * a(p)\} = \Lambda_\phi(\nu)A(\nu) = F(\mathbf{n}\nu)A(\nu) \equiv \hat{F}(\mathbf{n}\nu) \ , \tag{37}$$

where $\hat{F}(\boldsymbol{\rho})$ is the 2D Fourier transform of $\hat{f}(\mathbf{r})$. In other words, the effect of the apodization is to yield a reconstructed estimate of $F(\boldsymbol{\rho})$ that is equal to the true value multiplied by the apodizing function $A(\rho)$, where $\rho$ is the magnitude of the vector $\boldsymbol{\rho}$. Thus, even though $A(\nu)$ was introduced as a 1D weighting function, it now plays the role of the 2D modulation transfer function (MTF) of the reconstruction system. As a result of the central-slice theorem, weighting the projection in the 1D frequency domain with a function $A(\nu)$ is equivalent to weighting the original function $F(\boldsymbol{\rho})$ *with the same apodizing function,* but now regarded as a 2D, rotationally symmetric function.

To gain further insight into the behavior of this important algorithm, let us trace the fate of a point object. In other words, let us calculate the point spread function (PSF) of the system. Since all objects can be described as a superposition of points and since the algorithm is linear, the PSF is a complete description of the system.

A point object in 2D is represented by a 2D delta function $\delta(\mathbf{r}-\mathbf{r}_0)$, where $\mathbf{r}_0$ is the location of the point. The projection of this object is a 1D delta function given by

$$\mathcal{R}_2\{\delta(\mathbf{r}-\mathbf{r}_0)\} = \lambda_\phi^\delta(p) = \int_\infty d^2r\, \delta(\mathbf{r}-\mathbf{r}_0)\, \delta(p-\mathbf{r}\cdot\mathbf{n})$$

$$= \delta(p-\mathbf{r}_0\cdot\mathbf{n}), \tag{38}$$

where the superscript $\delta$ denotes the special case of a point object. If we use this projection in Eq. (36), we find easily that

$$\text{PSF} = \hat{f}^\delta(\mathbf{r}) = \int_0^\pi d\phi \int_{-\infty}^\infty dp\, [\delta(p-\mathbf{r}_0\cdot\mathbf{n}) * h(p)]_{p=\mathbf{r}\cdot\mathbf{n}}$$

$$= \int_0^\pi d\phi\, h[(\mathbf{r}-\mathbf{r}_0)\cdot\mathbf{n}]. \tag{39}$$

Thus the PSF for the combined operations of projection, filtering, back-projection, and summation over projection angles is simply the back-projection of the filter function. We see also that these operations taken together constitute a linear shift-invariant system since the PSF is a function only of $\mathbf{r}-\mathbf{r}_0$. The filter function $h(p)$ is of the form of Eq. (35), where $a(p)$ is the inverse 1D Fourier transform of some apodizing function $A(\nu)$; if $A(\nu)$ is broad and slowly varying, then the PSF is sharply peaked, approximating a 2D delta function.

Of course, the PSF is also the 2D Fourier transform of the MTF, and we have already seen that $A(\rho)$ serves as the MTF, so it must also be true that

$$\int_0^\pi d\phi\, h(\mathbf{r}\cdot\mathbf{n}) = \mathcal{F}_2^{-1}\{A(\rho)\}, \tag{40}$$

but we shall leave the explicit proof of this result as an exercise for the student.

The steps in the convolution algorithm, listed below Eq. (36), may also be executed in a different order. We may:
 1. back-project;
 2. sum over projection angles;
 3. filter the resulting 2D back-projected image.

From Eq. (37), it is straightforward to show that the reconstruction obtained with this sequence is the same as that obtained by the earlier algorithm, in which filtering is done on the 1D projections before back-projection, provided the 1D and 2D filter functions have the same functional form in their respective frequency domains. However, once again it is instructive to trace the fate of a delta function. We define a new function $b(\mathbf{r})$ that is the result of back-projection and summation without filtering, i.e.

$$b(\mathbf{r}) = \int_0^\pi d\phi \, \lambda_\phi(\mathbf{r} \cdot \mathbf{n}) \, . \tag{41}$$

In the specific case of a point object located at $\mathbf{r}_0$, we have

$$b^\delta(\mathbf{r}) = \int_0^\pi d\phi \, \delta[(\mathbf{r}-\mathbf{r}_0) \cdot \mathbf{n}] \, . \tag{42}$$

To evaluate this expression, we define $\mathbf{R} = \mathbf{r}-\mathbf{r}_0$ and let $(R, \theta_R)$ be its polar coordinates. Then

$$b^\delta(\mathbf{r}) = \int_0^\pi d\phi \, \delta[R\cos(\theta_R-\phi)] \, . \tag{43}$$

To proceed, we make use of an identity involving delta functions:

$$\delta[q(x)] = \frac{\delta(x-x_0)}{q'(x_0)} \, , \tag{44}$$

where it is assumed that $q(x)$ has only one zero within the range of integration of whatever integral this identity is to be used in, so that $q(x_0) = 0$, and the derivative $q'(x_0)$ does not vanish at this point. These conditions are satisfied in the present problem where $\phi$ plays the role of $x$ and $R\cos(\theta_R-\phi)$ plays the role of $q(x)$. The zero of this function occurs at $\phi = \phi_0$, where $\theta_R-\phi_0 = \pi/2$, and the derivative of the function at this point is just R. Therefore

$$b^{\delta}(\mathbf{r}) = \int_0^{\pi} d\phi \; \frac{\delta(\phi-\phi_0)}{R} = \frac{1}{R} = \frac{1}{|\mathbf{r}-\mathbf{r}_0|} \; . \qquad (45)$$

For a more general object, we find

$$b(\mathbf{r}) = f(\mathbf{r}) ** \frac{1}{r} \; , \qquad (46)$$

where the double asterisk denotes 2D convolution, and r is the magnitude of **r**.

The PSF for back-projection and summation, without any filtering, is thus the cusp-like function $1/r$. Although it does have a sharp central peak, $1/r$ is actually a terrible function for a PSF because of its long wings. When the PSF is convolved with some broad, extended function, the wings build up and cause severe quantitative errors and loss of contrast in the reconstruction. It is for this reason that a further filtering operation is needed. To discover the required form for this 2D filter, let us take the 2D Fourier transform of Eq. (46); since it is well known that

$$\mathscr{F}_2\{1/r\} = 1/\rho \; , \qquad (47)$$

we find that

$$B(\boldsymbol{\rho}) = \frac{F(\boldsymbol{\rho})}{\rho} \; , \qquad (48)$$

where B and F denote the transforms of b and f as usual, and of course $\rho$ is the magnitude of $\boldsymbol{\rho}$. The desired expression for $f(\mathbf{r})$ follows by an inverse 2D Fourier transform:

$$f(\mathbf{r}) = \mathscr{F}_2^{-1}\{\rho B(\boldsymbol{\rho})\} \; . \qquad (49)$$

Although this is an exact expression, and in fact yet another form of the 2D inverse Radon transform, in practice it is necessary to use an apodizing function to control noise at high spatial frequencies. Thus we obtain an estimate of $f(\mathbf{r})$ given by

$$\hat{f}(\mathbf{r}) = F_2^{-1}\{\rho A(\rho)B(\rho)\} \ . \tag{50}$$

To summarize, the steps in this algorithm are:
1. back-project each projection;
2. sum over projection angles;
3. filter the resulting 2D image b(r).

Note that step 3 may be carried out either by multiplying $B(\rho)$ in the 2D frequency domain by $\rho A(\rho)$, or by convolving $b(\mathbf{r})$ with the 2D inverse Fourier transform of that function. If the transforms are performed with the fast Fourier transform algorithm (FFT), the frequency domain is much faster in practice.


## 5. THE INVERSE RADON TRANSFORM IN THREE DIMENSIONS

The derivation of the 3D inverse Radon transform from the central-slice theorem follows exactly the path of the corresponding 2D derivation, as given in sec. 4, but there are some surprises along the way. The forward transform is given by Eq. (5), repeated here for convenience

$$\lambda_{\mathbf{n}}(p) = \mathscr{R}_3\{f(\mathbf{r})\} = \int_\infty d^3r \ f(\mathbf{r}) \ \delta(p - \mathbf{r}\cdot\mathbf{n}) \ , \tag{5}$$

where $\mathbf{n}$ is the 3D unit vector with spherical polar coordinates $(1, \theta_n, \phi_n)$. The general form of the 3D inverse Fourier transform is

$$f(\mathbf{r}) = \mathscr{F}_3^{-1}\{F(\sigma)\} = \int_{-\infty}^{\infty} \sigma^2 d\sigma \int_{2\pi} d\Omega_\sigma \ F(\sigma) \ e^{2\pi i \sigma \cdot \mathbf{r}}, \tag{51}$$

where $\sigma$ is the 3D spatial frequency vector with Cartesian coordinates $(\xi, \eta, \zeta)$ and polar coordinates $(\sigma, \theta_\sigma, \phi_\sigma)$, so that $\sigma\cdot\mathbf{r} = \xi x + \eta y + \zeta z$, and $d\Omega_\sigma = \sin\theta_\sigma \ d\theta_\sigma \ d\phi_\sigma$ is the element of solid angle in Fourier space. The angular integral runs over any hemisphere ($2\pi$ steradians).

The central-slice theorem, Eq.(10), requires that $\sigma=\nu$, $\theta_\sigma=\theta_n$, and $\phi_\sigma=\phi_n$, so that $\Omega_\sigma=\Omega_n$ and

$$f(\mathbf{r}) = \mathcal{R}_2^{-1}\{\lambda_\mathbf{n}(p)\} = \int_{-\infty}^{\infty} \nu^2 d\nu \int \frac{d\Omega_n}{2\pi} \Lambda_\mathbf{n}(\nu) \, e^{2\pi i \nu \mathbf{r} \cdot \mathbf{n}} \,, \tag{52}$$

This is one form of the 3D inverse Radon transform. Regrouping terms (cf. Eq. (15)), we find

$$f(\mathbf{r}) = \int \frac{d\Omega_n}{2\pi} \left[ \int_{-\infty}^{\infty} d\nu \, \nu^2 \Lambda_\mathbf{n}(\nu) \, e^{2\pi i \nu p} \right]_{p=\mathbf{r}\cdot\mathbf{n}} . \tag{53}$$

The substitution $p = \mathbf{r}\cdot\mathbf{n}$ in this equation is the operation of 3D back-projection, which means smearing the 1D function uniformly over the original plane of integration. Thus the inverse Radon transform in 3D is accomplished by:

1. taking the 1D Fourier transform of each projection;
2. filtering it by multiplication by $\nu^2$;
3. performing the inverse 1D Fourier transform;
4. back-projecting each filtered projection;
5. integrating over all projection directions.

With the Fourier derivative theorem, we may also write

$$f(\mathbf{r}) = -\frac{1}{4\pi^2} \int \frac{d\Omega_n}{2\pi} \, [\lambda_\mathbf{n}''(p)]_{p=\mathbf{r}\cdot\mathbf{n}} = -\frac{1}{4\pi^2} \int \frac{d\Omega_n}{2\pi} \, \lambda_\mathbf{n}''(\mathbf{r}\cdot\mathbf{n}) \,, \tag{54}$$

where $\lambda_\mathbf{n}''(p)$ is the second derivative of the projection. Thus the reconstruction may also be accomplished by:

1. taking the second derivative of each projection;
2. back-projecting it;
3. integrating over all projection directions.

This version of the algorithm shows the first surprise in the 3D case. In place of the convolution with a rather ill-behaved filter function as in Eq. (17), we have a simple second derivative in Eq. (54). This difference arises because $|\nu|$ has a slope discontinuity at the origin, while $\nu^2$ and all of its derivatives are continuous. Of course, the second derivative may also be regarded as a convolution with the second derivative of a delta function, but there is an important practical difference. The second derivative of a delta function is a compact function, defined only in a small neighborhood, while the filter function in the 2D case, $-1/p^2$, has long wings extending to infinity, and truncation of these wings can result in serious error. In 3D, the filtering operation is more computionally efficient and, moreover, less sensitive to missing data due to truncation of the measurement.

As in the 2D case, we may again reverse the order of the operations in this algorithm. From the definition of the Laplacian operator, it is straightforward to show that

$$\ddot{\lambda}_n(r \cdot n) = \nabla^2 \lambda_n(r \cdot n) \,, \tag{55}$$

from which it follows that

$$f(r) = -\frac{1}{4\pi^2} \nabla^2 \int_{2\pi} d\Omega_n \, \lambda_n(r \cdot n) \,. \tag{56}$$

Thus we may:
1. back-project each projection;
2. integrate over all projection directions;
3. take the Laplacian of the resulting 3D function.

Let us denote the output of step 2, (the unfiltered, back-projected image) as $b(r)$, which is given by (cf. Eq. (41))

$$b(r) = \int_{2\pi} d\Omega_n \, \lambda_n(r \cdot n) \,. \tag{57}$$

It is instructive to ask what the relationship is between $b(r)$ and the original function $f(r)$. To answer this question, let us again consider a point object $\delta(r-r_0)$. If we use a superscript $\delta$ to denote the special case of a point object, the delta-function identity given by Eq. (44) allows us to write

$$b^\delta(r) = \int_{2\pi} d\Omega_n \, \delta[(r - r_0) \cdot n] = \frac{\pi}{|r - r_0|} \,. \tag{58}$$

Thus the second surprise: Even though back-projection now involves smearing over a plane rather than a line as in 2D, the form of the PSF after back-projection and integration over projection directions is exactly the same as in the 2D case (cf. Eq. (45)).

For the general input object $f(\mathbf{r})$, we have

$$b(\mathbf{r}) = f(\mathbf{r}) ** \frac{\pi}{r} ,  \qquad (59)$$

where $r = |\mathbf{r}|$.

Again taking liberties with divergent functions, we may show that

$$\mathcal{F}_3\{1/r\} = 1/(\pi\sigma^2) . \qquad (60)$$

Thus Eq. (59) transforms to

$$B(\boldsymbol{\sigma}) = F(\boldsymbol{\sigma})/\sigma^2 , \qquad (61)$$

so that the 3D MTF for the operations of back-projection and integration over directions is just $1/\sigma^2$, compared to $1/\rho$ in the 2D case.

If we solve Eq. (61) for $F$ (or, equivalently, inverse-filter $B$ by multiplication with the reciprocal of the MTF) and take the inverse 3D Fourier transform, we obtain

$$f(\mathbf{r}) = \mathcal{F}_3^{-1}\{\sigma^2 B(\boldsymbol{\sigma})\} = -\frac{1}{4\pi^2}\nabla^2 b(\mathbf{r}) = -\frac{1}{4\pi^2}\nabla^2 \int_{2\pi} d\Omega_\mathbf{n} \, \lambda_\mathbf{n}(\mathbf{r}\cdot\mathbf{n}) , \qquad (62)$$

where the last step makes use of the Fourier derivative theorem, which in 3D says that multiplication by $-4\pi^2\sigma^2$ in the frequency domain is equivalent to taking the Laplacian in the space domain. This alternate route has thus again led to Eq. (56).

A very interesting difference between 2D and 3D arises when the object has circular or spherical symmetry, so that $f(\mathbf{r}) = f(r)$, where $r = |\mathbf{r}|$. Then the subscript $\mathbf{n}$ or $\phi$ on $\lambda$ is unnecessary since all projection directions are equivalent. In 2D the Radon transform reduces to another well-known integral transform, the Abel transform, in this case. In 3D, however, we have

$$\lambda(x) = \int_{-\infty}^{\infty} dy \int_{-\infty}^{\infty} dz\, f(r) = 2\pi \int_{|x|}^{\infty} r dr\, f(r) \, . \tag{63}$$

The inverse is readily obtained by differentiating with respect to x, yielding

$$f(r) = -\frac{1}{2\pi r}\lambda'(r) \, . \tag{64}$$

Thus the next surprise: For spherically symmetric objects the inverse Radon transform is not an integral transform at all, but just a simple derivative.

As one last surprise, we note that the mathematics outlined above is also encountered in elementary electrostatics; indeed, it was just this connection that led Radon to investigate the problem in the first place. The function $f(\mathbf{r})$ corresponds to the charge density and $b(\mathbf{r})$ corresponds to the potential; Eq. (62) is then Poisson's equation and (59) is its usual integral solution, since $1/r$ is just the Green's function or potential of a point charge.

6. THE GENERAL mD CASE

For completeness, we include the results for the general m-dimensional problem. The derivations are omitted since they parallel those given in the 2D and 3D cases. For details, see Barrett (1984).

The obvious generalization of Eq. (5) expresses the forward Radon transform in mD, while Eq. (6) still works for the central-slice theorem. The counterpart of Eq. (14) or (52) is

$$f(\mathbf{r}) = \mathcal{R}_m^{-1}\{\lambda_{\mathbf{n}}(p)\} = \int_{-\infty}^{\infty} |\nu|^{m-1} d\nu \int_h d\Omega_n \Lambda_{\mathbf{n}}(\nu)\, e^{2\pi i \nu \mathbf{r} \cdot \mathbf{n}} \, , \tag{65}$$

where the angular integral runs over a hemisphere (denoted by the subscript h) in the mD space. Thus the filter function in mD has the form $|\nu|^{m-1}$.

At this point a distinction must be made between even and odd m since $|\nu|^{m-1}$ has a slope discontinuity for m even but not for m odd. For the former case we find

$$f(\mathbf{r}) = \frac{i}{\pi} \left[\frac{1}{2\pi i}\right]^{m-1} P \int_h d\Omega_n \left[\frac{1}{p} * \frac{d^{m-1}\lambda_\mathbf{n}(p)}{dp^{m-1}}\right]_{p=\mathbf{r}\cdot\mathbf{n}} . \qquad (66)$$

(m even)

For m odd, the result is

$$f(\mathbf{r}) = \left[\frac{1}{2\pi i}\right]^{m-1} \int_h d\Omega_n \left[\frac{d^{m-1}\lambda_\mathbf{n}(p)}{dp^{m-1}}\right]_{p=\mathbf{r}\cdot\mathbf{n}} . \qquad (67)$$

(m odd)

Thus the filtering operation is local (a simple derivative) for odd m and global (because of the convolution with 1/p) for m even.

REFERENCES

Barrett, H.H. (1984). The Radon Transform and Its Applications, *Progress in Optics XXI*, E. Wolf (ed.).
Lighthill, M.J. (1962). *Fourier Analysis and Generalized Functions*, Cambridge University Press.

SUGGESTED READING

Barrett, H.H. and Swindell, W. (1981). *Radiological Imaging: Theory of Image Formation, Detection and Processing*, Academic Press, New York.
Deans, S.R. (1983). *The Radon Transform and Some of Its Applications*, Wiley, New York.
Helgason, S. (1980). *The Radon Transform*, Birkhaüser, Boston.
Herman, G.T. (1980). *Image Reconstruction from Projections, The Fundamentals of Computerized Tomography*, Academic Press, New York.
Radon, J. (1917). Ueber die Bestimmung von Funktionen durch ihre Integralwerte längs gewisser Mannigfaltigkeiten, *Ber. Saechs. Akad. Wiss.* (Leipzig) 69, pp. 262-278.
Swindell, W. and Barrett, H.H. (1977). Computerized Tomography: Taking Sectional X-rays, *Physics Today* 30, pp. 32-41.

# REGULARIZATION TECHNIQUES IN MEDICAL IMAGING

Frank Natterer

*Universität Münster*
*West Germany*

ABSTRACT

*We give a very short account of ill-posed problems and the method of regularization. We then show how this method is being used in various problems from tomography, such as incomplete problems and the problem of attenuation correction in emission computed tomography.*

1. THE CONCEPT OF REGULARIZATION. ILL-POSED PROBLEMS

A problem is called ill-posed if it suffers from either of the following deficiencies:

(i)   It has no solution.
(ii)  Its solution is not unique.
(iii) Its solution does not depend continuously on the data.

The problems of medical imaging are notoriously ill-posed, even to a very different degree. There is a vast and rapidly growing literature on ill-posed problems, see e.g. Tikhonov et al. (1977), Lavrent'ev et al. (1983), Groetsch (1983).

More specifically, let us consider a linear bounded operator $A: H \to K$ where $H, K$ are Hilbert spaces. We want to solve the equation

$$Af = g \qquad (1)$$

for f. Many problems in medical imaging can be formulated in this way.

Equation (1) is ill-posed if $A^{-1}$ does not exist, or if $A^{-1}$ is not defined on all of K, or if $A^{-1}$ is unbounded. The practical difficulty in the latter case is that the solution of $Af_\delta = g_\delta$ need not be close to f even if $g_\delta$ is close to g. Thus, Eq. (1) cannot be solved in the presence of modelling or measurement errors.

In the sequel we shall remove step by step the deficiencies of an ill-posed problem. First we define a substitute for the solution if there is none. We simply take a minimizer of $\|Af - g\|$ as such a substitute. Then we dispose of a possible non-uniqueness by choosing among all those minimizers that one which has minimal norm. We denote this element by $A^+g$. $A^+$ is known as the generalized inverse in the sense of Moore and Penrose. It is defined in a dense subset of K.

$A^+$ may still be an unbounded operator. Any method for evaluating $A^+g$ approximately in a stable way is called a regularization method. More precisely, a family $(T_\gamma)_{\gamma>0}$ of linear bounded operators from K into H is called a regularization of $A^+$ if

$$T_\gamma g \to A^+g \quad \text{as} \quad \gamma \to 0 \tag{2}$$

whenever $A^+g$ makes sense. We give some examples.

a. *The truncated singular value decomposition*

By a singular value decomposition we mean a representation of A in the form

$$Af = \sum_{k=1}^{\infty} \sigma_k (f,f_k) g_k \tag{3}$$

where $(f_k)$, $(g_k)$ are normalized orthogonal systems in H,K resp. and $\sigma_k$ are positive numbers, the singular values of A. We always assume the sequence $\{\sigma_k\}$ to be bounded. Then, A is a linear continuous operator from H into K with adjoint

$$A^*g = \sum_{k=1}^{\infty} \sigma_k (g,g_k) f_k \tag{4}$$

and the operators

$$A^*Af = \sum_{k=1}^{\infty} \sigma_k^2 (f,f_k) f_k \tag{5}$$

$$AA^*g = \sum_{k=1}^{\infty} \sigma_k^2 (g,g_k) g_k \qquad (6)$$

are selfadjoint operators in H,K respectively. The spectrum of $A^*A$ consists of the eigenvalues $\sigma_k^2$ with eigenelements $f_k$ and possibly of the eigenvalue 0 whose multiplicity may be infinite. The same is true for $AA^*$ with eigenelements $g_k$. The two eigensystems are related by

$$A^* g_k = \sigma_k f_k, \qquad A f_k = \sigma_k g_k. \qquad (7)$$

Vice versa, if $(f_k)$, $(g_k)$ are normalized eigensystems of $A^*A$, $AA^*$ respectively such that Eq. (7) holds, then A has the singular value decomposition (3). In particular, compact operators always admit a singular value decomposition.

In terms of the singular value decomposition, $A^+$ can be written as

$$A^+ g = \sum_{k=1}^{\infty} \sigma_k^{-1} (g,g_k) f_k. \qquad (8)$$

We see from Eq. (8) that $A^+$ is unbounded if and only if $\sigma_k \to 0$ as $k \to \infty$. In that case we can construct a regularization of $A^+$ by the truncated value decomposition

$$T_\gamma g = \sum_{k \leq 1/\gamma} \sigma_k^{-1} (g,g_k) f_k. \qquad (9)$$

It follows from Eq. (8) that $T_\gamma g \to A^+ g$ as $\gamma \to 0$, and $T_\gamma$ is bounded with

$$\|T_\gamma\| \leq \sup_{k \leq 1/\gamma} \sigma_k^{-1}. \qquad (10)$$

More generally we can put

$$T_\gamma g = \sum_{k=1}^{\infty} F_\gamma(\sigma_k) (g,g_k) f_k \qquad (11)$$

where $F_\gamma(\sigma)$ approximates $\sigma^{-1}$ for $\sigma$ large and tends to zero as $\sigma \to 0$. $F_\gamma$ is called a filter and regularization methods based on $F_\gamma$ are referred to as digital filtering.

The singular value decomposition gives much insight into the character of the ill-posedness. Let $g_\delta$ be an approximation to $g$ such that $\|g - g_\delta\| \leq \delta$. Knowing only $g_\delta$, all we can say about the expansion coefficients for $A^+g$ in (8) is

$$|\sigma_k^{-1}(g,g_k) - \sigma_k^{-1}(g_\delta,g_k)| \leq \delta/\sigma_k. \tag{12}$$

We see that for $\sigma_k$ small, the contribution of $g_k$ to $A^+g$ cannot be computed reliably. Thus looking at the singular values and the corresponding elements $g_k$ shows which features of the solution $f$ of Eq. (1) can be determined from an approximation $g_\delta$ to $g$ and which ones can not.

b. *The method of Tikhonov-Phillips*

Here we put

$$T_\gamma = (A^*A + \gamma I)^{-1} A^* = A^*(AA^* + \gamma I)^{-1}. \tag{13}$$

Equivalently, $f_\gamma = T_\gamma g$ can be defined to be the minimizer of

$$\|Af - g\|^2 + \gamma \|f\|^2. \tag{14}$$

If $A$ has the singular value decomposition (3) it is readily seen that $f_\gamma$ has the form (11) with

$$F_\gamma(\sigma) = \frac{1}{1 + \gamma/\sigma^2} \frac{1}{\sigma}. \tag{15}$$

Thus the method of Tikhonov-Phillips is a special case of digital filtering.

c. *Iterative methods*

As most simple case we consider the iteration proposed by Landweber (1951). Here, a sequence $f^t$, $t = 0,1,\ldots$ is computed from

$$f^{t+1} = f^t + \gamma^2 A^*(g - Af^t) \tag{16}$$

$\gamma > 0$ being a parameter. We analyse (16) with the help of the singular value decomposition (3) of $A$. Let $f^0 = 0$. Then,

$$f^t = \sum_{k=1}^{p} c_k^t f_k \tag{17}$$

where the $c_k^t$ satisfy the recursion

$$c_k^{t+1} = (1 - \gamma^2 \sigma_k^2) c_k^t + \gamma^2 \sigma_k (g, g_k). \tag{18}$$

From $c_k^o = 0$ we get

$$c_k^t = F_t(\sigma_k) \sigma_k^{-1} (g, g_k) \tag{19}$$

$$F_t(\sigma) = 1 - (1 - \gamma^2 \sigma^2)^t. \tag{20}$$

Hence

$$f^t = \sum_{k=1}^{p} F_t(\sigma_k) \sigma_k^{-1} (g, g_k) f_k. \tag{21}$$

We see that Landweber's iteration can be viewed as a special case of digital filtering. It converges to $A^+ g$ provided that $\gamma \sigma_1 < 1$. We also note that the speed of convergence of the contribution of $f_k$ depends on the size of $\sigma_k$. For $\sigma_k$ large, the convergence is fast, but for small $\sigma_k$'s, the convergence is slow. Thus in the early iterates, the contribution of the large singular values are well represented, while the contribution of the small singular values appear only later in the iteration process. This shows that stopping the iteration after a finite number of steps has the same effect as regularization. Carrying out too many steps may destroy the accuracy already obtained since for t large, the iterates pick up the contributions of the small singular values which are very sensitive with respect to data errors. This semi-convergence of iterative methods for ill-posed problems in the presence of noise is very typical.

## 2. REGULARIZATION IN STANDARD CT. FILTER SELECTION

In CT we have to invert the Radon transform

$$Rf(\theta, s) = \int_{x \cdot \theta = s} f(x) dx \tag{22}$$

where $\theta \in S^{n-1}$, the surface of the unit sphere in $\mathcal{R}^n$, and $s \in \mathcal{R}^\wedge$. Of course in CT we are only interested in $n = 2$, but we treat the general case.

Minimizing

$$\|Rf - g\|^2 + \gamma \|f\|^2 \tag{23}$$

in $L_2(\mathcal{R}^n)$ leads to

$$f = R^*(RR^* + \gamma I)^{-1} g \tag{24}$$

with

$$R^*g(x) = \int_{S^{n-1}} g(\theta, x \cdot \theta) d\theta \tag{25}$$

the backprojection operator, which happens to be the adjoint of R. Taking Fourier transforms it can be shown that

$$(RR^* + \gamma I)^{-1} g = h * g \tag{26}$$

with

$$h(\sigma) = \frac{1}{2} (2\pi)^{1/2-n} |\sigma|^{n-1} F_\gamma(\sigma),$$

$$F_\gamma(\sigma) = 1/(1 + \gamma' |\sigma|^{1-n}), \quad \gamma' = \gamma/2(2\pi)^{n-1/2}, \tag{27}$$

see e.g. Natterer (1986, chapt. II). From Eqs. (24), (26), (27) we see that computing the Tikhonov-Phillips regularized solution is equivalent to a filtered backprojection method with a special filter depending on the regularization parameter $\gamma$. Thus, not much is gained using regularization techniques for standard CT.


3. A REGULARIZATION ALGORITHM FOR CT IN THE CASE OF ROTATIONAL INVARIANCE

If the scanning geometry is rotational invariant, a technique quite different from filtered backprojection can be used. Assume that $L_{j,\ell}$, $\ell = 1,...,q$, $j = 0,...,p-1$ model the strip or (2D or 3D) cones which we integrate on, i.e. our data are now

$$R_{j\ell}f = \int_{L_{j,\ell}} f(x)dx = g_{j\ell}. \qquad (28)$$

We want to compute a regularized solution of (28), i.e. the minimizer of

$$\sum_{j,\ell} |R_{j\ell}f - g_{j\ell}|^2 + \gamma\|f\|^2 \qquad (29)$$

in $L_2(\Omega)$, $\Omega \subseteq \mathcal{R}^n$. Putting

$$R_j = \begin{pmatrix} R_{j1} \\ \vdots \\ R_{jq} \end{pmatrix}, \qquad g_j = \begin{pmatrix} g_{j1} \\ \vdots \\ g_{jq} \end{pmatrix} \qquad (30)$$

and

$$R = \begin{pmatrix} R_0 \\ \vdots \\ R_{p-1} \end{pmatrix}, \qquad g = \begin{pmatrix} g_0 \\ \vdots \\ g_{p-1} \end{pmatrix} \qquad (31)$$

We have $R: L_2(\Omega) \to \mathcal{R}^{pq}$. Minimizing (29) is now equivalent to solving

$$(RR^* + \gamma I)h = g, \qquad f = R^*h. \qquad (32)$$

The adjoint $R^*$ is easily seen to be

$$R^*h = \sum_{j=0}^{p-1} R_j^*h_j, \qquad R_j^*h_j = \sum_{\ell=1}^{q} h_{j\ell}\chi_{j,\ell} \qquad (33)$$

with $\chi_{j,\ell}$ the characteristic function of $L_{j,\ell}$. Thus, $R^*$ is simply a discrete version of the backprojection operator.

The matrix $RR^* + \gamma I$ is a qp×qp-matrix. In general this is by far too large for inverting this matrix on present day computers. However we shall see that in the case of rotational invariance of the $L_{j,\ell}$, the linear system in (32) can be solved very efficiently by means of FFT. By rotational invariance we mean the following: $L_{j+1,\ell}$ is obtained from $L_{j,\ell}$ by a rotation with angle $2\pi/p$ around the origin. It is clear that the parallel scanning geometry as well as fan-beam scanning satisfy this assumption. In 3D the cone-beam geometry, with the sources spinning around the object on one or on two parallel circles provide examples for rotational invariance. In that case, we have for the $k,\ell$-element of the $q \times q$-matrix $R_i R_j^*$

$$\begin{aligned}(R_i R_j^*)_{k\ell} &= \int_\Omega \chi_{i,k} \chi_{j,\ell}\, dx \\ &= \int_\Omega \chi_{o,k} \chi_{j-i,\ell}\, dx \\ &= (R_o R_{j-i}^*)_{k\ell}\end{aligned} \qquad (34)$$

where $i - j$ has to be taken mod $p$. Thus, $R_i R_j^* = S_{i-j}$, and the matrix $RR^*$ of (32) assumes the form

$$RR^* = \begin{pmatrix} S_o & S_1 & \cdots\cdots & S_{p-1} \\ S_{p-1} & S_o & \cdots\cdots & S_{p-2} \\ & & \vdots & \\ S_1 & & & S_{p-1}\ S_o \end{pmatrix}. \qquad (35)$$

We see that we end up with a block-cyclic matrix which is amenable to FFT methods. Properly implemented, (32) can be done in about the same number of operations as filtered backprojection, see Natterer (1980), Buonocore et al. (1981).


4. REGULARIZATION METHODS FOR INCOMPLETE DATA PROBLEMS

The great avantage of (32) over filtered backprojection is that it can be used also for incomplete data, if only the assumption of rotational invariance can be satisfied. This is the case for the interior and for the exterior problem. In the exterior problem, only lines in an annulus $0 < a \leqslant |x| \leqslant 1$ are measured, while in the interior problem, only lines in $|x| \leqslant a < 1$ are measured. Of course the reconstructions have to be done in the respective scanned regions only.

A general approach to incomplete data problems is data completion, i.e. computing the missing data from the available ones. We describe the procedure, as suggested by Louis (1980), Peres (1979), for the limited angle problem. Here, $g^I = Rf(\Theta,s)$ is given only for $\Theta = \begin{pmatrix}\cos\varphi\\ \sin\varphi\end{pmatrix}$, $|\varphi| \leq \phi < \pi/2$, while for $|\varphi| > \phi$, only a rough approximation $g^*$ to $Rf$ is available, e.g. $g^* = 0$. The idea is to compute a 'consistent' data set $g^C$ which coincides with $g^I$ on $|\varphi| \leq \phi$ and is not too far from $g$ in $|\varphi| > \phi$. By 'consistent' we mean that $g^C$ satisfies the Helgason-Ludwig consistency conditions for the range of the Radon transform, see Helgason (1980). For $f$ supported in the unit disk, these conditions read ($U_m$ are the Chebyshev polynomials of the second kind)

$$g_m^C(\varphi) = \frac{2}{\pi}\int_{-1}^{+1} U_m(s) g^C(\Theta,s) ds = \sum_{|\ell|\leq m} c_{m\ell} e^{i\ell\varphi} \quad (36)$$

with constants $c_{m\ell}$ which are 0 for $m + \ell$ odd. The idea is now to determine the $c_{m\ell}$ from (36) in $|\varphi| \leq \phi$ and then computing $g_m^C(\varphi)$ from (36) for $|\varphi| > \phi$.

For a finite number of angles $\varphi_j$, $j = 1,\ldots,p$, distributed in $[-\phi,\phi]$ the determination of the constants $c_{m\ell}$ calls for the solution of the linear system

$$g_m^I(\varphi_j) = \sum_{|\ell|\leq m} c_{m\ell} e^{i\ell\varphi_j} \quad (37)$$

which is overdetermined for $m < p$ and underdetermined for $m > p$. In any case it is highly unstable. This is to be expected since the limited angle problem is severely ill-posed, see Davison (1983). Therefore we replace (37) by the regularized problem: Minimize

$$\sum_{j=1}^{p}\left| \sum_{|\ell|\leq m} c_{m\ell} e^{i\ell\varphi_j} - g_m^I(\varphi_j)\right|^2 + \gamma \int_{-1}^{+1}\int_{|\varphi|>\phi} |g^C - g^*|^2 ds d\Theta. \quad (38)$$

We investigate the character of the ill-posedness of (37) more closely. Denoting the matrix of (37) by $A_m$, we have

$$(A_m^* A_m)_{k\ell} = \sum_{j=1}^{p} e^{i(\ell-k)\varphi_j}, \quad k,\ell = -m,\ldots,m;\ k+m,\ \ell+m\ \text{even.} \quad (39)$$

For equally spaced angles $\varphi_j$ we have

$$\lim_{p\to\infty} \frac{2\phi}{\pi p} (A_m^* A_m)_{k\ell} = \frac{1}{\pi}\int_{-\phi}^{\phi} e^{i(\ell-k)\varphi} d\varphi. \quad (40)$$

This matrix has been studied by Slepian (1978). For $m$ large, roughly $\frac{2\phi}{\pi} m$ of its $m + 1$ eigenvalues are close to 1, while the other ones are exponentially small. That means that (37) can be solved, even approximately, for very modest values of m only, and regularization is mandatory here. In our practical examples we used regularization by the truncated singular value composition and by Tikhonov-Phillips. Both methods performed satisfactorily when properly tuned. The resulting artifacts in the reconstructions are discussed in Natterer (1986, chapt. VI), for general incomplete data problems.

In some applications, in particular in industrial tomography, $Rf(\Theta,s)$ is known for very few directions $\Theta$ only, see Rangayan, Dhawan and Gordon (1985). In this case we cannot expect f to be determined uniquely. However, there is a result due to Gardner and Mc Mullen (1980), saying that f can be determined from 4 directions, provided f describes a convex homogeneous object, i.e. f is constant within a convex set and zero outside. A stability result for this reconstruction problem has been obtained by Volčič (1983).

All these results encouraged us to try the following approach. Assume f to be constant in a star-shaped domain whose boundary is given in polar coordinates by

$$r = r(\Theta), \qquad \Theta = \begin{pmatrix} \cos\varphi \\ \sin\varphi \end{pmatrix}, \qquad 0 \leq \varphi < 2\pi. \tag{41}$$

Starting out from the so-called 'projection theorem'

$$\hat{f}(\sigma\omega) = (2\pi)^{-1/2}\, \hat{g}(\omega,\sigma) \tag{42}$$

where

$$\hat{f}(\xi) = \frac{1}{2\pi} \int e^{-i \cdot \xi}\, f(x)\, dx,$$

$$\hat{g}(\omega,\sigma) = (2\pi)^{-1/2} \int e^{-i\sigma s}\, g(\omega,s)\, ds \tag{43}$$

are the Fourier transforms of f,g, respectively, we derive an integral equation for the function r. We have

$$\begin{aligned}
\hat{f}(\sigma\omega) &= \frac{1}{2\pi} \int_0^{2\pi} \int_0^{r(\Theta)} \rho e^{-i\sigma\rho\omega \cdot \Theta}\, d\rho\, d\varphi \\
&= \frac{1}{2\pi} \int_0^{2\pi} \int_0^{r(\Theta)} \rho e^{-i\sigma\rho\cos(\varphi-\psi)}\, d\rho\, d\varphi
\end{aligned} \tag{44}$$

where $\omega = \begin{pmatrix} \cos\psi \\ \sin\psi \end{pmatrix}$. The inner integral is evaluated by the formula

$$\int_0^r \rho e^{-ia\rho} d\rho = r^2 K(iar)$$

$$K(u) = \frac{1 - e^{-u}}{u^2} - \frac{e^{-u}}{u}, \quad u \neq 0 \tag{45}$$

which is easily established by integration by parts. We obtain

$$\hat{f}(\sigma\omega) = \frac{1}{2\pi} \int_0^{2\pi} r^2(\Theta) K(i\sigma r(\Theta) \cos(\varphi - \psi)) d\varphi. \tag{46}$$

Note that $K$ is a $C^\infty$ function. This makes the nonlinear integral equation (46) seriously ill-posed. We solved it by a Tikhonov-Phillips-type regularization, which consists in minimizing

$$\sum_{j=1}^p \|\hat{g}(\omega_j, \cdot) - A_j r\|_{L_2(\mathcal{R}^1)} + \gamma \|r\|^2_{L_2(0, 2\pi)} \tag{47}$$

where, by (42)

$$A_j r(\sigma) = (2\pi)^{-1/2} \int_0^{2\pi} r^2(\Theta) K(i\sigma r(\Theta) \cos(\varphi - \psi_j)) d\varphi, \tag{48}$$

$\psi_j$ being the directions for which $g$ is known. This is done by a Gauss-Newton method. We got quite satisfactory results even for non-convex objects, using only 2-4 directions, see Natterer (1986, chapt. VI).

## 5. ATTENUATION CORRECTION IN EMISSION COMPUTED TOMOGRAPHY

In emission computed tomography one has to solve the integral equation

$$g = e^{-R\mu} Rf \tag{49}$$

(PET = positron emission tomography), or

$$g(\Theta, s) = \int_{x \cdot \Theta = s} e^{-D\mu(x, \Theta^\perp)} f(x) dx$$

$$D\mu(x, \Theta) = \int_0^\infty \mu(x + t\Theta) dt \tag{50}$$

(SPECT = single particle emission computed tomography), see Budinger and Gullberg (1974). In both cases, f is the distribution of a radiopharmaceutical inside the body whose absorption is $\mu$. f is the quantity we are looking for, while $\mu$ plays the role of a parameter. If $\mu$ is known, (49) reduces to the standard tomography problem, while (50) becomes an integral equation very similar to the Radon integral equation which can be solved by iterative methods (Heike, 1984) or by a filtered backprojection type algorithm if $\mu$ is constant inside the body, Tretiak and Metz (1980).

In general, $\mu$ is unknown, and has to be determined somehow. The usual procedure is to do a transmission scan prior to the emission scan in order to determine $\mu$. This is a time-consuming procedure. Therefore, attempts have been made to determine $\mu$ from the data. One method, suggested by Censor, Gustafson, Lent and Tuy, tries to compute $\mu$ and f simultaneously from g by an iterative procedure. Another method consists in computing $\mu$ from g prior to solving (49) or (50) by using the consistency conditions for g, see Natterer (1983). We describe this method for (49), which is the simpler case.

Let f be a density with compact support, e.g. f = 0 outside the unit disk. Then, Rf satisfies the Helgason-Ludwig consistency conditions

$$\int_{-1}^{+1} s^m Rf(\Theta,s) ds = P_m(\Theta), \qquad m = 0,1,\ldots \tag{51}$$

where $P_m$ is a homogeneous polynomial of degree m. If we apply this to (49), we get

$$\int_{-1}^{+1} s^m e^{R\mu(\Theta,s)} g(\Theta,s) ds = P_m(\Theta). \tag{52}$$

This is a system of nonlinear equations for $R\mu$, and we may try to solve (52) for $R\mu$.

To begin with, let us assume that $\mu$ is rotationally invariant, i.e. $\mu(x)$ depends on $|x|$ only. Then, $R\mu$ is a function of s only. Putting $h = \exp(R\mu)$, Eq. (52) reads for m = 0

$$\int_{-1}^{+1} h(s) g(\Theta,s) ds = c_o \tag{53}$$

$c_o$ being a constant. Incidentally, $c_o$ is the total mass of f.

Equation (53) is a first kind integral equation for h. Since $c_o$ is not known, it permits (provided it is uniquely solvable) to determine $R\mu$ up to an additive constant which is easily determined from $\mu = 0$ outside the body. Unfortunately, a singular value decomposition of (53) reveals that (53) is severely ill-posed to such an extent that its solution is virtually impossible. To demonstrate this — and to show how we overcame this difficulty — we describe a numerical experiment in a detailed way.

Our source distribution f is 1 in the circle around $(0,-0.2)$ with radius 0.4 and zero elsewhere. The attenuation $\mu$ is 1.74 in $|x| \leq 0.675$, 2.08 in $0.675 < |x| \leq 0.75$ and 0 elsewhere. This corresponds roughly to a model of the human head at 100 KeV, see Herman (1980, chapt. 4). We took 64 directions with 41 equally spaced line integrals each. The first 6 singular values are given in the second column of Table 1. The remaining ones are less than 0.001. Because of the high noise level in emission tomography, it is completely out of question to use these small singular values in a truncated singular value expansion. Even the use of the second singular value is questionable.

The situation becomes much more favourable if we add an exterior point source of strength $f_o = 0.1$ at $(1,0)$. Then we get the singular values in the third column of Table 1. In contrast to the first column we now have a gentle decrease of the singular values, indicating that Eq. (53) is only modestly ill-posed in the presence of an exterior point source.

In order to explain what happened we study (53) with f replaced by $f_o \delta_a + f$ where $\delta_a$ is Dirac's $\delta$-function at a, obtaining

$$\int_{-1}^{+1} h(s) \int_{-1}^{+1} \left\{ f_o \delta_a (s\theta + t\theta^\perp) + f(s\theta + t\theta^\perp) \right\} dt\, ds = c_o. \tag{54}$$

Putting $x = s\theta + t\theta^\perp$ we get

$$\int h(x \cdot \theta) \left\{ f_o \delta_a(x) + f(x) \right\} dx = c_o \tag{55}$$

or

$$f_o h(a \cdot \theta) + \int h(x \cdot \theta) f(x) dx = c_o. \tag{56}$$

| k | $\sigma_k$ ($f_o = 0$) | $\sigma_k$ ($f_o = 0.1$) |
|---|---|---|
| 1 | 0.104 | 0.304 |
| 2 | 0.021 | 0.250 |
| 3 | 0.007 | 0.245 |
| 4 | 0.003 | 0.204 |
| 5 | 0.002 | 0.200 |
| 6 | 0.001 | 0.143 |
| 10 | - | 0.027 |
| 21 | - | 0.007 |

*Table 1. Singular values of (53) without and with exterior point source.*

Thus, even though (53) looks like an (ill-posed) integral equation of the first kind it is in fact a (well-posed) integral equation of the second kind, due to the presence of the point source.
The condition number of this second kind integral equation can be made arbitrarily close to 1 by choosing $f_o$ sufficiently large.

Our numerical experiments confirm that rotationally symmetric attenuation coefficients can be recovered reliably with the help of an exterior source. Unfortunately, human beings are not rotationally symmetric. Therefore we have studied the general problem, again with the help of the singular value decomposition. It is not yet clear if this approach can be used in the clinical practice.

REFERENCES

Budinger, T.F. and Gullberg, G.T. (1979). Emission computed tomography. In: *Image reconstruction from projections*, G.T. Herman (ed.), Springer-Verlag, Berlin.

Buonocore, M.H., Brody, W.R., and Macowski, A. (1981). A natural pixel decomposition for two-dimensional image reconstructions, *IEEE Trans. on Biom. Eng.* BME-28, pp. 69-77.

Censor, Y., Gustafson, D.E., Lent, A., and Tuy, H. (1979). A new approach to the emission computerized tomography problem: simultaneous calculations of attenuation and activity coefficients, *IEEE NS, Special Issue*, April 1979.

Davison, M.E. (1983). The ill-conditioned nature of the limited angle tomography problem, *SIAM J. Appl. Math.* 43, pp. 428-448.

Gardner, R.J. and Mc Mullen, P. (1980). On Hammer's X-ray Problem, *J. London Math. Soc.* 21, pp. 171-175.

Groetsch, C.W. (1983). *The theory of Tikhonov regularization for Fredholm equations of the first Kind*, Pitman, Boston.

Heike, U. (1984). *Die Inversion der gedämpften Radon-Transformation, ein Rekonstruktionsverfahren der Emissionstomographie*, Thesis, Universität Münster.

Helgason, S. (1980). *The Radon transform*, Birkhäuser, Boston.

Herman, G.T. (1980). *Image reconstruction from projections*, Academic Press, New York.

Lavrent'ev, M.M., Romanov, V.G., and Sisatskij, S.P. (1983). *Problemi non ben posti in Fisica matematica ed Analisi*, Conzilio Nazionale delle Ricerche, Publicazioni dell'Instituo di Analisi Globale e Applicazioni, Serie 'Problemi non ben posti ed inversi', n. 12, Florence.

Louis, A.K. (1980). Picture reconstruction from projections in limited range, *Math. Meth. in the Appl. Sci.* 2, pp. 209-220.

Natterer, F. (1980). Efficient implementation of 'optimal' algorithms in computerized tomography, *Math. Meth. in the Appl. Sci.* 2, pp. 545-555.

Natterer, F. (1983). Computerized tomography with unknown sources, *SIAM J. Appl. Math.* 43, pp. 1201-1212.

Natterer, F. (1986). *The mathematics of computerized tomography*, Wiley-Teubner, Stuttgart.

Peres, A. (1979). Tomographic reconstruction from limited angular data, *J. Comput. Assist. Tomogr.* 3, pp. 800-803.

Rangayan, R.M., Dhawan, A.P., and Gordon, R. (1985). Algorithms for limited-view tomography: an annotated bibliography and chalange, *Appl. Optics* 24, pp. 4000-4012.

Slepian, D. (1978). Prolate spheroidal wave functions, Fourier analysis, and uncertainty V: The discrete case, *Bell. Syst. Techn. J.* 57, pp. 1371-1430.

Tikhonov, A.N. and Arsenin, V.Y. (1977). *Solution of ill-posed problems*, Wiley, New York.

Tretiak, O. and Metz, C. (1980). The exponential Radon transform, *SIAM J. Appl. Math.* 39, pp. 341-354.

Volčič, A. (1983). Well-posedness of the Gardner-Mc Mullen reconstruction problem, *Proceedings of the conference on measure theory*, Springer-Verlag, Berlin, Lecture Notes in Mathematics 1089, pp. 199-210.

STATISTICAL METHODS IN PATTERN RECOGNITION

C. Robert Appledorn

*Indiana University Medical Center*
*Indianapolis, USA*

ABSTRACT

*When measurements group together and begin to form clusters in some measurement feature space, one tends to remark that a pattern is developing. Furthermore, the size and shape of the pattern can be provided a statistical description. In a variety of applications, one is faced with the problem of using these statistical descriptions to classify a particular measurement to a specific cluster; that is, to make a decision regarding which pattern group generated the measurement.*
*This paper presents an overview of the mathematical considerations that statistical pattern recognition entails. Topics that receive emphasis include normalizations based upon covariance matrix eigen-factorizations, eigen-expansion feature extraction methods, linear classifier functions, and distance measurements. Particular emphasis is given to linear algebraic techniques that lead to simple computational implementations. Lastly, estimation in a noisy environment is briefly discussed.*

1. INTRODUCTION

Imagine for the moment a machine or some other process generating a sequence of numbers that an investigator is able to observe and measure. During repeat measurements, the investigator observes different sequences of numbers although they are related to the previously acquired measurements from the generator. After a period of time, the investigator is able to describe statistically the ensemble of measurements from this machine in terms of a mean value (or mean vector) and a variance (or a covariance matrix).

Imagine now a number of machines or processes, each generating a sequence of numbers. Each generator then has a statistical description associated with it. We assume each machine has some distinguishing quality about it so that the measurements will be different in some sense. Thus the statistical descriptions (mean and covariance) will be different also.

The investigator is often presented with the following situation which we refer to as the pattern recognition problem: Given an arbitrary sequence of measurements from a single source, decide which generator process produced it.

Complicating what appears to be a simple classification problem solved using a series of statistical hypothesis tests are a few issues the investigator needs to remain cognizant of. For example, the measurement sequences may consist of such a high dimensionality that the covariance matrix becomes numerically singular, i.e. non-invertible. A reduction in this dimensionality requires the investigator to choose judiciously a new basis onto which the measurements are projected.

In the monograph that follows, we will be discussing the statistical nature of the descriptions for the different class measurements. Reduction of measurement dimensionality will be discussed as well as other topics. Particular attention will be paid to maintaining separation between the class distributions and, in effect, minimizing classification errors. The emphasis of the paper will be upon linear methods that are easily implemented in most computational environments (Fukunaga, 1972; Swain and Davis, 1978).

## 2. DEFINITIONS AND NOTATION

A class $w_j$ (j = 1,...,M) is defined as the generator process that produces the observed set of measurements. The measurements are organized as an N-vector $\mathbf{x} = [x_1, x_2, ..., x_N]^T$ and are treated as random vectors. The individual elements $x_i$ (i = 1,...,N) are random variables and frequently are correlated with each other. Probability density functions (pdf) describe the statistical distribution of the random vectors.

The joint pdf over all measurements and all classes is denoted as $p(\mathbf{x},w)$. It can be either continuous or discrete in $\mathbf{x}$, but is discrete in w. More useful will be the marginal pdf, $p(\mathbf{x})$ and $p(w)$, and the conditional pdf, $p(\mathbf{x}|w)$ and $p(w|\mathbf{x})$. Although no restrictions are placed upon the pdf structure, one often will assume a normal or Gaussian structure for the conditional pdf $p(\mathbf{x}|w)$.

The random vectors $\mathbf{x}$ within a given class $w_j$ have a mean vector $\mathbf{m}_j$ and a covariance matrix $C_j$ that describe the expected value statistics for that class. We denote these expected values as

$$\mathbf{m}_j = E\{\mathbf{x} | w_j\} \tag{1}$$

$$C_j = E\{(\mathbf{x} - \mathbf{m}_j)(\mathbf{x} - \mathbf{m}_j)^T | w_j\}, \qquad (j = 1,\ldots,M). \tag{2}$$

The expected value operator $E\{\cdot\}$ is defined as the integral of the argument weighted by its pdf. If a Gaussian pdf is assumed for a particular class $w_j$, this is denoted as $\mathbf{x} \sim N[\mathbf{m}_j, C_j, P_j]$ $(j = 1,\ldots,M)$ where $P_j = p(w_j)$, the probability that the class $w_j$ occurs. Frequently, the more compact notation $\mathbf{x} \sim N[\mathbf{m}_j, C_j]$ will be employed.

The population statistics, without regard to class, are given by

$$\mathbf{m}_0 = E\{\mathbf{x}\} \tag{3}$$

$$C_0 = E\{(\mathbf{x} - \mathbf{m}_0)(\mathbf{x} - \mathbf{m}_0)^T\}. \tag{4}$$

The mean vector for the entire population of measurements $\mathbf{m}_0$ is related to the individual class mean vectors by

$$\mathbf{m}_0 = \Sigma_j P(w_j) E\{\mathbf{x}|w_j\} = \Sigma_j P_j \mathbf{m}_j. \tag{5}$$

The total population covariance $C_0$ is more interesting. If the above expression for $C_0$ is expanded, one finds

$$C_0 = B + W \tag{6}$$

where the between-class covariance $B$ represents the separation between class mean vectors and is given by

$$B = \Delta\Delta^T \tag{7}$$

$$\Delta = \frac{1}{2}(\mathbf{m}_1 - \mathbf{m}_2). \tag{8}$$

The within-class covariance $W$ represents the distribution scatter about a given class mean vector and is given by

$$W = \frac{1}{2}(C_1 + C_2). \tag{9}$$

Here, we have assumed for simplicity of illustration a two-class problem ($M = 2$) with equal class probabilities ($P_1 = P_2$). Of course, the notion of between-class covariance and within-class covariance can be generalized to higher order multi-class problems with unequal class probabilities.

Often it is necessary (or desirable) to express a covariance matrix ($C_0$, $C_j$, or $W$) as its eigenvector-eigenvalue expansion, i.e.,

$$C = V\Lambda V^T \tag{10}$$

where $V = [v_1|v_2|\ldots|v_N]$ is the matrix of column eigenvectors and $\Lambda = \mathrm{diag}(\lambda_i)$ is a diagonal matrix of the corresponding eigenvalues. The eigenvectors are normalized to unit length thus satisfying $V^T = V^{-1}$. We will assume for the time being that $\lambda_i > 0$ ($i = 1,\ldots,N$) thus assuring the existence of the inverse of the covariance matrix.

The importance of the eigenvalue-eigenvector expansion of the covariance matrix is to determine the orientation of the class distribution within a measurement or feature space (Fukunaga, 1972). The class distribution can be imagined as being described as a generalized ellipsoid distributed about the class mean vector **m**. The direction of each principal axis of this ellipsoid is given by each of the eigenvectors. The length of each ellipsoid principal axis is given by the square root of the corresponding eigenvalue. Thus, the mean vector and the covariance eigen-expansion provides one with information regarding the position, size, shape, and orientation of the class distribution in the multi-dimensional measurement space.


3. NORMALIZATIONS

The topic of normalization needs to be approached very carefully. Before an arbitrary normalization procedure is applied to a data set of measurements, one must always ask the question, "What is this doing to my hard-earned data?" Such procedures might include unit-length and unit-area normalizations. Inappropriately applied normalizations could possibly destroy features that might have led to the successful classification of a measurement pattern. The normalizations that are considered here are restricted to those which preserve classification features and which recast the classification problem into a simpler mathematical form. In particular, the aim is to reduce the off-diagonal structure of the covariance matrix, which in turn reduces the computational complexity of the classifier functions considered in the material that follows.

a. *Covariance diagonalization — one-class problem*

It is well-known that if a sequence of random numbers is used to produce a new sequence by subtracting the mean value and dividing by the standard deviation, the result has zero mean and unit variance. This procedure is easily generalized to the case involving random vectors if the vector elements are decorrelated - the resulting statistics are the zero vector and the identity matrix.

For the case of correlated measurement vector elements (i.e., the covariance matrix has non-zero off-diagonal structure), a more complete transformation of the data vector is required (Fukunaga, 1972). For $x \sim N[m,C]$, the covariance matrix is expressed as its eigen-expansion, $C = V\Lambda V^T$. The normalizing transformation

$$T_a = V\Lambda^{-1/2} \quad \text{or} \quad T_a^T = \Lambda^{-1/2}V^T \tag{11}$$

is applied to the measurement vector

$$y = T_a^T(x - m) \tag{12}$$

to generate a new normalized measurement. The resulting statistics are $y \sim N[0,I]$ where $0$ is a zero vector and $I$ is an identity matrix, both of appropriate dimension.

Descriptively, the transformation has three distinct stages:
  1. Shift the distribution to the origin by subtracting the mean vector $m$. This step results in a zero mean vector.
  2. Rotate the distribution about the origin to align the distri-

bution's principal axes parallel to the coordinate axes. This step diagonalizes the covariance matrix and equivalently, results in decorrelated random variables.

3. Divide the measurement by the square root of the corresponding eigenvalue. This step rescales the distribution's principle axes and results in unit variance random variables.

b. *Simultaneous diagonalization — two-class problem*

In this case, measurements arise from a two-class problem in which two distributions are involved. The object is to diagonalize (decorrelate) both covariance matrices using a sequence of normalizing transformations (Fukunaga, 1972). As before, the eigen-expansions of the covariance matrices are required: $C_1 = V_1 \Lambda_1 V_1^T$ and $C_2 = V_2 \Lambda_2 V_2^T$. The normalizing transformation

$$T_b = V_1 \Lambda_1^{-1/2} V_2 \quad \text{or} \quad T_b^T = V_2^T \Lambda_1^{-1/2} V_1^T \tag{13}$$

is applied to the measurement vector

$$y = T_b^T (x - m) \tag{14}$$

resulting in a new measurement vector with decorrelated elements (i.e., diagonal covariance matrices). For this transformation, the normalized class 1 covariance is the identity matrix and the normalized class 2 covariance is diagonal but not necessarily the identity matrix.

The mean vector that is subtracted, denoted as $m$, can be $m_1$, $m_2$, or $m_0$. If $m = m_1$ is chosen, then the class 1 distribution is shifted to the origin resulting in $y \sim N[0,I]$ and $N[-2\delta,D]$ where $\delta = T_b^T \Delta$. A similar result is obtained if $m = m_2$. If $m = m_0$, the population is shifted to the origin resulting in $y \sim N[+\delta,I]$ and $N[-\delta,D]$. This author prefers the choice $m = m_0$ only because it results in a symmetrical distribution of the mean vectors about the origin so that $m_0 \rightarrow 0$.

In the (unlikely) event that $C_1 = C_2$, the simultaneous diagonalization transformation produces new covariance matrices both of which are equal to the identity matrix. In a higher order multi-class problem ($M > 2$) with all classes having the same covariance matrices $C_j = C$, all distributions are transformed to the identity covariance matrix. Unfortunately, this usually is not the case.

## 4. CLASSIFICATION

Although classification of a pattern into one of many ($M > 2$) classes is of general interest, the emphasis of this section will be restricted to the two-class classification problem. A further restriction will be to develop classifiers that are linear (and/or affine) in structure. Once this subject is treated, one extension to the general multi-class problem is to test a pattern pairwise between the multiple classes.

The general two-class classification problem is to test a measurement vector $x$ with a function $L_x(x)$, the value of which is used to select the class $w_j$ that most likely produced the measurement:

$$L_x(x) \gtrless \alpha \rightarrow \begin{matrix} x \in w_1 \\ x \in w_2 \end{matrix} \qquad (15)$$

If the value of the classifier function $L_x(x)$ exceeds the threshold, we select class 1; otherwise, we select class 2. The threshold parameter $\alpha$ is chosen to minimize a classification error criterion. Under appropriate conditions, the assignment of this parameter value degenerates to letting $\alpha = 0$.

### a. *Linear classifier — minimum mean-square error*

We require the classifier function $L_x(x)$ to be linear (actually, affine):

$$L_x(x) = \alpha_0 + \alpha_1 x_1 + \ldots + \alpha_N x_N = \alpha_0 + \alpha^T x \qquad (16)$$

where the classifier parameter vector $\alpha_i$ ($i = 0, \ldots, N$) will be determined using a mean-square error minimization technique. In order to set up the 'right-hand side' of this least-squares problem, target function values for the classifier must be specified. We make the following arbitrary choice:

$$E\{L_x(x) | x \in w_1\} = +1$$
$$E\{L_x(x) | x \in w_2\} = -1. \quad (17)$$

This is sufficient information to specify the least-squares problem for the linear classifier parameters:

$$\begin{bmatrix} 1 & x^T \\ 1 & x^T \end{bmatrix} \begin{bmatrix} \alpha_0 \\ \alpha \end{bmatrix} = \begin{bmatrix} +1 \\ -1 \end{bmatrix} \begin{matrix} \leftarrow \text{class 1} \\ \leftarrow \text{class 2.} \end{matrix} \quad (18)$$

Premultiplying both sides by the larger matrix above and taking expectations (first, over the random vectors, then over classes 1 and 2), the resulting normal equation for this problem becomes

$$\begin{bmatrix} 1 & m_0^T \\ m_0 & C_0 + m_0 m_0^T \end{bmatrix} \begin{bmatrix} \alpha_0 \\ \alpha \end{bmatrix} = \begin{bmatrix} 0 \\ \Delta \end{bmatrix}. \quad (19)$$

The left-hand side clearly is full rank, therefore, its inverse exists and a solution for the parameter values can be determined. After some tedious matrix algebra and an application of the matrix inversion lemma, one determines

$$L_x(x) = \Delta^T W^{-1} (x - m_0). \quad (20)$$

This is the desired result.

b. *Linear classifier — Bayes Gaussian*

We assume a two-class problem with the classes each normally distributed with pdf $N[m_j, C_j]$, $(j = 1, 2)$. We further assume that each class has equal covariance; i.e., $C_1 = C_2 = W$. The classifier function is taken to be the log likelihood function given by

$$L_x(x) = \log [p(x|w_1)/p(x|w_2)]. \quad (21)$$

In solving this Bayesian-type problem, one observes that due to the equal covariance assumption, the quadratic terms $(x^T W^{-1} x)$ cancel. This leaves a linear expression that reduces to

$$L_x(x) = \Delta^T W^{-1} (x - m_0). \tag{22}$$

This is in fact the same linear classifier that was obtained in the previous section, which is not particularly surprising - just reassuring.

c. *Linear classifier — other forms and remarks*

In the previous problems involving the determination of the linear classifier function, it was implicitly assumed that the class probabilities $P_j = p(w_j)$, $(j = 1,2)$ were equal. If this is not the case, then the threshold value $\alpha$ must be changed to a non-zero value in order to minimize classification errors. The Bayes optimal choice should be taken as

$$\alpha = \log \frac{P_2}{P_1} = \log \frac{1 - P_1}{P_2} \tag{23}$$

although some fine-tuning about this value may be required in practice.

One can make an interesting observation regarding the normalizing transformations of Section 3. For the case of equal covariances $W = C_1 = C_2$, note that $W^{-1} = T_a T_a^T = T_b T_b^T$. The linear classifier can be rewritten as

$$\begin{aligned} L_x(x) &= \Delta^T T_a T_a^T (x - m_0) \\ &= (T_a^T \Delta)^T y \\ &= \delta^T y \\ &= L_y(y) \end{aligned} \tag{24}$$

where $y = T_a^T (x - m_0)$ is simply the normalized measurement vector. We observe that $L_x(x) = L_y(y)$; thus, the normalization has absolutely no effect upon the performance of the linear classifier function. In fact, this statement is true for all invertible linear transformations. The benefit gained by using the normalized coordinate system is to reduce the computational complexity of the linear classifier function. The classification function becomes that of performing a simple vector dot product because the covariance matrix is the identity matrix and the matrix inversion is no longer required.

The linear classifier performs best when there exists separation between the class mean vectors ($\Delta \neq 0$). If this is not the case, then one must rely on separability due to unequal class covariances ($C_1 \neq C_2$). The Bayes Gaussian classifier then is no longer linear but takes on a quadratic form:

$$L_x(x) = \frac{1}{2} x^T (C_2^{-1} - C_1^{-1}) x + \frac{1}{2} \log [\det(C_2)/\det(C_1)]. \tag{25}$$

It can be seen that when the two-class problem degenerates to a single-class problem (equal means and equal covariances), no decision regarding classification can be made because $L_x(x) = 0$ always.

If the two-class distributions exhibit both separability of the means and the covariances, the Bayes classifier function is again quadratic. However, if the investigator wishes to employ a (sub-optimal) linear classifier then a minimum mean-square error approach applies. The linear classifier function has the same structure as before, except the within-class covariance is now taken as

$$W = \frac{1}{2} (C_1 + C_2). \tag{26}$$

## 5. FEATURE EXTRACTION

It is not an uncommon problem that the dimensionality of the measurement vectors **x** becomes so large (N large) that one begins to encounter numerical difficulties. Of course, there can be other motivating factors to reduce the measurement dimension to a more tractable size. This process is referred to by a number of names, a few of which include feature extraction, pattern representation, dimensionality reduction, and so forth (Fukunaga, 1972; Swain and Davis, 1978).

A number of procedures, quite different, exist and approach this general problem with different optimization goals as the criteria. These different optimization criteria can vary widely. One may wish to represent the measurement vector so that the approximation error is minimized or that the class separability is maximized. Both situations need to be examined and similarities identified if one is to gain insight into the feature extraction problem.

a. *Karhunen-Loève transformation*

The Karhunen-Loève transform (KLT) is a favorite among the feature extraction techniques. It involves the eigenvalue-eigenvector expansion of the covariance matrix and exhibits a great many error minimization properties. A number of forms of the KLT exist and space limitations do not permit an exhaustive overview of its variations. Therefore, this author must be satisfied with presenting one form of the KLT (naturally, his favorite).

Consider a single-class problem with mean vector **m** and covariance matrix **C**. The covariance matrix has an eigen-expansion $C = V \Lambda V^T$ in which the eigenvalues are ordered $\lambda_1 > \lambda_2 > \ldots > \lambda_N$ and the corresponding eigenvectors are organized similarly. We form the single-class normalization transformation (11). The complete Karhunen-Loève expansion **f** of a random vector **x** then is given by

$$f = T_a^T (x - m) \qquad (27)$$

which is invertible. The expansion coefficients $f = [f_1, f_2, \ldots, f_N]^T$ are referred to as factor weighings, factor loadings, as well as other labels that reflect the KLT's relationship to factor analysis and principle components.

We recognize that the vector **f** is identical to the normalized measurement vector **y** of the previous sections. We can infer then that the resulting pdf for the expansion coefficients has a zero mean vector and an identity covariance matrix. For the normally distributed case, we denote this as $f \sim N[0, I]$.

Thus far we have not extracted any features - we have only represented the measurements differently. To achieve a reduction in the dimensionality of the measurement vector, we select and save only L expansion coefficients where $L < N$. This new feature vector $f_L = [f_1, f_2, \ldots, f_L]^T$ is determined using

$$f_L = T_L^T (x - m) \tag{28}$$

where $T_L = V_L \Lambda_L^{-1/2}$. The non-square eigenvector matrix $V_L$ is given by $V_L = [v_1 | v_2 | \ldots | v_L]$ corresponding to the L largest eigenvalues $\Lambda_L = \text{diag}(\lambda_1, \ldots, \lambda_L)$. This linear transformation is not invertible.

An approximate inversion of this transformation is provided by

$$\hat{x}_L = m + V_L \Lambda_L^{1/2} f_L = m + T_L \Lambda_L f_L \tag{29}$$

where $\hat{x}_L$ is the estimate of $x$ using L expansion coefficients. We further note that $\hat{x}_N = x$ which results from the invertible normalizing transformation. The KLT is easily shown to minimize the approximation error

$$J = E\{\|\hat{x} - x\|^2\}. \tag{30}$$

Thus, for representing single-class distributions, we conclude that KLT is "best".

When the KLT is generalized to the multi-class problem of random vector representation, the origin is taken to be $m = m_0$. This selection is easily shown to minimize the representation (approximation) error over all classes. The difficult problem is with the choice of the covariance matrix from which the eigenvalue-eigenvector expansion is obtained; hence, the existence of the many variants of the KLT.

In this situation, the author prefers the following choice:

$$C = W + B \tag{31}$$

where $W$ is the average within-class covariance matrix and $B$ is the average between-class covariance matrix. Note that this latter choice is simply the population covariance matrix $C_0 = E\{(x - m_0)(x - m_0)^T\}$.

b. *Foley-Sammon transformation — two-class problem*

The Foley-Sammon transformation (FST) arises from an optimization of the Fisher discriminant ratio for a multi-class classification problem (Fisher, 1936; Sammon, 1970):

$$R(d) = (d^T B d)/(d^T W d). \tag{32}$$

The goal of the optimization is to determine a vector $d = [d_1, d_2, \ldots, d_N]$ such that the between-class separation ($d^T B d$) is a maximum while simultaneously the within-class scatter ($d^T W d$) is a minimum. Satisfying both conditions maximizes the Fisher discriminant ratio $R(d)$.

The two-class problem is considered first where $B = \Delta \Delta^T$. Differentiation of the scalar ratio $R(d)$ with respect to the vector $d$ and equating the result to zero leads directly to the generalized eigenvalue-eigenvector problem

$$[B - RW] \, d = 0. \tag{33}$$

We recognize that the Fisher discriminant ratio $R$ is an eigenvalue of this problem; therefore, maximizing $R$ is the same as identifying the maximum eigenvalue of the problem

$$[W^{-1}B - R_j I] \, d_j = 0. \tag{34}$$

This problem has exactly one non-zero eigenvalue because rank $(W^{-1}B)$ = rank $(B)$ = rank $(\Delta \Delta^T)$ = 1. The solution is

$$d_1 = W^{-1} \Delta \tag{35}$$

and

$$R(d_1) = (d_1^T B d_1)/(d_1^T W d_1) = R_{max} \tag{36}$$

which can be confirmed by direct substitution into the generalized eigen problem above.

An interesting observation is that this single feature extraction for the two-class problem corresponds identically to the linear classifier function

$$L_x(\mathbf{x}) = \Delta^T W^{-1}(\mathbf{x} - \mathbf{m}_0). \tag{37}$$

When the Foley-Sammon method is generalized to multi-class distributions (M classes), then M-1 discriminant vectors $\mathbf{d}_j$ (j = 1,...,M-1) corresponding to M-1 non-zero eigenvalues $R(\mathbf{d}_j)$ are found. This follows from rank $(W^{-1}B)$ = rank $(B)$ = M-1. However, the matrix $W^{-1}B$ is rarely symmetrical; therefore, its eigenvectors $\mathbf{d}_j$ will not be mutually orthogonal. This implies a redundancy in the resulting FST coefficients. One approach is to perform a Gram-Schmidt orthonormalization upon the FST discriminant vectors. Another approach is described next.

c. *Foley-Sammon transformation — multi-class problem*

The goal here is to optimize the Fisher discriminant ratio $R(\mathbf{d})$ with respect to the discriminant vector $\mathbf{d}$ subject to the additional constraint that $\mathbf{d}_k$ is orthogonal to $\mathbf{d}_j$ (j < k), see Foley and Sammon (1975). To pose this problem mathematically, we append the k-1 constraints to the Fisher ratio with the use of Lagrange multipliers:

$$R(\mathbf{d}_k) = (\mathbf{d}k^T B \mathbf{d}k)/(\mathbf{d}k^T W \mathbf{d}k) + \beta_1 \mathbf{d}_k^T \mathbf{d}_1 + \ldots + \beta_{k-1} \mathbf{d}_k^T \mathbf{d}_{k-1}. \tag{38}$$

Differentiating the scalar function $R(\mathbf{d}_k)$ with respect to the discrim-vector $\mathbf{d}_k$ and equating the result to zero yields one equation. To obtain the other k-1 equations, one also must differentiate the function with respect to the Lagrange multipliers $\beta_j$ which yields the orthogonality constraint equations. The entire system of equations must be solved simultaneously to obtain a solution for the discrim-vector $\mathbf{d}_k$ in terms of the previously identified $\mathbf{d}_j$ (j = 1,...,k-1).

One can see that the procedure described is iterative. Also, the dimension of the problem increases by one with each iteration because an additional constraint equation is added as each new discrim-vector is identified. Due to the complexity of the solution, it regrettably is not presented here in full detail. Instead, the reader is referred to the literature. Again, the alternative of Gram-Schmidt orthonormalization is suggested as a simpler means of implementation.

As a final comment, the Foley-Sammon methods for identifying basis vectors are related to that of maximizing the Hotelling criterion

$$J = \text{trace } \{W^{-1}B\} \tag{39}$$

which is equal to the sum of the M-1 non-zero eigenvalues.

## 6. DISTANCE AND CLASS SEPARABILITY

With any classification problem, one should be concerned with the frequency with which misclassification errors occur. A misclassification error occurs when an incorrect class selection is made for a pattern measurement vector. For example, assume a two-class problem in which a classifier function indicates that a particular measurement sample belongs to class 1 when in fact it actually belonged to class 2. The probability associated with these misclassification events is known as the probability of error or the Bayes error, $P_e$.

It is clear that the Bayes error must be related to the class conditional pdf's $p(\mathbf{x}|w_j)$. If the class pdf's great overlap one another, then the probability of error will be large ($P_e \to \frac{1}{2}$). On the other hand, if the distributions are well separated, then this error will become quite small ($P_e \to 0$).

The common approach employed in hypothesis testing for determining the Bayes error is to integrate the class conditional probabilities (actually, the a posteriori pdf's) over the regions of overlap. Of course, this assumes that analytical expressions for these pdf's are available.

The situation is different in a pattern classification problem involving feature extraction. The ultimate test of a set of selected features will be their contribution to the probability of classification error. Unfortunately, direct relationships between this Bayes error and the features used do not usually exist; therefore, analytical determination of the Bayes error by integration of the class pdf's is an approach not available to the investigator. Hence, various "distance" measures of class separation have been proposed for determining the effectiveness of a given set of features.

a. *Matusita distance*

Of the many distance or separability indices available, the Matusita distance measure is an easily determined value that can be used to estimate class separation (Kailath, 1967). It is given by

$$J_M = \left\{ \int_{S_x} [p(x|w_1) - p(x|w_2)]^2 \, dx \right\}^{1/2}. \tag{40}$$

Anticipating the material that follows, consider the two-class problem in which a measurement vector $x \sim N[m_1, C_1]$ and $N[m_2, C_2]$ is used to generate a new feature vector $y = A^T(x - m_0)$ where the linear transformation $A$ is, at this time, an arbitrary feature selection transform. The feature vector then has class conditional pdf's $y \sim N[+u, P_1]$ and $N[-u, P_2]$ where

$$\begin{aligned} u &= A^T \Delta \\ P_1 &= A^T C_1 A \\ P_2 &= A^T C_2 A. \end{aligned} \tag{41}$$

For the normal class distribution case, the Matusita distance reduces to

$$J_M = \frac{1}{2} u^T P^{-1} u + \frac{1}{2} \log [(\det P)/(\det P_1 \cdot \det P_2)^{1/2}] \tag{42}$$

with $P = \frac{1}{2}(P_1 + P_2) = A^T W A$.

The above information measure obeys most of the rules that one would expect of a distance measure. Although a complete justification of using this measure will not be provided, a partial list of some of its properties is presented below:

(i) It is non-negative.
(ii) It becomes very large if $y$ permits almost perfect discrimination between classes.
(iii) It goes to zero if $y$ is useless for distinguishing between classes.
(iv) It will never decrease if $A$ is augmented with an additional column.
(v) It has the same value for equivalent choices of $A$ that are related by invertible transformations.

One can see that feature selection can now be interpreted as a problem involving the choice of an appropriate transformation **A** that optimizes (maximizes) the Matusita distance measure $J_M$. By determining this distance-maximizing transform, we are equivalently minimizing the probability of classification error. Thus, we are attacking directly the class separation (discrimination) problem.

b. *Déja vu — feature selection revisited*

In this section we wish to re-examine the feature selection problem from the point of view of the distance measure (Henderson and Lainiotis, 1969). To facilitate the ensuing discussion, it is convenient to separate the Matusita distance into

$$J_M = J_I + J_{II} \tag{43}$$

where $J_I = \frac{1}{2} u^T P^{-1} u$ and $J_{II} = \frac{1}{2} \log [...]$.

CASE I
We assume separability of the means ($\Delta \neq 0$) and equal class covariances (**W** = $C_1$ = $C_2$). Under these conditions $J_{II} = 0$ and the feature selection problem depends upon $J_I$ only. This problem has been examined in previous sections and has a solution provided by the two-class Foley-Sammon discrimvector. It leads directly to the single feature linear classifier; hence,

$$A = d_1 = W^{-1}\Delta. \tag{44}$$

CASE II
We assume the mean vectors are equal ($\Delta = 0$) and the class covariances are not equal ($C_1 \neq C_2$). Under these conditions, $J_I = 0$ and the feature selection depends upon $J_{II}$ only. This problem has not been fully examined previously and is explored here. Optimization of $J_{II}$ with respect to a column vector $a_k$ of the transformation **A** yields a generalized eigenvalue-eigenvector problem

$$C_2 a_k = \gamma_k C_1 a_k \tag{45}$$

where the eigen-solutions $(\gamma_k, a_k)$, $(k = 1,\ldots,N)$ are ordered so that

$$\gamma_1 + 1/\gamma_1 > \gamma_2 + 1/\gamma_2 > \ldots > \gamma_N + 1/\gamma_N. \qquad (46)$$

Based upon this ordering of the eigen-solutions, we select the feature transformation $A = [a_1|a_2|\ldots|a_L]$ $(L < N)$.

CASE III

We assume mean separability $(\Delta \neq 0)$ and class covariance separability $(C_1 \neq C_2)$. The problem is to select $K < N$ feature vectors that optimize $J_M = J_I + J_{II}$. We note that neither term of this sum is zero. Unfortunately there is no general solution to this problem. Gradient techniques could be used to identify the optimum $A$ but this method is impractical when the measurement sample dimension N is large.

A suboptimal approach is suggested here. First, we observe that $J_{II}$ is independent of $u$ (and $\Delta$). Thus the eigen-solution method of Case II applies to the optimization of $J_{II}$ leading us to select the K eigenvectors for $A = [a_1|\ldots|a_K]$. With this choice, $J_{II}$ is maximum but $J_I$ may not be the maximum it can attain. To maximize $J_I$, it is necessary to have a feature vector of the form of the linear classifier function. Thus, we append a K+1 feature vector $a_{K+1} = W^{-1}\Delta$ to A. This procedure for generating an Nx(K+1) feature selection transformation will result in a classifier that will perform at least as well as the truly optimum NxK matrix. This is achieved at very little expense; we exceed our design goal of K feature vectors, but only very slightly.

c. *Comments on distance measures*

The primary utility of distance measures (Matusita, Bhattacharyya, Chernoff, etc.) is for theoretical investigations. The net result of the many papers published during the 1960's and 1970's has been to conclude that while distance measure bounds for estimating $P_e$ are interesting, one should try to estimate the error probability in some direct manner (Kanal, 1974).

A more direct approach for measuring the performance of a classifier function is provided by the resubstitution techniques, also known as the sample-partitioning methods. Of the many methods proposed, the widely held conclusion appears to be that the 'leave-one-out' method is the method of choice.

In this performance test, given Q measurement samples, a classification function is designed using Q-1 samples. (Estimation of statistical matrices is discussed in the next section). The classifier is tested with the 'left-out' sample and the outcome of the classification (correct or incorrect) is recorded. This procedure is repeated Q times for all samples in the training/testing set. The classification outcomes are averaged to obtain an unbiased estimate of $P_e$.

Foley has determined that if the ratio of the sample size to the number of features Q/L exceeds three, then one obtains a reasonable estimate of $P_e$ using the 'leave-one-out' method. The error estimates compared well with those obtained when the linear classifiers were tested against truly independent data sets (Foley, 1972).

## 7. ESTIMATION OF CLASS STATISTICAL MATRICES

Consider for the moment a single class problem where $x \sim N[m,C]$. A sample set $\{x_k, k = 1,2,\ldots,K\}$ is available and is to be used to obtain estimates of the mean vector $m$ and the covariance matrix $C$. The estimator equations are well-known and are given by

$$\hat{m}_K = \frac{1}{K} \sum_{\ell=1}^{K} x_\ell \tag{47}$$

and

$$\hat{C}_K = \frac{1}{K-1} \sum_{\ell=1}^{K} (x_\ell - m_K)(x_\ell - m_K)^T. \tag{48}$$

The estimate of the mean vector $\hat{m}_K$ is unbiased ($E\{\hat{m}_K\} = m$) and consistent ($Var\{m_K\} = K^{-1}C$). Similar statements are true for the estimate of the covariance matrix $\hat{C}_K$, provided the normalization factor K-1 is used.

The biased estimator for covariance which uses a normalization factor of K is easily shown to underestimate the true covariance $C$. However, an eigenvalue-eigenvector expansion of the biased estimate will still yield the correct eigenvectors. Hence, the normalization transformation of Section 3 will properly rotate the class distribution. The effect of the biased normalization factor is upon the eigenvalues - they are too small. Their relative magnitude with respect to each other, however, is correct. We conclude that while use of the unbiased estimate for covariance (K-1) is preferred over that of the biased estimate (K), the difference is very small. Clearly, if K is large enough, the difference becomes inconsequential.

Depending upon the environment, it may be necessary to calculate the estimates in 'real-time', i.e., update a current estimate with a new measurement. This problem is known as sequential estimation. The sequential estimator for the mean vector becomes

$$\hat{m}_K = \hat{m}_{K-1} + \frac{1}{K}(x_K - \hat{m}_{K-1}) \quad (49)$$

which is exact when initialized with $\hat{m}_0 = 0$ (Robins and Monroe, 1951).

An alternative form of this sequential estimator equation is obtained by defining an error vector $\tilde{x}_{K,K-1} = x_K - \hat{m}_{K-1}$ which leads to

$$\hat{m}_K = \hat{m}_{K-1} + \frac{1}{K}\tilde{x}_{K,K-1}. \quad (50)$$

This equation can be interpreted as a current estimate $\hat{m}_{K-1}$ that is sequentially updated with an error term $\tilde{x}_{K,K-1}$ which is appropriately weighted. The weighting sequence $\{1/K\}$ is recognized as the harmonic sequence and has the following properties:

$$\begin{aligned}&(i) \quad K \to \infty \quad 1/K = 0 \\ &(ii) \quad \sum_{k=1}^{\infty} 1/K = \infty \\ &(iii) \quad \sum_{k=1}^{\infty} (1/K)^2 < \infty.\end{aligned} \quad (51)$$

These three properties are actually necessary and sufficient conditions for error-driven estimators to converge to unbiased results (Dvoretsky, 1956). (These conditions, known as the Dvoretsky conditions, are well-known in the field of stochastic environment self-organizing control theory). Basically the conditions state the following: (i) the estimate converges, (ii) corrective action is always provided, and (iii) the variance of the estimate diminishes.

The sequential estimate for the covariance matrix can take on a variety of forms. This variation depends upon whether a biased or an unbiased estimate is generated and whether the error vector update uses $\tilde{x}_{K,K}$ or $\tilde{x}_{K,K-1}$. Most numerical difficulties can be avoided if the following biased estimate of the covariance matrix is updated

$$\hat{C}_K = \hat{C}_{K-1} + \frac{1}{K}\left[\frac{K-1}{K} \tilde{x}_{K,K-1} \tilde{x}_{K,K-1}^T - \hat{C}_{K-1}\right] \qquad (52)$$

which is exact with the initialization of the covariance as $\hat{C}_0 = 0$, the zero matrix. Of course, an unbiased estimate is always available by simply multiplying the matrix by $K/(K-1)$.

The probability density functions associated with these estimates are known and have been studied. The estimate of the mean vector has a normal (Gaussian) multivariate density, $\hat{m}_K = N[m, k^{-1}C]$, as is well-known. The estimate of the covariance has a Wishart density which is, in a sense, a generalized multidimensional Chi-square density.

Kaleyeh and Landgrebe (1983) studied this Wishart density and developed a criterion for predicting the number of measurements in a training set for estimating the covariance matrix. They suggested that the required number of training samples be at least five times the number of features in the measurement vector. Naturally the quality of the estimate improves as additional samples are used.

## 8. CONCLUDING REMARKS — NOISE

It is not possible within the space available to provide a comprehensive survey of a field as broad as statistical pattern recognition. That task must be left to texts and other reference materials such as Fukunaga (1972). We must be content with restricting the presentation to a limited set of topics and, hopefully, providing sufficient mathematical detail as to allow the reader to gain some insight into these selected subjects. Thus, we arrive at the final topic - noise.

One frequently (always?) obtains measurements that consist of two components: a true underlying signal which is contaminated with a random noise process. This problem can be modelled as

$$z = x + n \qquad (53)$$

where the signal $x \sim N[m,C]$ has the usual normal distribution and the noise $n \sim N[0,D]$ is a zero mean process. The resulting observed measurement has pdf $z \sim N[m,C+D]$.

If one wishes to obtain an estimate of the underlying signal, given the measurement, the linear minimum mean-square error estimate is appropriate:

$$\hat{x} = m + C(C + D)^{-1}(z - m) \qquad (54)$$

or equivalently,

$$\hat{x} - m = C(C + D)^{-1}(z - m). \qquad (55)$$

This result is well-known from filtering/estimation theory and is referred to as the optimal estimate (Gelb, 1974; Kalman, 1960). A significant property of this estimator is that it minimizes the error criterion $E\{\|\hat{x} - x\|^2\}$ in a noisy environment.

When the Karhunen-Loève transform is extended to this stochastic case, one does not use the eigenexpansion of $C + D$ for calculating the expansion coefficients $f$. Instead the eigenvectors of $C = V\Lambda V^T$ are used. To minimize the effects of noise that propagate into the KL coefficients, it is necessary to apply a "noise handling" modification to the feature extracting eigenvectors. As it turns out, this procedure is equivalent to performing the KL expansion on the optimal estimate $x$ rather than the actual measurement $z$.

If this KL expansion is explored a little further, it is possible to develop an interesting analogy with signal system theory. For simplicity, assume the noise covariance matrix is diagonal with all variance terms being equal, i.e., $D = \text{diag}(d) = dI$. The minimum error feature extraction vector then becomes

$$a = \frac{\lambda^{1/2}}{\lambda + d} v = \frac{\lambda^{-1/2}}{1 + d/\lambda} v \tag{56}$$

where $v$ is an eigenvector of $C$ and $\lambda$ is its corresponding eigenvalue. The multiplicative term is the "noise handling" term and is recognized as a Wiener filter expression (Wiener, 1949).

We finally address the issue of how the linear classifier function must be modified to perform properly in a noisy environment. If one assumes equal class covariances $C = C_1 = C_2$, then the classifier function becomes

$$L_z(z) = \Delta^T (C + D)^{-1} (z - m_0) \tag{57}$$

as one would expect. The more interesting result is obtained if one investigates the form the linear classifier function assumes if the optimal estimate $\hat{x}$ is used.

The covariance matrix for $\hat{x}$ is given by $\text{cov}(x) = C(C + D)^{-1} C$; hence, the classifier becomes

$$L_{\hat{x}}(\hat{x}) = \Delta^T C^{-1} (C + D) C^{-1} (\hat{x} - m_0) \tag{58}$$

because the population mean $m_0$ and the difference between the class means $\Delta$ are unaffected by the optimal filtering. Note, however, that

$$(C + D) C^{-1} (\hat{x} - m_0) = z - m_0 \tag{59}$$

leading us to conclude that $L_{\hat{x}}(\hat{x}) = L_z(z)$. The classifier functions are identical. The classification can be performed with either the optimal estimate or the noisy measurement - the result is the same.

We close with that final demonstration of internal mathematical consistency.

# REFERENCES

Fukunaga, K. (1972). *Introduction to Statistical Pattern Recognition*, Academic, New York.

Swain, P.H. and Davis, S.M. (1978). *Remote Sensing: The Quantitative Approach*, McGraw-Hill, New York.

Fisher, R.A. (1936). The use of multiple measurements in taxonomic problems, *Ann. Eugenics* 7-II, pp. 179-188.

Sammon, J.W. (1970). An optimal discriminant plane, *IEEE Trans. Comput.* C-19, pp. 826-829.

Foley, D.H. and Sammon, J.W. (1975). An optimal set of discriminant vectors, *IEEE Trans. Comput.* C-24, pp. 281-289.

Kailath, T. (1967). The divergence and Bhattacharyya distance measures in signal selection, *IEEE Trans. Commun. Technol.* COM-15, pp. 52-60.

Henderson, T.L. and Lainiotis, D.G. (1969). Comments on linear feature extraction, *IEEE Trans. Inform. Theory* IT-15, pp. 728-730.

Kanal, L. (1974). Patterns in pattern recognition: 1968-1974, *IEEE Trans. Inform. Theory* IT-20, pp. 697-722.

Foley, D.H. (1972). Considerations of sample size and feature size, *IEEE Trans. Inform. Theory* IT-18, pp. 618-626.

Robins, H. and Monroe, S. (1951). A stochastic approximation method, *Ann. Math. Stat.* 22, p. 400.

Dvoretsky, A. (1956). On stochastic approximation, *Proc. 3rd Berkeley Symposium on Math. Stat. Prob.*, Univ. of California Press, Los Angeles.

Kalayeh, H.M. and Landgrebe, D.A. (1983). Predicting the required number of training samples, *IEEE Trans. Pattern Anal. Mach. Intell.* PAMI-5, pp. 664-667.

Gelb, A. (ed.) (1974). *Applied Optimal Control*, MIT Press, Cambridge.

Kalman, R. (1960). A new approach to linear filtering and prediction problems, *Trans. ASME J. Basic Eng.* 82, pp. 34-35.

Wiener, N. (1949). *The Extrapolation, Interpolation, and Smoothing of Stationary Time Series*, Wiley, New York.

# IMAGE DATA COMPRESSION TECHNIQUES: A SURVEY

A. Todd-Pokropek

*University College London*
*United Kingdom*

ABSTRACT

*Data compression methods as applied to medical images are essentially grouped into reversible (redundancy reducing) and irreversible (entropy reducing) methods. Firstly, this paper describes some methods for determining the limits and efficiency of data compression, in particular the definitions of entropy, rate distortion, and some commonly used definitions of error. The remainder of the paper discusses various data compression methods such as transform coding, block and run-length coding, DPCM, transform coding, the use of multi-resolution techniques. Some indication of possible future extensions are given and some results of the use of data compression for medical images are cited.*

1. THE NEED FOR DATA COMPRESSION: NETWORKS

As the use of images in digitial form increases in Medical Diagnosis, so there will naturally follow an increased use of Image Networks. At present, all images obtained in CT and MRI (magnetic resonance imaging), DSA (digital subtraction angiography), and most images in nuclear medicine and ultrasound exist in digital form. It is anticipated that when appropriate captors have been developed, most conventional radiology will also have the capability of being performed digitally. Image processing for enhancement, expert system aided diagnosis, and digital long term storage will then be possible on most medical images. However, such facilities do not in themselves improve medical care or decrease costs, and may be considered to be of limited value. To be exploited properly, it is necessary that access to such digital images, and the means to manipulate them, be widely available. This supposes an image network (with appropriate image workstations). Descriptions of particular systems are given in, amongst many, Ackerman et al (1984) and Maguire (1987), and many other general references may be found in the publications of SPIE (1983-1987) on Picture Archiving and Communication Systems (PACS). Many different technologies have been employed; ethernet, coaxial broadband, fibre-optic cable, etc. However the major problem, whatever support is used, appears to be in managing the very large amounts of data involved.

A major limitation in most conventional PACS networks is that they tend to be limited to a small geographical area. For example the ACR/NEMA interface specifications (1986) for an industry wide common hardware and software interface for radiological use specify data rates of about 80Mbits/sec. Such rates cannot be maintained over long distances, and therefore cannot be used for a wide area network. Networks should therefore be designed such that images can be transmitted via a high speed network, throughout an extended 'campus', but also available via low speed bridges, for example, to clinics or individual General Practitioners. This implies the use of data compression.

A typical raw image is about 0.5Mbytes in size, although in conventional radiology images might be up to 16 times larger. At a data transmission rate of 19.2Kbaud then the time to transmit this typical image would be about 200 seconds. A highly desirable data compression ratio would be to reduce this time by a factor of 100, corresponding to about 0.1-0.2bits/pixel on average, if this could be achieved.

One approach has been attempted by the U.S. National Institutes of Health to provide remote radiological cover at a number of Veterans Administration Hospitals over (9600 baud) telephone links by transmitting radiographs to a central site after image compression for interpretation (Gitlin, 1986). Such studies indicate that a data compression ratio of about 10 (to achieve about 1bits/pixel) can be and must be achieved to make suitable use of data communication channels. Such techniques require considerable processing power at both transmitting and receiving station.

Data compression therefore is needed for transmission of medical data, in certain circumstances. It is also appropriate with respect to reducing the storage requirements of an archive (see for example Maguire, 1987). In addition it may be required as part of the data acquisition process to reduce the data rates to an amount such that they can be handled by current storage systems (see Peters et al., 1987). Finally, they are of interest in their own right, in that there is a close link between data compression and classification. Data compression has the aim of reducing redundancy in the data, and, in certain cases, of producing a more compact description of the data. Thus certain advanced data compression systems employ the methods of feature extraction (Kunt, 1985). There is considerable disagreement about the choice of appropriate method to be used for data compression, much of which is essentially a disagreement about the purpose of data compression. This is discussed further.

This paper will firstly consider data compression techniques in general, and then give some example of some specific methods considered to be of potential interest. Excellent general reviews are given by Netravali and Limb (1980), Jain (1981), Kunt et al (1985) and also the chapter on data compression in Rosenfeld and Kak (1982). These reviews contain many further references to the data compression literature. Three valuable books in the area or those of Held (1983), Lynch (1986), and Wade (1987), while an older book but with useful references is Huang and Tretiak (1972).

## 2. THE LIMITS OF DATA COMPRESSION- ENTROPY

Data compression may be considered from the point of view of information theory, that is, considering the transmission of a series of symbols over some (noisy) line in order to transmit information about the **source** to the **receiver**. A source paper in this area is by Shannon (1948), but see also Wolfowitz (1978). Consider a symbol $a_j$, in 'alphabet' of symbols **a**, being received. Information theory suggests that, for continuous additive functions (which are linear and homogeneous) a suitable measure of the information $I(a_j)$ that has been gained when a symbol has been received is

$$I(a_j) = - \log(\mathbf{P}[a_j]) \tag{1}$$

where **P**[] is the probability of occurrence of the symbol within the alphabet. This is actually the self-information. Its expected value is called the entropy **H** such that

$$\mathbf{H} = - \Sigma \, \mathbf{P}[a_j] \, \log(\mathbf{P}[a_j]) \tag{2}$$

over all possible symbols.

Important mathematical concepts associated with these ideas are those of stationarity, ergodicity, the notion of a Markov chain, and of course, sampling associate with the transformation of a continuous signal into a discrete representation (see for example Berger, 1971, or any textbook on information theory, for example, Khinchin, 1957), which will not be defined here for lack of space.

Bmax, the MAXIMUM number of bits needed to be transmitted for an image (excluding identification information etc) is the number of pixels N x N assuming the image to be square multiplied by the number of bits per pixel Nmax. Thus a conventional CT image might have an upper limit Bmax of 512x512x16 bits, i.e. 0.5Mbytes or 4Mbits.

One value for the LOWER limit is defined by the entropy (see Jain, 1981 and Rosenfeld and Kak, 1982). Let $P[i]$ be the probability that a given pixel takes the value i then zero order entropy $H_o$ is defined as

$$H_o = \sum_i P[i] \log_2(P[i]) \tag{3}$$

where effectively we have restricted our alphabet to integers between 0 and Nmax.

A typical value reported for a CT image is about 6bits per pixel (Rosenfeld and Kak, 1982). Thus a data compression technique which achieves the limit supposed by zero order entropy could compress a CT image from 4M bits to about 1.5M bits, that is, achieve a compression of just over a factor of 2. Huffman coding can come very close to achieving such compression (see Huffman, 1952, and also Rosenfeld and Kak, 1982).

For the rest of this paper, the efficiency of a data compression technique will be considered only in terms of the average number of bits per pixel after compression (Npix). If a compression technique were to achieve Npix=0.5 on a CT image, then typically the total amount of data would be reduced by about 32 (Nmax/Npix). However, such use of 'compression ratios' can sometimes be misleading since they depend rather arbitrarily on the upper limit of size for the image, and will therefore be avoided.

Shannon's noiseless coding theorem states that it is possible to code without distortion a source (of many pixels) of entropy H using H + epsilon data bits. Thus zero order entropy would appear to be the absolute lower limit for compression, but **only** under certain constraints. Firstly, the data were assumed to be formed from independent Gaussian variables, and therefore completely decorrelated. Secondly, it was assumed that the compression needed to be completely reversible, that is, the original image can IDENTICALLY be reconstructed after expansion of the compressed image. Methods designed to achieve this limit are called redundancy reducing or reversible techniques.

However, firstly, most medical images have very considerable correlation between pixel values. Secondly, if the compressed image after expansion cannot be 'distinguished' from the original image, even though it might not be identical, such a compression method would normally be considered 'acceptable', the definitions of distinguishable and acceptable being questions of 'perception'. Both these facts lead to the exploitation of data compression methods which can give a value of bits per pixel very much less than that suggested from zero order entropy. Such methods are called entropy reducing, or irreversible, techniques. The aim of such methods is to reduce entropy by throwing away noise or variance in the image with is uncorrelated with the (desirable) signal.

However, zero order entropy is only a limit when considering individual pixels and not 'neighbourhoods'. For example, first order entropy is defined similarly as

$$H_1 = \sum_{i_1} \sum_{i_2} P[i_1|i_2] \log_2(P[i_1|i_2]) \tag{4}$$

where $P[i_1|i_2]$ is the conditional probability of state $i_1$ given state $i_2$. Higher order entropies $H_n$ are defined correspondingly using n states. Typical values quoted for $H_1$ and $H_2$ are 4 and 2 bits correspondingly for CT scan like data. This gives some indication of the potential for compression by exploiting the correlation between pairs of pixels, i.e. a 'neighbourhood' of two.

It is also of interest to see how zero order entropy changes as a function of $\sigma^2$ (the variance) and Nmax (the maximum number of bits per pixel) in an image. By simulating image data for independent Gaussian distributed data, it may be shown that entropy is, to a good approximation, independent of Nmax, although it is of course very dependent on $\sigma$, only reaching useful values when $\sigma$ is very small. This provides support for expressing compression in terms of bits/pixel (Npix) rather than as a compression ratio.

A basic tool of data compression techniques is to eliminate the co-variance between pixels, and to transmit only the independent components, which are essentially the random variations between the expected image and the observed image. As might be expected, in medical images, such random variations are mostly (but not entirely) noise, Thus, if noise may be suppressed (which is therefore non-reversible), very efficient data compression can be achieved. An important issue is how noise can be recognised, and what threshold should be placed with respect to noise reduction.

## 3. THE RATE-DISTORTION FUNCTION

In the previous discussion, we have considered only that a given **pixel** can take on one of n states, and then noted that the information content is dependent on correlation between that pixel and its surrounds. Consider an image in general as being one symbol in (a very large) alphabet. Let us define also the mutual information $H(X;Y)$ between a source X and receiver Y as being

$$H(X;Y) = \sum_{ij} P[a_i,b_j] \log(P[a_i,b_j]/(P[a_i]Q[b_j])) \qquad (5)$$

where

$$Q[b_j] = \sum_j P[a_i,b_j] . \qquad (6)$$

If the self-information of the source X was $H(X)$ then

$$H(X;Y) = H(X) - H(X|Y) \qquad (7)$$

where

$$H(X|Y) = - \sum P[a_i,b_j] \log(P[a_i|b_j]) \qquad (8)$$

is the conditional entropy, $P[a_i|b_j]$ is the conditional probability of $a_i$ given $b_j$. The channel capacity can be defined as $\max H(X;Y)$. Now assume that a succession of symbols $a_i$, which here may be considered as images, are independent and obey the same distribution function. We need to define a distortion function $\delta[a_i,b_j]$ such that the expected value for the distortion d is given by

$$\begin{aligned} d &= \sum_{ij} \delta[a_i,b_j] \, P[a_i,b_j] \\ &= \sum_{ij} \delta[a_i,b_j] \, P[a_i] \, C[b_j|a_i] \end{aligned} \qquad (9)$$

where $C[b_j|a_i]$ is the probability of receiving $b_j$ given $a^i$. Let the set $C_c$ be the set of all the conditional probabilities $C[b_j|a_i]$ and define a set $C_d$ such that the distortion d is less the D for all $P[a_i]$. Then the rate-distortion function $R[D]$ is defined as

$$R[D] = \min_{C_d \in C_c} H(X;Y) \qquad (10)$$

for some given function $\delta[a_i,b_j]$. (For more details see Berger, 1971) This can be explained in words.

Some function giving the 'cost' of distortion is defined. We need to restrict the channel so as to be 'D-admissible' (i.e. giving acceptable distortion). We are only interested in mutual information, that is, the information actually conveyed by the receipt of the signal. The rate distortion function is the minimum value for the mutual information as a function of the distortion that is accepted as giving acceptable distortion.

Obviously R[D] is only defined for positive D. As D increases so R[D] will decrease, in fact R[D] is positive, decreasing, continuous and convex.

Unfortunately, while the rate distortion function provides a good conceptual handling of the limits of data compression, it is almost impossible to compute and therefore remains primarily of theoretical interest. Rosenfeld and Kak (1982) who give a good description of this function state that in order to compute the function we need to measure of the order of $2^{524288}$ images!.

Thus we need to look at the problem from a more empirical point of view, for example, by considering individual pixels and their co-variance.

## 4. MEASURES OF FIDELITY

If an image is transmitted in a non-reversible manner, we must have some measure of how far away it is from the original. In general, when images are transmitted over noisy lines where not all errors can be corrected, we still need this distance measure. No distance measure that is currently used appears to be completely satisfactory for medical images.

Let x(i,j) be a pixel in the original image, and x'(i,j) be a pixel in the image after compression and restoration (which we will call the 'transmitted' image). Most fidelity measures are based on some kind of L2 norm. For example, the average mean squared error $e^2_{ms}$ has been used, and defined such that

$$e^2_{ms} = 1/(N^2) \sum_i \sum_j E\{(x(i,j)-x'(i,j))^2\} \tag{11}$$

where as before E{} is the expected value and the image is assumed to be square of side length N. The sample mean square error $e'^2_{ms}$, which is in fact often used, is exactly the same as for $e^2_{ms}$ but using the observed squared difference rather than its expectation.

Related to these two error estimates are definitions of signal to noise ratios (SNRs) which may be expressed either in dBs or as pure ratios. For example peak-to-peak SNR may be expressed as

$$SNR = 10\log_{10}((MAX-MIN)^2/e^2_{ms}) \quad dB \tag{12}$$

where MAX-MIN is the peak to peak difference in the original image. A commonly used form for MAX-MIN is just the total range of values in the image, e.g. the largest possible integer. A better estimate of signal might be considered to be the variance $\sigma^2_x$ from which an alternative definition of SNR can be defined as

$$SNR' = 10\log_{10}(\sigma^2_x/e^2_{ms}) \quad dB \tag{13}$$

All these least square error based estimates have the usual problem of being sensitive to gross changes in the image and insensitive to local changes.

Some attempts have been made to use measures based on some visual perception criterion. For example Mannos and Sakirson (1974) have proposed a weighted mean square contrast error as follows. Let f() be some non-linear function. Let c(i,j) be an estimate of contrast change at the point i,j such that

$$c(i,j) = f(x(i,j) - x'(i,j)) \tag{14}$$

and

$$C(u,v) = F\{c(i,j)\} \tag{15}$$

where F{} is the 2-D Fourier transform, and let $e^2_c$ be the mean squared error in contrast defined as

$$e^2_c = \iint |C(u,v) H(u,v)|^2 \, du \, dv \tag{16}$$

where H(u,v) is the chosen weighting function. It is interesting to note that Mannos and Sakirson suggest using

$$f(u) = u^{1/3}$$
$$H(u,v) = H(w) = A(a + (w/w_0)^\alpha) \exp(-((w/w_0)^\beta) \tag{17}$$

with appropriate constants for $A, a, \alpha, \beta$ and $w_0$. Note that this is a separable exponential model which will be observed again used with KL transform. It may also be noted that such a model is often hypothesised when performing Wiener type filtering operations.

Another measure often used is based on the $Chi^2$ distance, where the least squares difference is normalized with respect to the variance, with a general form of the type

$$Chi^2 = [\sum_i \sum_j |(x(i,j) - x'(i,j)|^2 ]/\sigma^2 \qquad (18)$$

where $\sigma$ is defined in an appropriate (local or global manner) with respect to the original image. The variance (expected at a point) is sometimes included within the summation.

In summary, it is clear that, as compression increases, so fidelity is likely to become worse, in exactly the same manner as indicated for the Rate distortion function. No single fidelity measure has been found to be ideal. A few more empirical methods are helpful, or even essential. It is useful to display the difference image, i.e. $d(i,j) = x(i,j) - x'(i,j)$ as a function of compression. It is also helpful to use ROC methods comparing data before and after compression to check whether any (significant) loss of information has occurred, or if any difference in interpretation of the image would result. For a further discussion see, for example, Lynch (1986).

Quantization methods, and the associated error estimates, have not been considered here, but can be found discussed in Jain (1981), Lynch (1986) and also in Arnstein (1975) for DPCM coders. However, one result will be stated. When independent Gaussian variable of variance $\sigma^2$ are quantized using an optimal uniform encoder (called a Shannon quantizer) then, where D is the average mean square distortion per sample, if $D<\sigma^2$, the average bit rate n is given by

$$n = (1/2) \log_2 (\sigma^2/D), \qquad (19)$$

or

$$D = \sigma^2 2^{-2n}, \quad n \geq 0. \qquad (20)$$

This result will be used later. D is a lower bound on the distortion given the number of quantization levels.


## 5. DECORRELATION OF PIXELS- THE KL TRANSFORM

Many different forms of transform coding have been used, see for example the reviews by Wintz (1972) and Jain (1981). However, the Karhunen-Loeve (KL) transform is fairly fundamental in this respect. In the discrete case the KL transform is defined by the relationship

$$F(u,v) = \sum_m \sum_n f(m,n) \, \Phi^{(u,v)*}(m,n) \tag{21}$$

and its inverse by

$$f(m,n) = \sum_u \sum_v F(u,v) \, \Phi^{(u,v)}(m,n) \tag{22}$$

where $f(m,n)$ is the original image, $F(u,v)$ is the transformed image, where the basis functions $\Phi^{(u,v)}$ are a set of orthonormal matrices or eigenmatrices, of which $\Phi^{(u,v)}(p,q)$ is the p,qth element. These eigenmatrices are given from the solution of

$$\sum_p \sum_q R(m,n,p,q) \, \Phi^{(u,v)}(p,q) = \Gamma_{uv} \, \Phi^{(u,v)}(m,n) \tag{23}$$

where R is the autocorrelation function

$$R(m,n,p,q) = E\{f(m,n) \, f^*(p,q)\} \tag{24}$$

and $\Gamma$ is defined by

$$\Gamma_{uv} = E\{|F(u,v)|^2\}. \tag{25}$$

In this case it may be shown that

$$E\{F(u,v) \, F^*(u',v')\} = 0 \quad \text{for} \quad u \neq u' \;\; v \neq v' \tag{26}$$

since $E\{F(u,v)\} = 0$ provided that $E\{f(m,n)\} = 0$, i.e. that the original images had a mean value of zero. This is the desired result, in that the correlations of the transformed image are expected to be zero.

In this sense, the Karhunen-Loeve (KL) transform is therefore optimal in generating uncorrelated values from an image with correlated pixels. Thus the KL transform can be used to generate the set of values $F(m,n)$ with minimum entropy, which is, in this sense, optimal for data compression. However, the use of the KL transform itself is image dependent. In fact, of all the transform coding methods, it is the only technique which, in itself, is adaptive. Strictly, it must be recomputed for every image, and therefore is computationally not very convenient, requiring $N^4$ operations for $N^2$ coefficients. It is desirable to look for alternatives or approximations.

One approximation has been employed to generate a so-called Fast KL transform. This is based, as mentioned above, on the hypothesis that some image model can be supposed. Here, as before, let us assume that the autocorrelation function is separable and suppose some model of form

$$E\{f(m,n)\,f(p,q)\} = \exp(-\alpha|m-p|)\,\exp(-\beta|n-q|)$$

$$= r_1^{|m-p|}\, r_2^{|n-q|} \qquad (27)$$

Let $r = r_1 = r_2 < 1$. It may be shown that the eigenvalues can be derived analytically, such that

$$\Phi^{(u)}(m) = \sqrt{(2/(N+\Gamma_u))}\, \sin[w_u(m-(N-1)/2)+\pi(u+1)/2] \qquad (28)$$
$$\text{for } m=0 \ldots N-1$$

$$\Gamma_u = (1-r^2)/(1 - 2r\cos w_u + r^2) \qquad (29)$$

and $w_u$ are the N roots of the equation

$$\tan Nw = ((1-r_2)\sin w) / (\cos w - 2r + r^2\cos w) \qquad (30)$$

This one dimensional form can be extended trivially into two dimensions.

The use of a fast KL transform is well described by Rosenfeld and Kak and is given in summary here. Let the f(m,n) be scaled so that its variance is unity, and the autocorrelation function is separable. Let v(m,n) be formed excluding edge elements which are assumed to be known, as the estimation error equal to $f(m,n) - \hat{f}(m,n)$ where $\hat{f}(m,n)$ is a linear mean square estimate

$$\hat{f}(m,n) = \sum_{k,l \neq (0,0)} \sum a(k,l) \, f(m-k,n-l) \tag{31}$$

with the coefficients a(k,l) as derived from the model of the autocorrelation function. Then f(m,n), where m and n are not on the edge of the image, can be determined from the equation

$$f(m,n) = v(m,n) + \alpha \sum \sum f(k,l) \tag{32}$$

where the summation is over all the pixels adjacent to the point m,n and $\alpha$ is as in the model (see Eq. 27). Let [Q] be the tridiagonal Toeplitz matrix

$$[Q] = \begin{bmatrix} 1 & -\alpha & & & & & \\ -\alpha & 1 & -\alpha & & & & \\ & -\alpha & 1 & -\alpha & & & \\ & & \ddots & & & & \\ & & & \ddots & & & \\ & & & & \ddots & & \\ & & & & -\alpha & 1 & -\alpha \\ & & & & & -\alpha & 1 \end{bmatrix} \tag{33}$$

then, in matrix notation, a difference equation may be set up where

$$[Q][f][Q] = [v] + [B], \tag{34}$$

and

$$[B] = \alpha[B_1][Q] + \alpha[Q][B_2] + \alpha^2[B_3] \tag{35}$$

is a collection of edge elements of [f]; $[B_1]$ being the top and bottom rows, $[B_2]$ being the left and right edges, and $[B_3]$ the four corners. Now a new matrix [h] may be established such that

$$[h] = [Q]^{-1} [v] [Q]^{-1} \tag{36}$$

and therefore

$$[h] = [f] - [Q]^{-1} [B] [Q]^{-1} \tag{37}$$

Create a vector **h** from the matrix [h]. Note that there may be better methods than just reading the matrix into the vector element by element, and some kind of zig-zag path may give better results (Di Paola and Todd-Pokropek, 1981). Let [H] be the KL transform of [h], and **H** be the vector derived from [H]. Now, let the matrix [ϕ] be defined by

$$[Q]^{-1} = [\phi] [\Gamma] [\phi] \tag{38}$$

where [Γ] is a diagonal matrix of eigenvalues. Then

$$[H] = [\phi] [h] [\phi] \tag{39}$$

or

$$H(u,v) = (2/(N+1)) \sum_m \sum_n h(m,n) \sin(\pi um/(N+1)) \sin(\pi vn/(N+1)) \tag{40}$$

which can be implemented using an FFT. The whole operation can thus be performed firstly by calculating $[Q]^{-1} [B] [Q]^{-1}$ which can be performed by an FFT, secondly generating [h] and thirdly generating **H**, which being largely decorrelated may be used for data compression. When reconstructing, the reverse procedure is applied, which requires a knowledge of [B] that therefore must be transmitted separately.

## 6. THE DISCRETE COSINE TRANSFORM

An approximation to the use of the KL transform which has been described by Ahmed et al (1974), being easier to implement than the KL, is the use of the discrete cosine transform (DCT), defined as

$$F(u,v) = \frac{4 c(u,v)}{N^2} \sum_n \sum_m f(m,n) \cos((2m+1)\pi u /2N) \cdot \cos((2n+1)\pi v /2N) \quad (41)$$

where $c(u,v) = 2^{-1}$ for $u=v=0$, $2^{-1/2}$ for $u=0$, $v \neq 0$ or $u \neq 0$, $v=0$, and 1 elsewhere.

It may be demonstrated that, for a small matrix size of 16x16, the basis functions of the DCT transform are very close to those of the KL transform (Ahmed et al., 1974), where, as before, the autocorrelation matrix is assumed to be separable and obeying an exponential model.

The advantage of using the DCT over the KL transform is primarily that of computational efficiency. The DCT can also be implemented using a conventional FFT with appropriate weights, and there is a fast form of the transform in addition (Ahmed et al., 1974). When the image is decomposed into blocks of size nxn (see below) the DCT also has a symmetry property over 2n which tends to reduce the so called blocking artefact.


## 7. DISCRETIZATION, ENCODING AND BLOCKING

Having performed some transformation, one is left with a series of real (or possibly complex) values which it would be inefficient, as such, to transmit. These values must be turned into symbols, that is, located with respect to a set of discrete intervals, a process known as 'discretization'. The simplest form of such an operation is to convert the real numbers into integers. However, the range of values found after use of the transform have typically very different (dynamic) ranges. The fundamental is usually very large, but higher order coefficients fall off very rapidly. The choice of an appropriate encoding scheme can achieve significant gains in compression for given noise properties.

Normally, it is found very inefficient to manipulate the whole image (which might be of size 1Kx1K) in a single operations, and the image is usually treated as a series of blocks of size nxn, for example 16x16. Non square blocks may also be employed. However, it is important that after transmission the edges between blocks should not generate (blocking) artefacts; and that the selection of values be chosen to given good decorrelation after the transform.

Optimal codes may be selected by studying the performance of the transform for many samples of image data. Essentially Huang and Schultheiss (1963) have performed a constrained minimization and obtained the result that

$$b(u,v) = b_{aver} - 0.5 \log_2(\sigma^2(u,v))$$
$$- (1/2n^2) \sum_u \sum_v \log_2(\sigma^2(u,v)) \qquad (42)$$

where $\sigma^2(u,v)$ is the expected variance at the position u,v in the nxn transform. $b(u,v)$ is the number of bits to use to encode the data at that position and $b_{aver}$ is the desired average number of bits/pixel.

Such coding processes can be made adaptive, for any transform method, and many papers have been published discussing appropriate strategies, for example Habibi (1977) gives a general description of adaptive block encoding.

An example of an interesting, and simple, block coding method is called block truncation compression and was originally proposed by Delp and Mitchell (1979). For some block size, for example a 4x4 array of pixels, the (local) sample mean m and first moment σ are computed. All the pixels with values greater or equal to m are set to 1 while those below are set to zero. Two additional numbers A and B are computed such that, if q is the number of pixels ≥ m then

$$A = m - \sigma\sqrt{q/(m-q)}$$
$$B = m + \sigma\sqrt{q/(m-q)} \quad . \qquad (43)$$

Thus the block is reconstructed by replacing pixels marked with a 1 by the value A, and those marked with a 0 by B. The values of A and B are such that the first moment of the block is preserved. Only the bit pattern, plus the values for m and σ and transmitted. The method is clearly irreversible, but has been shown to have good properties in certain situations, for example in remote sensing. For a 4x4 block, if 8bits are used to transmit m and σ, the average number of bits per pixel is 2.

# 8. RUN-LENGTH CODING, PREDICTIVE DPCM CODING, AND NON-REDUNDANT CODING

Run length coding (RLC) has often been suggested as a suitable data compression technique. In essence, when a pixel value is found to be repeated, RLC consists of recording the value of the pixel and the number of times that it occurs, rather than storing the complete sequence. Thus edges of images when uniformly zero may be largely eliminated, that is, replaced by a zero value and a repeat count. Many improvements can be made, for example by using variable length bit codes to store changes rather than absolute values etc. Unfortunately, the run lengths observed in medical images tends to be very short, when edges are excluded (Chan, 1984). Much of this is a result of random fluctuations associated with statistical noise. The limits of run length data compression have been discussed by Gray and Simpson (1972), and in general, it has been estimated that the efficiency of run length coding (in achieving the entropy limit) is about 60%, as compared to the efficiency of the Huffman code which is typically about 95%. See also the discussion in Lynch (1986), p. 121.

Predictive coding is, in a way, a related technique. It appears to have been proposed originally by Elias (1955). Essentially, if one knows the expected form of the image, then one needs record only the differences between that image, and the observed image (see also Jain, 1981; Rosenfeld and Kak, 1981). Such differences might be expected to be very uncorrelated, and thus very efficient data compression could be achieved.

The simplest form of predictive coding is one using a one step delay, with a one bit quantizer, to indicate <greater> or <less> in the comparison between estimation and observation. This form of coding is called delta modulation originally patented by Deloraine et al. (1946) but later undergoing many extensions (see for example Abate, 1967).

Other simple techniques can and have been employed, for example, by predicting the value of the jth pixel horizontally in an image in terms of linear interpolation (or extrapolation) of previous values (see Davisson, 1968). A zero order predictor estimates the next (pixel) value to be received as being just the previous value, and thus only the difference is transmitted. A first order predictor estimates $\hat{x}_j$ from

$$\hat{x}_j = x_{j-1} + (x_{j-1} - x_{j-2}). \tag{44}$$

More generally, Differential Pulse Code Modulation (DPCM) may be defined by some linear combination of the previous i pixels such that

$$\hat{x}_j = a_1 x_{j-1} + a_2 x_{j-2} + \ldots + a_i x_{j-i} . \tag{45}$$

In all these cases, only the difference between the predicted value and the observed value is recorded, and compression results for the decrease in the number of bits required to store such a difference, with respect to that of the original pixel value. The coefficients $a_1..a_i$ may be defined in terms of a suitable filter, or in terms of some polynomial interpolation. Common implementations that have often been employed are first to third order linear predictors.

In one dimension i 'previous' sample points are used to estimate the next value. In two dimensions, some scheme must be used to define 'previous' pixels. Predictors can be horizontal, vertical, linear, non-linear, and include various numbers of terms (previous picture elements). In general, a reasonable assumption is to consider the image developing from top to bottom primarily, and from left to right secondarily. Thus for a pixel at (i,j), any pixel at (i',j') may be considered either if i'<i for any j', or j'<j for i'=i. These definitions may be mirrored or inverted. The problem remaining is to define appropriate values for the coefficients, the optimal solution depending on the image being encoded, and indeed, an adaptive solution, with varying values of the coefficients dependent on local image statistics, is likely to perform better than the use of fixed values.

Let $\hat{f}(m,n)$ be the estimated value, using some rule, at the point m,n and f(m,n) be the actual value. The error e(m,n) is given by

$$e(m,n) = f(m,n) - \hat{f}(m,n) \tag{46}$$

and is the value that needs to be transmitted. For a random field $E\{e(m,n)\} = 0$. If a similar model for the autocorrelation functions is assumed as in eqn 27. then

$$E\{e(m,n)e(p,q)\} = 0, \quad m \neq p \; n \neq q \tag{47}$$

and the transmitted values are actually decorrelated as desired.

Let $\delta(m,n)$ be the error in predicting $e(m,n)$. Its mean value should be zero, and its variance $\Omega$ ($= E\{\delta(m,n)\}$) will be the mean square distortion created by the coding (see Eq. (11)). Thus from Eq. (19),

$$n = (1/2) \log_2 (\sigma^2/\Omega) \qquad (48)$$

where n is the lower limit on the achievable compression. Using this expression, signal to noise ratios can be calculated, and various coding schemes compared. These have been evaluated for example by Ehrman (1967) and Lippmann (1976), amongst others.

The techniques previously described are generally termed predictive coding. A further development is called non-redundant sample coding. In predictive coding a difference between prediction and observation is sent for every sample point. In non-redundant coding, a value is only transmitted when this difference is greater than some threshold. Thus the transmission of values and the sequence of inspecting and reconstructing values are asynchronous, and a buffer must be set up with appropriate flow control. Thus such methods are more sensitive with respect to errors. As for DPCM, nth order and polynomial predictor may be used. However, an important difference between these two classes of method is that, in this case, an error estimate must also be established. The key point of this method is that rather, like run length coding, it is asynchronous, and that the threshold can be chosen (adaptively if necessary) as a function of the noise properties of the image.

A commonly employed algorithm (see Ehrman, 1967) is called the fan algorithm where, for each successive value, the tolerance of the error estimate for the next value is established from the previous values (in the form of a fan) where the spread of the fan is reduced if the successive values fall within the fan, and vice versa. If the observed value falls outside the fan (and only then), an error term is transmitted. The reconstructed signal is determined from interpolation between these values where such a signal was transmitted. This method is therefore not reversible as such. However, since the threshold can be determined with respect to knowledge of local noise levels, for example using a Kalman type filter, it may prove very useful for medical images. A suitable statistical predictor may be devised.

## 9. MULTI-RESOLUTION CODING

Pyramids and stacks (Koenderink, 1987) are structures in 'scale' space, i.e. at multiple resolutions, and are defined (here) by a set of images $G_L$ where each image is defined by means of some rule, for example summation of adjacent pixels, from the previous image in the pyramid $G_{L-1}$. The initial image $G_0$ is the raw data or original image.

Burt (1984) has described a novel form of data compression using pyramids, and examples are also shown in Kunt (1985). A Gaussian pyramid can be constructed in terms of the **REDUCE** operation defined as

$$G_L = \textbf{REDUCE} [ G_{L-1} ] \qquad (49)$$

such that

$$G_L(i,j) = \sum_{m,n=-2}^{+2} w(m,n)\, G_{L-1}(2i+m, 2j+n) \qquad (50)$$

where $G_L$ is the result of performing the operation L times and $G_0$ is the original image, w(m,n) is a suitable weighting function which is separable, normalised and symmetric. After each operation the size of $G_L$ is reduced linearly by a factor of 2 (1/4 the number of pixels). Basically w(m,n) is a Gaussian weighting function.

A similar operation **EXPAND** can be defined as

$$G_{L,K} = \textbf{EXPAND} [ G_{L,K-1} ] \qquad (51)$$

such that

$$G_{L,K}(i,j) = 4 \sum_{m,n=-2}^{+2} w(m,n)\, G_{L,K-1}((i+m)/2, (j+n)/2) \qquad (52)$$

where only those elements for which (i+m)/2 and (j+n)/2 are integer are used.

A Laplacian pyramid **LAP** is defined as the set of images over L such that

$$\text{LAP}_L = G_L - \text{EXPAND}[G_{L+1}] \qquad (53)$$

Using the subscript L,L to indicate that the corresponding Lth image $\text{LAP}_L$ in **LAP** has been expanded L times, then **LAP** has the property that

$$G_0 = \sum_L \text{LAP}_{L,L} \qquad (54)$$

where $G_0$ is the original image. If $G_0$ is of size NxN then the entire set of data in **LAP** is only about $(4/3)N^2$. Thus the operation of creating the Laplacian pyramid will slightly increase the amount of data (by 1/3).

However, on 'discretization', or encoding, the amount of data can be considerably reduced. The lowest level of the Laplacian pyramid contains mostly noise and can be encoded very coarsely. Higher order pyramids are each reduced in size by a factor of 4 and because of this are therefore less critical. They may then be, and indeed should be, encoded more finely. Burt (1984) has suggested using bin sizes of 19,9,5,3 for the four lowest levels of the pyramid for an image with a maximum grey value of 255. Each level of the pyramid corresponds to the equivalent of the result of band pass filtering at progressively lower spatial frequencies. The image can be built up in reverse order, starting with the highest order (smallest) Laplacian pyramid levels, and progressively adding in lower orders containing finer detail. Thus provided the **EXPAND** operation can be performed rapidly by the display hardware, the image can be recovered in coarse resolution, and resolution progressively improved, while the observer is watching. This assumes that the **EXPAND** operation can be performed rapidly in hardware.

Perhaps one of the most attractive features of the pyramid coding method is the similarity between the processing of such multi-resolution image sequences, and one possible mechanism for visual perception. It may be that such a coding scheme is a 'natural' coding scheme in that there is a better mapping between the data structures and the manner in which they are handled by an observer. The pyramid method has some relationship with an old method known as the 'synthetic high' technique, where an image is split into low frequency and high frequency components, the low frequencies encoded very coarsely, the high frequencies being treated essentially as edges (Kunt, 1985).

For this method to be efficient, then the lowest level which contains the most pixels must be compressed efficiently. This level contains mostly noise, and therefore for this method to be efficient, it must be used as a non-reversible method, where noise is eliminated.

An alternative method called hierarchical interpolation has been described more recently by Peters et al. (1987) which seems to be better adapted for use in a reversible manner. This method also has the advantage of being computationally very convenient and rapid, and has therefore been used for data compression during acquisition of angiograpic images. Consider one level of a pyramid containing nxn pixels. The next level at finer resolution can be estimated by linear interpolation between each of the pixels at the higher level. This can also be computed as a two stage operation by selecting appropriate neighbourhoods (for example squares and diamonds in succession). At each stage an error can be computed between the predicted value and the correct value. Thus the transmitted image comprises this ensemble of error terms, where the linear interpolation has essentially been used as a predictor to reduce the co-variance. The reconstruction process proceeds from the top of a pyramid by generating the next level (higher resolution) in the pyramid by linear interpolation and correcting with suitable error terms. Thus the bulk of the work occurs at the lowest level (regenerating the original picture) where, it is to be hoped, the error terms are very small.

10. THE 'S' TRANSFORM

A transform which does not appear to be directly related to compression is the 'S' transform (Lux, 1977). However, there is an interesting application. Consider a vector of 4 input values [a,b,c,d] multiplied by a matrix [s] to give a vector of 4 output values such that

$$[s] = \frac{1}{2} \begin{bmatrix} 1 & 1 & 1 & 1 \\ 1 & 1 & -1 & -1 \\ 1 & -1 & 1 & -1 \\ 1 & -1 & -1 & 1 \end{bmatrix}, \quad \begin{bmatrix} e \\ f \\ g \\ h \end{bmatrix} = \frac{1}{2} \begin{bmatrix} 1 & 1 & 1 & 1 \\ 1 & 1 & -1 & -1 \\ 1 & -1 & 1 & -1 \\ 1 & -1 & -1 & 1 \end{bmatrix} \begin{bmatrix} a \\ b \\ c \\ d \end{bmatrix} \quad (55)$$

then the following operation may be performed. Let the original image be treated as a set of 2x2 subregions. The 'S' transform is applied to each subregion to give 4 separate output values. The data may be regrouped such that the complete original NxN image is transformed into 4 N/2 x N/2 images, which may be called the e,f,g and h sub-images. This may be repeated recursively to generate 16 N/4 x N/4 etc. Note that the so-called 'e' image is merely the 2x2 sum of the image used to generate it, and the set of 'e' images form a very conventional pyramid (quad tree). Alternatively, the 'S' transform may be applied, recursively, only to the 'e' image obtained at each stage.

The amount of data after each 'S' transform is constant, although the entropy may in fact be reduced. However, the 'e' image is a low resolution copy of the original image. Thus, as described in the previous section, the technique may be employed to transmit firstly a low resolution copy of the original image, followed by additional data to restore the image (at leisure) to its full resolution. This is of considerable value in terms of improving the apparent response time of image transmission. The particular value and interest of the 'S' transform is that the kernel [s] is its own inverse, and can be implemented by addition and subtraction only, if necessary by a simple logic array. The 'S' transform is closely related to the Hadamard transform, and is in fact a reordered, incomplete (one step of a) Hadamard transform.


## 11. OTHER METHODS- DIRECTIONAL DECOMPOSITION

Some other potential methods for improving compression are discussed in Kunt (1985) where the use of a general contour-texture model for coding is recommended as a method for establishing the a-priori model of the image based on aspects determined from a knowledge of the human visual system.

Let an ideal high pass filter $H_i(u,v)$ be defined such that

$$H_i(u,v) = 1 \quad \text{if} \quad \theta(i) < \arctan(v/u) < \theta(i+1)$$
$$\text{and} \quad \sqrt{(u^2+v^2)} < w$$
$$= 0 \quad \text{else} \tag{56}$$

where $\theta(i)$ are a series of angular increments, and w is the cut-off frequency. A low frequency image is also generated. Thus after filtering (using an appropriate window function) the set of i directional images are input into a zero crossing detector to find edges (by looking for negative values in the product of adjacent pixels) and their magnitudes. These filtered images may be sampled at coarser intervals, and the positions of edges Huffman encoded. The image is reconstructed by interpolating between edges and adding the low frequency component.

An improvement to the above can be obtained (it is suggested) if pixels are classified into 'contour' pixels, those pixels defining an edge, and 'texture' pixels, those within an edge in some region. This implies, first of all, the use of a preprocessor to remove structure (granularity) within edges, then detecting edges, and then eliminating artefacts. Having segmented the image in that manner, the segments themselves can be encoded, using the raw image, in terms of some texture parameter (such as average run-length). This process can also be performed using a split-merge type of algorithm, or using pyramid type algorithms. Very efficient coding has been claimed by Kunt (1985). Further developments of these ideas generate classes of algorithms where the image is decomposed into some set of objects with an appropriate syntax, as employed in feature extraction methods, and the compression of the data results from the transmission of the list of features and their inter-relationships.


## 12. 3-D DATA COMPRESSION

A dramatic improvement in data compression efficiency can be achieved when considering not just single images but sets of images either in 3-D, or in time. The compression techniques previously discussed were achieved by predicting structure in 2-D. Such methods have commonly be employed on video signals where interframe coding schemes are very natural. It appears that elimination of co-variance in the 3rd dimension can, in principle, produce results of up to a factor of ten better, but at the expense of a considerable increase in computational complexity and storage requirements.

One method which has been used is to perform a 3-D DCT block transform over a 16x16x16 volume, in a manner equivalent to that used for the 2-D 16x16 transform. Some work in this area has been described by Roese et al (1977).

The major drawback with this method occurs when the number of slices is limited, or resolution is very different along the 3 spatial axes. One partial solution is to encode data over other volumes, for example 16x16x4, with some loss of efficiency.

If a 3-D pyramid is constructed, reducing each 2x2x2 volume to a 1x1x1 element, the total number of voxels in the entire image is only about $(8/7)N^3$ for a 3-D image of side N. Thus there is even less difference in the total number of voxels than for a 2-D pyramid. Laplacian 3-D pyramids can be constructed using an extension of the algorithm given previously, and data compressed correspondingly. The main limitation of 3-D coding schemes to date has been the rather large memory requirements needed for implementing them efficiently.

## 13. SOME RESULTS

Various data compression methods have been tested at UCL, and applied to CT scans, and radioisotope scans. The size of the CT scans was 256x256, but the radioisotope scans were only 64x64. The total number of bits required after compression was noted, and converted to a average value per pixel. The image was then 'decompressed' and compared visually with the original image. A difference image was also generated, together with an estimate of the difference before and after compression. Various distance measures where used: an $L_2$ norm (in fact the RMS error), the Chi-squared distance, the maximum error, and signal to noise ratio (SNR). The following general results were obtained.

1.  Reversible noiseless data compression was only able to achieve useful data compression down to about 2-3bits/pixel. Irreversible data compression can achieve much greater compression without significant (i.e. noticeable) changes in compressed images.

2.  The use of run length coding was not helpful, with the exception of edge stripping, which, on its own, could at almost no cost remove 50% of the data. The best data compression found was similarly about 2.5 average bits/pixel.

| Ave. bits/pixel | | RMS difference |
|---|---|---|
| CT images | NM scans | |
| 2.44 | 4.24 | 6.22 |
| 2.31 | 3.1 | 6.4 |
| 2.21 | 2.95 | 6.59 |
| 2.09 | 2.93 | 6.9 |
| 1.9 | 2.91 | 7.51 |
| 1.62 | 2.89 | 9.63 |

*Table 1 shows a typical result (at UCL) for the amount of compression achieved using the DCT in average bits per pixel, together with the corresponding RMS error.*

3. The DCT transform can give good results, down to about 1.5bits/pixel, but some blocking artefacts were generated (see Table 1).

4. The Laplacian pyramid coding approach is encouraging, can give similar compression as the DCT, and does not generate blocking artefacts. It is rather compute intensive. Using Huffman coding data compression of from 1.0 to about 1.5 average bits/pixel can be achieved. This amount of data compression is irreversible. When used for a reversible compression, it is likely to be inefficient.

Huang (1986) has reported experiments on data compression using a DCT and Huffman coding of chest x-rays and other medical images where acceptable data compression ratios of up to 16:1 are reported. Quinn et al (1983) reported results on CT scans with data compression typically of about 40%. Dunham et al. (1983) reported compression ratios of 1-4. Kunt (1985) suggested that compression could be achieved below the 1 bit/pixel level. It is interesting to note not just the different units in which these and other workers have reported their results, but also the considerable differences in compression that is claimed. The primary difference in these results is the difference between reversible and irreversible methods. The very high compression ratios reported by Huang and Kunt result from the suppression of noise, and the fact that the resulting pictures are 'nearly indistinguishable' from the originals. Whether this criterion will prove acceptable to clinicians is still under debate. It is the author's opinion in that some kind of threshold compression may prove to be the most satisfactory where some noise reduction is achieved, which should give somewhat intermediate compression ratios.

## 14. COMPUTATIONAL REQUIREMENTS

The compression techniques employed often tend to be rather heavy on computation time, for example requiring of the order of several seconds of time (about 30 for a 256x256 matrix) on a PDP11/44, or about a factor of 10 better on a SunIII. They can be made more efficient, for example by the use of appropriate hardware such as an array (vector) processor and it is estimated that sub-second compression techniques can be achieved for use with high speed links. Both the DCT and the multi-resolution techniques are suitable for handling in this manner. The interest in multi-resolution image processing makes it possible that special hardware for creating and manipulating pyramids may become available. Sub-optimal, but computationally efficient algorithms have also been described, for example that of Ziv and Lempel (1978) which may be more efficient for data transmission, or for incorporation into the hardware of a disk drive.

For low speed bridges, where the computational requirement is less, the use of 80X86 or 680X0 (where X>1) CPUs should be quite adequate.

## 15. CONCLUSION

The use of digital information in medical imaging can be exploited to a much greater extent if image data can be widely distributed. This seems to imply the use of low speed bridges for some time into the future such that compression techniques will be essential for their exploitation. A (near) loss-less coding scheme, or one which only removes noise with respect to a suitable statistical threshold, even if it achieves a tenfold reduction in the amount of data in images, is only desirable if the computational overheads can be made acceptable.

## ACKNOWLEDGEMENT

The author would like to thank Bob Appledorn for providing much of the information about the S transform, and J.H. Peters for the information about the method of hierarchical interpolation (which was added during the process of revising this text).

REFERENCES

Abate, J.E. (1967). Linear and adaptive delta modulation, *Proc. IEEE* 55, pp. 298-308.
Ackerman, L.V., Flynn, M.J., Froelich, J.W., Zerwkh, J.P., Lund, S.R., and Block, R.W. (1984). Implementation of a broadband network in a diagnostic radiology department and large hospital, *Radiology* 153(P), abstract p. 249.
ACR/NEMA (1985). Digital imaging and communications standard.
Arnstein, D.S. (1975). Quantization error in predictive coders, *IEEE Trans. on Comm.* COM-23, pp. 423-429.
Ahmed, N., Natarajan, T. and Rao K.R. (1974). Discrete cosine transform, *IEEE Trans Comput.* C-23, pp. 90-93.
Berger, T. (1971). *Rate distortion theory, a mathematical basis for data compression*, Prentice-Hall, Englewood Cliffs.
Burt, P.J. (1984). The pyramid as a structure for efficient computation. In: *Multiresolution image processing and analysis*, A. Rosenfeld (ed.), Springer-Verlag, Berlin, pp. 6-37.
Chan, C.A. (1984). *Data compression of medical images*, MSc Dissertation, University College London.
Davisson, L.D. (1968). Data compression using straight line interpolation, *IEEE Trans. on Info. Theory* IT-14, pp. 390-394.
Delp, E.J., adn Mitchell, O.R. (1979). Image truncation using block truncation coding, *IEEE Trans. on Comm.* COM-27, pp. 1335-1342.
Deloraine, E.M., Derjavitch, B., and van Mierlo, S. (1946). *Delta modulation*, French Patent 932 140.
Di Paola, R., and Todd-Pokropek, A.E. (1981). New developments in techniques for information processing in radionuclide imaging. In: *Medical radionuclide imaging*, IAEA, Vienna, 1, pp. 287-312.
Dunham, J.G., Hill, R.L., Blaine, G.J., Snyder,D.L., and Jost R.G. (1983). Compression for picture archiving and communication in radiology. In: *Picture Archiving & Communications Systems (PACS II) for Medical Applications,* SPIE. 418, pp. 201-208.
Elias, P. (1955). Predictive coding. Part I and II, *IRE Trans. Info. Theory* IT-1, pp.16-33.
Ehrman, L. (1967). Analysis of some redundancy removal bandwidth compression techniques, *Proc. IEEE* 55, pp. 278-287.
Gitlin, J.N. (1985). Teleradiology. Presented at *RSNA/AAPM symposium. Electronic radiology: Image archiving, PACS and Teleradiology, Radiology* 157(P) p. 254.
Gray, K.G. and Simpson, R.S. (1972). Upper bound on compression ratio for run length coding, *Proc. IEEE* 60, p. 148.
Habibi, A. (1977). A survey of adaptive image coding techniques, *IEEE Trans. on Comm.* COM-25, pp. 1275-1284.
Held, G. (1983) *Data compression*, Wiley, New York.
Huang, J.J.Y. and Schultheiss, P.M. (1963). Block quantization of correlated Gaussian random variables, *IRE Trans Commun. Syst.* CS-11, pp. 289-297.
Huang, H.K. (1986). Recent developments in digital radiology. In: *Pictorial information systems in Medicine*, K.H. Hoehne (ed), Nato ASI Series F19, Springer-Verlag, Berlin.
Huang, T.S. and Tretiak, O.J. (1972). *Picture bandwidth compression,* Gordon and Beach, New York.
Huffman, D.A. (1952). A method for the construction of minimum redundancy codes, *Proc. IRE.* 40, pp. 1098-1101.
Jain, A.K. (1981). Image data compression: a review, *Proc IEEE* 69, pp. 349-406.

Khinchin, A.I. (1957). *Mathematical foundations of information theory*, Dover, New York.
Koenderink, J.J. (1987). Image structure, these proceedings.
Kunt, M. (1978). Source coding of X-ray pictures, *IEEE Trans. on Biomed. Eng.* BME-25, pp. 121-138.
Kunt M., Ikonomopoulos A., and Kocker, M. (1985). Second-generation image coding techniques, *Proc. IEEE* 73, pp. 549-574.
Lux, P. (1977). A novel set of closed orthogonal functions for image coding, *A.E.I.I.* 31, pp. 267-274.
Lippmann, R. (1976). Influence of channel errors on DPCM picture coding, *Acta Electronica* 19, pp. 289-294.
Lynch, T.J. (1986). *Data compression techniques and applications*, Lifetime Learning Publications, Belmont.
Maguire, G.Q.Jr. (1987). Introduction to PACS for those interested in image processing. In: *Information Processing in Medical Imaging*, De Graaf C.N. and Viergever M.A. (eds), Plenum Press, New York, in press.
Mannos, J.L. and Sakirson, D.J. (1974). The effects of a visual fidelity criterion on the encoding of images, *IEEE Trans Inform. Theory* IT-20, pp. 525-536.
Netravali, A.N. and Limb, J.O. (1980). Picture coding: a review, *Proc. IEEE* 68, pp. 366-406.
Peters, J.H., van Dijke, M.C.A., Roos, P., and Viergever, M.A. (1987). Loss-less interframe compression in digital angiography. In: *Information Processing in Medical Imaging*, De Graaf C.N. and Viergever M.A. (eds), Plenum Press, New York, in press.
Quinn, J.F., Rhodes, M.L., and Rosner, B. (1983). Data compression techniques for CT image archival. In: *Picture Archiving & Communications Systems (PACS II) for Medical Applications,* SPIE. 418, pp. 209-212.
Roese, J.A., Pratt, W.K., and Robinson, G.S. (1977). Interframe cosine transform image coding. *IEEE Trans. on Comm.* COM-25, pp. 1323-1329.
Rosenfeld, A. and Kak, A.C. (1982). *Digital Picture Processing.* 2nd Edition, Academic Press, New York, Vol 1, Chapter 5.
Shannon, C.E. (1948). A mathematical theory of communication, *Bell System Technical Journal* 27, pp. 379-423, 623-656.
SPIE (1983,1985,1986,1987). *Picture Archiving and Communication Systems (PACS II,III,IV and V) for Medical Applications.* SPIE, Bellingham Washington.
Wade, J.G. (1987). *Signal Coding and Processing: an introduction based on video systems*, Ellis Horward, Chichester.
Wintz, P.A. (1972). Transform picture coding, *Proc. IEEE* 60, pp. 809-820.
Wolfowitz, J. (1978). *Coding theorems of information theory*, A Series of Modern Surveys in Mathematics 31, Springer-Verlag, Berlin.
Ziv, J. and Lempel, A. (1978). Compression of individual sequences via variable-rate coding, *IEEE Trans. on Info. Theory* IT-24, pp. 530-536.

FROM 2D TO 3D REPRESENTATION

Gabor T. Herman

*University of Pennsylvania*
*U.S.A.*

ABSTRACT

*This tutorial describes methods that have been developed to obtain three-dimensional (3D) representations and rendering of medical objects based on a sequence of two-dimensional (2D) slices. The topics discussed are segmentation of 3D scenes, representation of segmented 3D scenes (binary arrays, segment-end-point representation, directed contours, octrees, objects and their surfaces), and display of objects in 3D scenes (projection onto a screen, hidden surface removal, visible surface rendering, stereoscopic presentation).*

1. SEGMENTATION OF 3D SCENES

Our aim is to provide the clinician with computer graphic tools which will enable him to visualize the three-dimensional internal structures of patients.

The necessary information can come from any medical imaging device which estimates the value of some physical parameter at each of a three-dimensional array of points: our prime example will be the CT (Computerized Tomography) scanner, which provides us with estimates of the x-ray attenuation of tissue in a sequence of parallel slices. The same graphic procedures can be applied to the output of other devices such as magnetic resonance imagers, positron or single photon emission tomography scanners, or even ultrasound scanners (provided that the data are collected appropriately).

The data from such devices contain detailed information regarding three-dimensional (3D) internal structures, but the visualization of the shapes of these structures based on the numbers, or even on the two-dimensional (2D) images of the individual slices, is nontrivial (see Fig. 1). The type of visualization that we have in mind is illustrated in Fig. 2 which shows

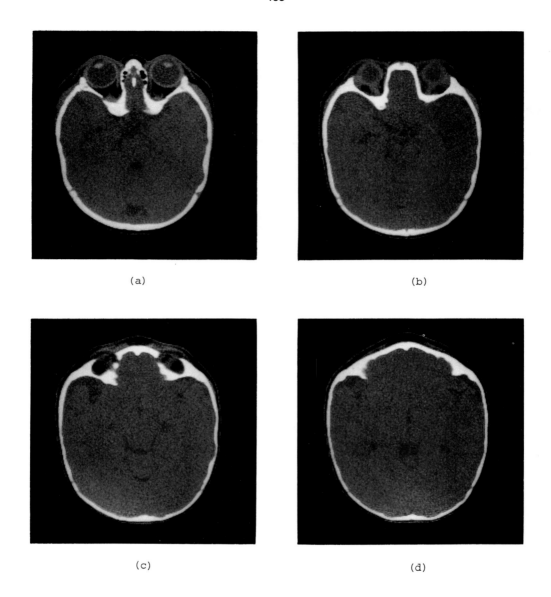

Fig. 1. Four out of the 41 cross-sectional slices of the CT scan of a patient prior to craniofacial surgery. The slices are 6 mm from each other and cover a region from the middle to the roof of the orbits.

images of the skull of a patient before and after craniofacial surgery.

In what follows we discuss computer graphic techniques for display of 3D imagery. We pay special attention to the data structures, that is to the way

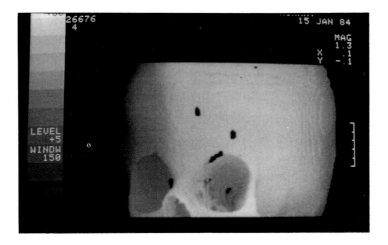

(a)

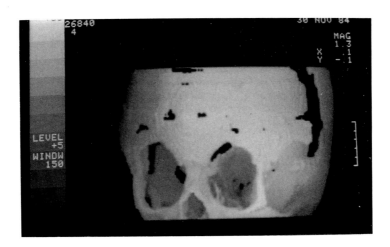

(b)

*Fig. 2. Displays of the skull of the patient whose CT slices are shown in Fig. 1. (a) 3D display of the skull prior to craniofacial surgery. (b) 3D display of the skull after supra-orbital reshaping to correct for a developmental anomaly.*

the information contained in the 3D data is stored in the computer. This is because different types of display operations are most efficiently performed using different types of data structures (Herman et al., 1986). In this article a *discrete 3D scene* (or *scene* for short) is considered to consist of a rectangular parallelepiped (referred to as the *region* of the scene) which is subdivided by three sets of parallel planes into smaller identical parallelepipeds (referred to as *voxels*) each one of which has a value assigned to it (referred to as the *density* of the voxel). In all applications, the 'density' represents some physical property; for example, in CT it is related to the x-ray absorbancy in the voxel. In practice the scene is *digitized* in the sense that the density must be one of the integers between a lower limit L and an upper limit U. If L = 0 and U = 1, then these are the only possible values, and we say that the scene is a *binary scene*.

We assume that together with the scene a rectangular coordinate system has been determined which assigns to each voxel a triple (i,j,k) of integers, where $1 \leqslant i \leqslant I$, $1 \leqslant j \leqslant J$, and $1 \leqslant k \leqslant K$. We refer to such a scene as an I×J×K scene. The I×J×1 scene formed by all voxels $(i,j,\bar{k})$, for a fixed $\bar{k}$, is called the $\bar{k}$'th *slice* of the scene.

*Scene processing* is the 3D analog of picture processing (Rosenfeld and Kak, 1982). Here we mention three scene processing operations: subregioning, interpolation, and segmentation. Each of these operations produces a new scene from another scene.

A scene produced by *subregioning* contains a subset of the voxels in the original scene, with densities unchanged. Taking a single slice out of a scene is an example of subregioning. Typically, subregioning is performed as an early step of scene processing: by isolating a region of interest the size of the data set that needs to be handled by further processing is reduced.

*Interpolation* produces a scene in which the voxel size is different from that in the original scene. The densities assigned to the new voxels are estimated based on the densities of the original voxels. For example, for reasons to be explained later, we desire to have a *cubic scene*, i.e. a scene in which each voxel is a cube. On the other hand, the voxels associated with a CT scanner are typically not cubic; e.g. 0.5 mm × 0.5 mm × 3 mm. In such a case, we use linear interpolation to estimate the densities assigned to 0.5 mm × 0.5 mm × 0.5 mm voxels.

*Segmentation* produces a binary scene. The original and the resulting scenes are of the same size and are subdivided into voxels in the same way. One method of segmentation is *thresholding*, in which a voxel in the binary scene has density 1, if the density in the original scene is above a predetermined (threshold) value. For the illustrations given earlier (Figs. 1 and 2) we were interested in the bony structures of our patient; we therefore chose the threshold so that it is below the densities of the voxels containing mostly bone, but it is above the densities of the other voxels. Thresholding is the simplest mode of producing from an arbitrary scene a binary scene of the same size. Other techniques are often required to deal with more complicated situations (these often arise when we wish to see soft tissue surfaces or when the CT scans contain artifacts such as those due to fillings in teeth or breathing motion). Some methods of segmentation in the medical environment are described in Udupa (1982) and Trivedi et al. (1986). We now give a brief description of the ideas expressed in the latter.

Sometimes the object we wish to image is surrounded by two different types of material, one with a higher density and one with a lower density than the object of our interest. Examples are: contrast carrying blood flowing past a calcified plaque and infarcted myocardium bordering both healthy myocardium and the outside of the heart. The problem with saying that a voxel belongs to the object of interest provided that its density is in a certain range is that those voxels which are on the border of the higher density and lower density material will often have a density lying in the same intermediate range. This way we may misclassify as blood the edge of the calcified plaque which is away from the blood and as infarct the edge of healthy myocardium.

One way of removing this effect is by the use of gradients. The idea is that voxels whose densities are in the intermediate range because they are on the border of (and are partially occupied by) tissues in the higher and lower density range can be identified by looking at their 3×3×3 neighborhood, because the rate of change around them (the gradient) will be high. In Trivedi et al. (1986) we demonstrated this approach by segmenting both a myocardical infarct based on MRI data and the contrast carrying blood in the presence of a calcified plaque.

## 2. REPRESENTATION OF SEGMENTED 3D SCENES

### a. *Binary arrays*

The binary scene produced by segmentation is a three-dimensional array of 0's and 1's. This is called the *binary array* representation of the scene. Geometrically, the relative location of an entry in the binary array matches the relative location of the corresponding voxel in the region of the scene.

For typical medical objects, the binary array representation seems to be unnecessarily 'detailed', since it contains large areas filled with only 0's or only 1's, each one of which is explicitly recorded. In the next three subsections we discuss more concise ways of representing those binary scenes which are typical in medicine.

### b. *Segment-end-point representation*

The *segment-end-point* (SE, for short) representation is a variant of the run length representation, which has been used for the efficient transmission of images; see, for example, Foley and van Dam (1983) and Quinn et al. (1984). SE is a slice-by-slice, row-by-row representation. Each row of each slice in the binary scene is examined for voxels with density 1. Each continuous run of 1-voxels in the row is referred to as a segment. The left and right end-points of all segments are recorded in an increasing order. This representation, due to its inherent simplicity, can be derived very easily from other representations of the binary scene; see Udupa (1982) and Merrill (1973).

It is clear that in this fashion every binary scence can be described by a sequence of integers, and that if the binary array representation is full of long sequences of 0's and 1's, then the SE representation is much more compact. The SE representation has been used for various scene processing operations such as region search, scene display, and scene manipulation as well as for quantitative analysis of scene regions (Trivedi, 1985).

*c. Directed contours*

In this representation, objects are described by their borders in a slice-by-slice fashion. In each slice, each border of a 1-region is represented by a sequence of voxels with the convention that as the border is traversed the 1-region lies to the left (see Fig. 3). We call such a sequence of voxels with a directionality assigned to it a *directed contour* (DC). The DCs are chain-encoded (Freeman, 1974) so that they can be stored very compactly. With each slice we associate a *containment tree* in which each node represents a contour, and its sons represent the contours which are 'immediately inside'. In the example in Fig. 3, the border C0 of the slice represents the root node, and C1, C2 are its sons representing contours which are 'immediately inside' C0.

In the above description, the contours in the DC representation are defined to be the borders of regions. There are situations in which the user has to supply this information by drawing contours; for example when the result of segmentation is unreliable, or when a part of the object has to be 'removed'

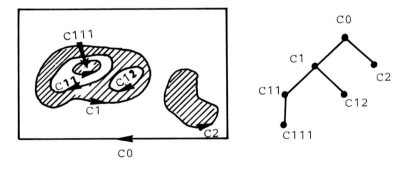

Fig. 3. Illustration of the DC representation. (a) In each slice 1-regions (shown hatched) are represented by directed contours such that the 1-region lies to the left of the contour. (b) A containment tree is associated with the contours to describe the relationship of containment of contours inside contours.

for better visualization. To be able to uniquely define the region represented by a set of contours in such situations, the contours should satisfy certain properties. Each contour should be 'closed and non-self-intersecting' and no two contours in the same slice can intersect each other (Udupa, 1982). Since the user-drawn contours do not always satisfy these properties, they should be appropriately processed to yield valid contours.

In producing the DC representation of a given binary array, one may choose to generate the nodes of the containment tree in either depth-first or breadth-first order. Both approaches require two basic processes: (i) searching for a contour within as well as outside the region enclosed by a given contour, and (ii) following a contour in a specified direction once its presence is detected. With reference to the example in Fig. 3 in the depth-first approach, first C1 is detected (assuming search for a contour is made row-by-row starting from the top), then search is restricted to the region inside C1 to detect C11, and then to the region inside C11 to detect C111, etcetera. The recursion stops when a leaf node (C111) is reached. Nodes C12 and C2 are then expanded similarly. Roughly speaking, the complexity of the algorithm for the depth- and breadth-first approaches is a function of the depth and breadth of the containment tree, respectively. Since, for medical objects, the depth of the tree is likely to be much smaller than its breadth, the depth-first algorithm is to be preferred; see Tuy and Udupa (1983) and Chen et al. (1985a).

Given a DC representation, we can easily produce the equivalent SE representation. For details see Herman et al. (1986).

The DC representation is extremely compact. For typical medical objects, it requires about 5-8 times less storage than that required by the binary array representation.

*d. Octrees*

We first explain the *quadtree* representation of a binary slice as follows (see Fig. 4).

> Look at the densities in the slice. If they are all 1, represent the slice by a terminal node labelled Full. If they are all 0,

represent the slice by a terminal node labelled Empty. Otherwise, label the node Partial and recursively subdivide the slice into four sub-slices (not necessarily equal) which may in turn be represented by nodes labelled Full, Empty, or Partial.

The subdivision ceases when all terminal nodes are either Full or Empty.

*Octrees* are the 3D analogs of quadtrees. The scene is recursively subdivided into eight sub-scenes until the nodes representing the sub-scenes are either Full or Empty (see Fig. 5).

The use of octrees results in substantial data-compression as compared to binary arrays (Meagher, 1982).

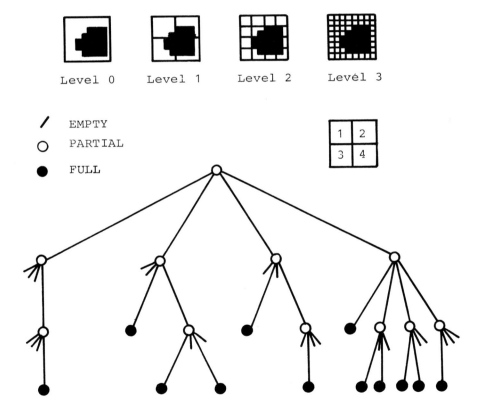

*Fig. 4. Quadtree representation of a binary slice.*

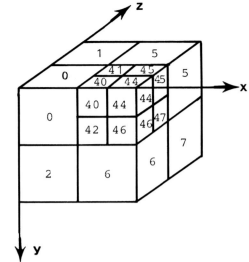

*Fig. 5. Octree encoding with octant 4 subdivided. For this viewpoint priorities in ascending order are: 3, 1, 2, 7, 0, 6, 5, 4; applied recursively to each octant.*

*c. Objects and their surfaces*

Even though human observers identify objects in 3D scenes without any apparent difficulty, the mathematically precise definitions of an 'object' and its 'surface' are far from trivial, and equally reasonable definitions can be (and have been) proposed giving rise to different types of entities being identified as objects.

In discussing objects and their surfaces one has to recognize that in most application areas the discrete 3D scene is considered only to be an approximation to the underlying domain of interest. Usually, we assume that there is a function $f(x,y,z)$ of three real variables and the density assigned to a voxel is an estimate of the average (or central) value of $f(x,y,z)$ in the voxel. For example, in computerized tomography $f(x,y,z)$ is the x-ray attenuation coefficient of the human body at the point $(x,y,z)$. The object of our interest in this underlying continuous space will occupy some voxels totally, will miss some others, and will partially intersect a third set of voxels. In the process of discretization some information regarding this 'true' object is irrevocably lost; nevertheless recovery of its surface (or a good approximation to it) may be our heuristic aim.

We now give a precise definition of an object in a discrete scene. We assume that a region is given and is already subdivided into voxels. We put the following restrictions on the notion of an *object in the scene*.

> *Property 1.* An object is the *union of voxels* (i.e., for each point p in the object, there is a voxel v entirely contained in the object and containing p).
>
> *Property 2.* An object is *connected*.

These properties, which make no reference to the densities of voxels, can be used to exclude certain subsets of the region as objects in the scene, but tell us nothing about how to find subsets which are objects. Many different approaches can be, and have been, taken.

We restrict our attention to binary scenes. A possible definition is the following. Let Q denote the union of all voxels with density 1. We may then define an object as a connected component of Q. We refer to objects defined this way as *0-objects*.

In our own work we have found it more convenient to use the following definition. A *1-object* is a 0-object which cannot be described as the union of two 0-objects U and V whose common points are isolated. (Geometrically this means that we do not allow 1-objects to be connected by vertices only, they must be connected at least by edges, which are 1-dimensional entities); hence the name 1-object.)

The mathematically precise definition of the surface of an object is quite complex (Herman and Webster, 1983), we do not reproduce it here. Instead we describe a surface tracking algorithm, which (when all details are provided) can be proven to produce surfaces which match the mathematically precise definition.

The input to the algorithm is a binary array and a face separating a 1-voxel from a 0-voxel (referred to as the *starting face*). The output is a list of voxel faces, each of which separates a 1-voxel from a 0-voxel. The faces in the list, when combined together, form a connected closed surface containing the starting face.

The surface to be tracked is defined as follows. The starting face has four edges. Each one of these edges is shared by another face which separates a 1-voxel from a 0-voxel. (There may be more than one such face, but a careful definition can be given which identifies exactly one of them.) We say that the four such faces are *adjacent* to the starting face. For each of these faces there are four faces adjacent to them (one of which is in fact the starting face) and so on. The surface is the set of all faces to which we can get from the starting face by repeatedly moving from a face to an adjacent face. Such a surface will be closed (separating a portion of 3D space from the rest of space) and connected (one can get from any point on it to any other point without leaving it).

The computational task of surface tracking is quite difficult. The reason for this is that in the medical environment we desire voxels to be small (for accuracy) and hence we need a very large number of voxels to describe an object (quite typically 10 million) and correspondingly quite a large number of faces in the tracked surface (quite typically 300,000). To avoid infinite loops during surface tracking we need to keep and check a record of the faces that have already been tracked, a significant task as the number of tracked faces increases. Nevertheless, a combination of powerful mathematical theorems (Herman and Webster, 1983) and sophisticated software techniques (Frieder et al., 1985) allows the tracking of such intricately detailed surfaces in approximately 15 minutes on the computer of the CT scanner. (For the GE 9800 CT scanner, this is a 16 bit minicomputer, the Data General Eclipse S/140).

This algorithm produces a surface based on the binary array representation. Alternative algorithms exist for producing surfaces from the SE, DC, and octree representations of a binary scene. For a published example, see Udupa (1982).

## 3. DISPLAY OF OBJECTS IN 3D SCENES

*a. Projection onto a screen*

In the previous section we discussed five different ways of representing the information in binary scenes: the binary array, SE, DC, octree, and as surface(s) of object(s) in the scene. Methods for displaying the information in the scene in a manner similar to that illustrated in Fig. 2 have been developed for all five representations. In this section we discuss such methods. Since techniques exist to translate from one representation into another, under suitable circumstances it may be best to do such a translation prior to display. In what follows we concentrate on direct display based on the different types of representation.

The two basic issues associated with displaying a view of (an opaque diffusely reflecting) 3D object on a 2D screen are: (i) ensuring that only that part of the object which is visible from the given view is displayed (hidden surface removal), and (ii) assigning a shade to the visible surface to impart an illusion of three-dimensionality (visible surface rendering).

These will be the subject matter of the next two subsections. Here we discuss how a point in the scene is to be projected onto the screen.

We assume that the coordinate system $(x,y,z)$ of the 3D scene is chosen in such a way that the axes are parallel to the planes which subdivide the region of the scene into voxels, and that the $x$, $y$, and $z$ directions correspond to the $i$, $j$, and $k$ indexing, respectively, of the voxels. Thus, for example, the $z$ coordinate is the same for all voxel centers with the same $k$ index. We think of the scene as movable in space and introduce a fixed coordinate system $(u,v,w)$ attached to the 2D screen on which we wish to display ($u$ horizontal on the screen, $v$ vertical on the screen, and $w$ perpendicular to the screen). We call this the *image space* coordinate system. For any desired view of the 3D scene, there is a 3×4 transformation matrix T (representing translation, rotation, and scaling of the scene) such that any point $(x,y,z)$ in the original scene can be expressed in terms of the image space coordinates according to

$$u = T_{11}x + T_{12}y + T_{13}z + T_{14}$$
$$v = T_{21}x + T_{22}y + T_{23}z + T_{24} \qquad (1)$$
$$w = T_{31}x + T_{32}y + T_{33}z + T_{34}$$

see Fig. 6. In our work we display a scene using orthographic projection; this means that the point (x,y,z) projects (if it is visible) onto the point (u,v) which is nearest to it on the screen and that its distance from the screen is w.

b. *Hidden surface removal*

We now describe how hidden surface removal can be achieved for each of the five representations discussed in Section 2.

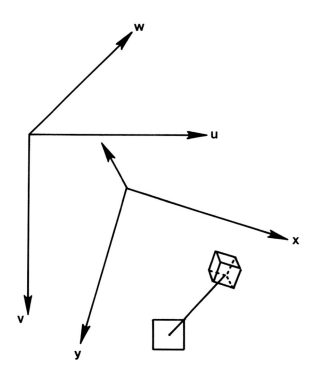

Fig. 6. *The 3D scene viewed in image space. A single 1-voxel and its 'projection' on the screen (w = 0) are shown.*

In many instances this is done by a so-called *back-to-front* method in which hidden surface removal is achieved by painting (and overpainting) on the screen images of individual 1-voxels in an order which is determined by their distance from the screen. To be more precise, eliminating x and y from Eq. (1) we get

$$w = \alpha_1 + \beta_1 z \qquad (2)$$

where $\alpha_1$ and $\beta_1$ are functions of the $T_{ij}$ and the position (u,v) of the point on the screen. Thus for a given point on the screen, the distance w from the screen either decreases or increases with the coordinate z. If the viewing direction is such that w increases with increasing z, then a point in a voxel in slice k cannot hide a point in a voxel in slice k-1 and thus painting voxels on the screen in order of decreasing k achieves hidden surface removal. Similar results can be shown for the i and j indices as well. Thus for the binary array representation, hidden surface removal can be achieved by three nested loops through the indices, either in increasing or decreasing order depending on the desired view. In our implementation the outer loop is through slices (k index), the middle loop is through rows (j index), and in the inner loop we paint onto the screen one-by-one all the 1-voxels in that row (i index).

The back-to-front method can be implemented in the SE representation, by simply changing the inner loop on the i index. The need for checking the individual voxels is eliminated by simply painting all voxels between the end-points of the segments. An interesting variation of this implementation arises from the fact that by an insignificant restriction on the viewing direction, we can insure that all voxels in a segment project onto the same scan lines in the image. This means that the geometrical transformation needs to be performed only on the end-points (Reynolds et al., 1985).

A further restriction on the viewing direction leads to extremely fast hidden surface removal with the DC representation. We assume that the viewing direction is obtained by rotation around the axis perpendicular to the slices, i.e., that z is parallel to v in Fig. 6. Under this assumption, only about half of a contour needs to be processed (the part lying between the points where the global maximum and the global minimum u-coordinates occur for the rotated contour). Since the 'inside' of the object lies to the left of the directed contour, only those voxels on the contour which are encountered while traversing in the direction of the contour starting from the voxel with the maximum u-coordinate and ending with the voxel with the minimum u-coordinate need to be displayed. Furthermore, only those contours which correspond to the sons of the root of the containment tree need to be considered for display (cf., Fig. 3). Since different slices project onto disjoint sets of scan lines, the object can be processed slice-by-slice, and at the same time, the image can be generated line-by-line. This nicety completely removes the need for time consuming memory management to handle input and output. These and other considerations lead to very efficient hidden surface removal; for details see Herman et al. (1986).

Hidden surface removal in the octree representation can be achieved by a back-to-front method which makes use of the recursive structure of the octree. For each view, a specific ordering of the eight descendents of each node labelled Partial will result in these descendents being ordered according to their distance from the screen (see Fig. 5). Hidden surface removal can thus be achieved by painting on the screen those subscenes which are labelled Full in the order determined by a depth-first traversal of the octree combined with the specific ordering of the descendents at each node.

In our approach to inexpensive but nevertheless realistic surface display in the medical environment we made use of the fact that the simple nature of the surface elements in a cubic scene allows us to prove powerful mathematical results. Such results lead to efficient computer procedures to handle the large amounts of data. An example of such a result, leading to efficient surface tracking, has been referred to in Subsection 2e. Another example is a theorem (Herman and Liu, 1979) which states that in a cubic scene the distances from an observer of two points A1 and A2 on the surface are ordered in the same way as the distances from the observer of the centers of the faces which contain A1 and A2, respectively. Such a theorem allows rapid removal of the hidden part of the surface.

The basic hidden surface removal technique used in displaying objects based on their surfaces is the so-called 'z-buffer' algorithm, which for our notation would be more appropriately called a w-buffer algorithm, since the buffer contains, for each screen point (u,v), the distance w from the screen of the point in object space currently displayed at (u,v). The surface is painted on the screen face-by-face. Whether or not the face currently being considered should overpaint what is on the screen is decided based on the contents of the w-buffer. It is here that the theorem mentioned above can speed things up considerably.

*c. Visible surface rendering*

We discuss in some detail how we render the visible surface for the binary array representation. We treat the SE, DC, and octree representations in a similar way.

Painting of a single 1-voxel is done as follows. First note that for a fixed view, the projections of any two voxels onto the screen have the same shape. Using Eq. (1), we work out the size of a rectangle (with edges parallel to u and v) which just encloses this shape. We paint a 1-voxel by painting on the screen such a rectangle located around the point (u,v) determined by Eq. (1) and the coordinates (x,y,z) of the center of the voxel (see Fig. 6). The brightness B assigned to the rectangular area is determined by the value of the distance w obtained from Eq. (2) according to the following rule (called *distance only shading*):

$$B = \frac{255 * (d_2 - w)}{d_2 - d_1} \tag{3}$$

where $d_2$ and $d_1$ are precalculated values such that for all views and all voxels $d_1 \leq w \leq d_2$.

This method of display does not require that the binary scene be cubic. In Fig. 7 we show the image of a spine obtained using this method. The CT data provided a 160×144×30 scene with voxel size 0.4 mm × 0.4 mm × 1.5 mm. The time required to display each 256×256 view of this scene is 111 seconds (84 seconds are needed for a 128×128 picture). Image quality is mediocre but acceptable for a preview of the scene. The same procedure can be used to display the object after interpolating the scene so that voxels are cubic. Figure 8 shows the same spine displayed after interpolation (scene size now is 160×144×109), demonstrating a considerable improvement. The time taken for this 256×256 picture is nearly four minutes (130 seconds are needed for a 128×128 picture).

The times quoted here refer to our current implementation on an Eclipse S/200 minicomputer. Since this is a program under development, we expect considerable improvements. For example, it was noticed that a large fraction of the time required to display the object was wasted by the painting of interior voxels (i.e., 1-voxels in the scene that are not visible from any direction). Many of the interior voxels can be easily identified by the fact that the six face-adjacent neighbors are also 1-voxels. By setting the density of such interior voxels to zero prior to displaying the scene speeds up the process without any loss of image quality. For example, by first altering the binary scene in this fashion reduced the time required to produce Fig. 8 from nearly four minites to less than one minute of which only 15 seconds was needed for the display, the rest was needed for the process of identifying interior voxels. The speed of the display process here approaches but does not quite reach the speed for special viewing directions of display using the DC representation.

The interpolation which resulted in the cubic scene in our medical examples has been used partly so that we can apply the rapid computer graphic display procedures which can be applied only to cubic scenes. The other reason is that we believe that interpolation results in more accurate approximation to the 'true' surface than what is obtained using the alternative of thresholding the original scene and then displaying the resulting surface. The display produced by this alternative method has a different 'resolution' in one direction from that in the other two. The use of original densities for interpolation and the thresholding of the interpolated scene is likely to result in a more accurate approximation to the true organ surface than the alternative. This is demonstrated for the binary array based display in Figs. 7 and 8.

At the early stages of our work the quality of the displays based on surfaces in cubic scenes was not quite as good as we desired, especially if they were displayed in a dynamic sequence. A number of techniques have been suggested to deal with this (Herman, 1980). Recently, we have proposed in Herman and Udupa (1981) and Chen et al. (1985b) an approach which improves the quality of the displays, but retains those properties of the approach which make it computationally efficient. In this new method, visibility of a point on the surface is determined in the same fashion as in the previously published approaches (Artzy, 1979 and Herman and Liu, 1979), that is

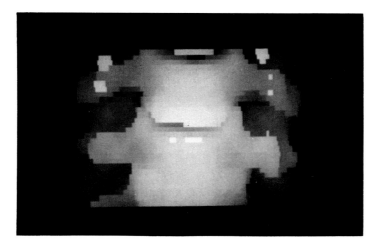

*Fig. 7. Display of a human spine from binary array representation derived from an uninterpolated scene. (Reproduced with permission from S.S. Trivedi, Interactive Manipulation of Three-Dimensional Binary Scenes, The Visual Computer, to appear.)*

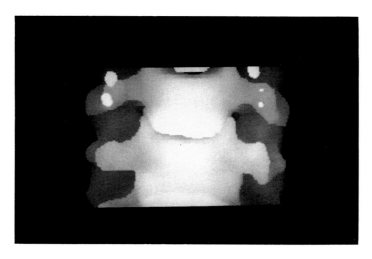

*Fig. 8. Display of a human spine from binary array representation derived from an interpolated scene. (Reproduced with permission from S.S. Trivedi, Interactive Manipulation of Three-Dimensional Binary Scenes, The Visual Computer, to appear.)*

using a variant of the z-buffer algorithm which is very rapid for displaying surfaces in a cubic scene. It is the shading of the visible points which is done differently. The convention of assigning a single shading value to all points of a face is retained, however the value depends on the orientation of all the faces adjacent to the given face. We call this *contextual shading*. Additional computation for contextual shading is minimal, since all necessary information regarding the faces adjacent to the given face can be gathered while the boundary surface is tracked and can be stored as attributes of the face in question. Striking improvements in the quality of displays due to contextual shading have been illustrated by Herman and Udupa (1981) and Chen et al. (1985b).

Figures 2(a) and 2(b) shown in Section 1 have been produced using contextual shading. A further example is shown in Fig. 9. The data for this figure are as follows. The 3D binary scene is 141×186×251. The surface consists of 296,970 elements. Surface tracking took 16 minutes of elapsed time on our Eclipse S/140 16 bit minicomputer. Displays of the detected surface, such as the one shown in Fig. 9, took 2 minutes per image to generate. (Images are generated at a 512×512 resolution, but are displayed at 256×256 using averaging to reduce sampling affects.)

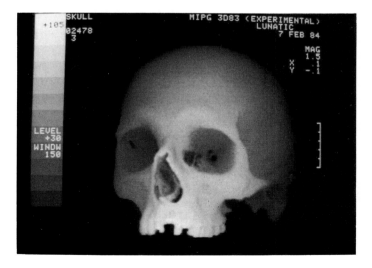

*Fig. 9. Surface-based display from x-ray CT data of a human skull specimen using the contextual shading method of Chen et al. (1985b).*

d. *Stereoscopic presentation*

By specifying a sequence of viewing directions, we can create a sequence of images similar to those that would be obtained if the specified surfaces rotated in front of a digitizing camera. By choosing a vertical axis of rotation and a small rotational increment, a pair of images can be produced which between them represent what the two eyes would see if they viewed the same stationary object. An example of such a stereo pair is shown in Fig. 10.

A sequence of such stereo pairs can be stored on a computer disk. If they are retrived in a rapid sequence (and viewed appropriately) they will provide a stereo 'movie' of the object. Three modes of viewing are discussed in Herman (1986):

(i) Side-by-side display to be merged either by crossing the eyes or by the use of mirrors or prisms to deliver the correct view to the correct eye.

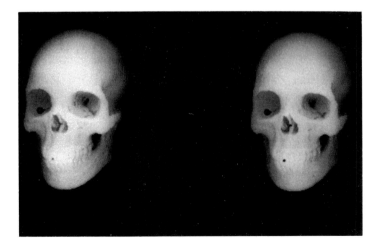

*Fig. 10. Slightly oblique frontal views of a specimen skull with a chipped incisor. Note that the morphology in the depth of cavities, such as the orbits, is clearly perceived and, in addition, the three-dimensional relationship of the teeth, mandibular rami, and zygomatic arches can be easily appreciated. Images are to be viewed cross-eyed; right-eye image is on the left, left-eye image is on the right.*

(ii) Alternating display of the two views with an active device blocking the vision of the inappropriate eye.

(iii) Overlapping displays of the two views with different optical properties (polarization or color) with a passive device in front of the eyes blocking the inappropriate view.

Illustrations of over 30 stereo pairs of medical objects based on CT and MRI can be found in Herman et al. (1985). The authors conclude with the following discussion.

Monoscopic viewing cannot give true depth perception. However, geometric perspective, relative size, texture shading, and interposition do provide monoscopic cues to depth in computer graphics. Motion parallax is another monoscopic depth cue used in computer graphics. True visual perception of depth can only be afforded by stereoscopic display. Stereoscopic viewing removes visual ambiguities inherent in monoscopic viewing. The advantages of stereoscopic viewing have to be seen to be truly appreciated. Stereoscopic viewing of certain anatomic areas consistently gives clinically relevant information not recognized in monoscopic viewing; these structures include the relationship of osteophytes to neural foramina, the volume and morphology of the orbits and cardic cavities, and the morphology of the skull. The addition of motion adds further information to stereoscopic display as it does to monoscopic viewing. The ease with which stereoscopic viewing can be added to present monoscopic display systems and the enhanced perception it affords make its use attractive. Clinical trials will determine its place in the imaging armamentarium.

*Acknowledgements*

*The research of the author is supported by NIH grants HL28438 and RR02456. He is grateful to Ms. M.A. Blue for typing, Mr. S. Strommer and Mr. K. Rude for photography, and Ms. M. Kirsch for drawing the illustrations. This work closely follows an earlier joint publication of the author with Drs. S.S. Trivedi and J.K. Udupa (Herman et al., 1986).*

# REFERENCES

Artzy, E. (1979). Display of three-dimensional information in computed tomography, *Comput. Graphics and Image Process.* 9, pp. 196-198.

Chen, L.S., Herman, G.T., Reynolds, R.A., and Udupa, J.K. (1985b). Surface shading in the cuberille environment, *IEEE Comput. Graphics and Appl.* 5 (12), pp. 33-43. Erratum (1986) 6 (2), pp. 67-69.

Chen, L.S., Hung, H.M., and Udupa, J.K. (1985a). Towards real-time interactive display of 3D objects, *Proc. 18th Hawaii Int'l. Conf. on System Sciences*, III, pp. 160-172.

Foley, J.D. and van Dam, A. (1983). *Fundamentals of Interactive Computer Graphics*, Addison Wesley Publishing Company, Inc., Reading.

Freeman, H. (1974). Computer processing of line-drawing images, *Comput. Surveys* 6, pp. 57-97.

Frieder, G., Herman, G.T., Meyer, C., and Udupa, J. (1985). Large software problems for small computers: an example from medical imaging, *IEEE Software* 2 (5), pp. 37-47.

Herman, G.T. (1980). *Image Reconstruction from Projections: The Fundamentals of Computerized Tomography*, Academic Press, New York.

Herman, G.T. (1986). Dynamic stereo display and interaction with surfaces of medical objects, *Proc. of SPIE* 671, to appear.

Herman, G.T. and Liu, H.K. (1979). Three-dimensional display of human organs from computed tomograms, *Comput. Graphics and Image Process.* 9, pp. 1-21.

Herman, G.T., Trivedi, S.S., and Udupa, J.K. (1986). Manipulation of 3D imagery, Technical Report MIPG106, Medical Image Processing Group, Dept. of Radiology, Univ. of Pennsylvania, January 1986. To appear in *Progress in Medical Imaging* 1, V.L. Newhouse (ed.), Springer-Verlag, Berlin.

Herman, G.T. and Udupa, J.K. (1981). Display of three-dimensional discrete surfaces, *Proc. of SPIE* 283, pp. 90-97.

Herman, G.T., Vose, W.F., Gomori, J.M., and Gefter, W.B. (1985). Stereoscopic computed three-dimensional surface displays, *Radiographics* 5, pp. 825-852.

Herman, G.T. and Webster, D. (1983). A topological proof of a surface tracking algorithm, *Comput. Vision, Graphics, and Image Process.* 23, pp. 162-177.

Meagher, D. (1982). Geometric modeling using octree encoding, *Comput. Graphics and Image Process.* 19, pp. 129-147.

Merrill, R.D. (1973). Representation of contours and regions for efficient computer search, *Comm. ACM* 16, pp. 69-82.

Quinn, J.F., Rhodes, M.L., and Rosner, B. (1984). A technique for locally optimal compression of CT image data, *Proc. of the 1984 Joint Alpine Symposium on Medical Computer Graphics, Image Communications and Clinical Advances in Neuro CT/NMR*, Innsbruck, pp. 31-36.

Reynolds, R.A., Gordon, D., and Chen, L.S. (1985). A dynamic screen technique for shaded graphics display of slice represented objects, Technical Report MIPG99, Medical Image Processing Group, Dept. of Radiology, Univ. of Pennsylvania, February 1985. To appear in *Comput. Vision, Graphics and Image Process.*

Rosenfeld, A. and Kak, A.C. (1982). *Digital Picture Processing*, 2nd edition, Academic Press, New York.

Trivedi, S.S. (1985). Representation of three-dimensional binary scenes, *NCGA '85 Techn. Sess. Proc.*, III, Dallas, pp. 132-144.

Trivedi, S.S., Herman, G.T., and Udupa, J.K. (1986). Segmentation into three classes using gradients, *IEEE Trans. on Med. Imag.* MI-5, pp. 116-119.

Tuy, H.K. and Udupa, J.K. (1983). Representation, display and manipulation of 3D discrete scenes, *Proc. of 16th Hawaii Int'l. Conf. on System Science*, II, pp. 397-406.

Udupa, J.K. (1982). Interactive segmentation and boundary surface formation for 3D digital images, *Comput. Graphics and Image Process.* 18, pp. 213-235.

VLSI-INTENSIVE GRAPHICS SYSTEMS

Henry Fuchs

*University of North Carolina
Chapel Hill, USA*

ABSTRACT

*This paper reviews some experimental and commercial graphics systems that intensively use VLSI (Very Large Scale Integration) technology. Described in some detail is the current state of one of these systems, our own system, Pixel-planes. The system renders about 30,000 full-screen, smooth-shaded, Z-buffered polygons per second, about 13,000 Z-buffered interpenetrating spheres per second.*

1. INTRODUCTION

Computer graphics has always been an expensive proposition and its users a demanding, unsatisfied lot — the picture never got onto the screen fast enough, or later, never moved fast enough once it got to the screen, and then, the picture was never sharp enough, never realistic enough. Many users have been tackling problems that could only be solved with the highest performance graphics systems — the crystallographer trying to understand the structure of a complex protein from noisy data, the radiologist trying to detect a possible tumor amid the clutter of healthy tissue, an architect trying to "walk" the client through the still-unbuilt house with only the images on a video screen. This paper describes the answers of various graphic system designers to this need. Regrettably, because the internal details of commercial systems are often not published and are revealed to outsiders only on a non-disclosure basis, some systems that the authors would like to have selected could not be described in this paper. Since this field moves very rapidly, it should also be noted that this paper was compiled during Fall, 1986.

## 2. HISTORICAL BACKGROUND AND OVERVIEW

Graphics displays have been a part of computers since at least 1950 with the point-plotting CRT on MIT's Whirlwind 1 computer (Everett, 1952). The random-deflection CRT, driven by a refresh list of point locations, was the basis for virtually all graphics displays through the 1960's. By 1968, at least one system had general 4×4 matrix transformations, clipping, and perspective divide (Sutherland, 1968; Sproull and Sutherland, 1968). This paved the way for commercial systems with real-time manipulation of 3D wire-frame models with perspective (Evans & Sutherland, 1971). These capabilities are the ones still found in today's high-performance random-scan vector systems.

The major drawback with all the random-scan vector systems has been their inability to produce realistically-rendered objects and scenes. Although algorithm development for generating continuous-tone renderings increased throughout the 1960's and early 1970's, the only readily-available output devices were film recorders. These took many seconds or minutes to expose a film in front of a CRT, on whose face was scanned out a sequence of hundreds of thousands of individual positions, each with a distinct intensity. Numerous users and system builders dreamed of dynamically interactive continuous-tone image generation, but until the mid-1970's, only very expensive systems, mostly visual subsystems of flight simulators, could afford this capability (Schumacker, Brand, Gilliland, and Sharp 1969; Watkins, 1970; Schachter, 1981). Continuous-tone digital images on video monitors were available, however, for image processing applications. These systems displayed the contents of an image buffer memory bank continuously on the video monitor. Unfortunately, these systems were of only limited use for computer graphics because their storage organization typically allowed only very restricted access to the pixel values. The medium was usually a disk track or a cyclical shift register that was only serially accessible. Thus, updating of random pixel values was unacceptably slow. In some systems, for example, it was only possible to change a pixel when its screen refresh time came around, once every 30 **milliseconds.**

The major turning point in graphics systems development came when RAM's (random-access memory chips) became sufficiently affordable so that a bank big enough to store an entire video image could be built at reasonable cost

(Kajiya, Sutherland, and Cheadle, 1974). Once random-access frame buffers became affordable, in the late 1970's, they quickly became the system of choice for many users. They produced continuous tone, usually color images, and, through various techniques with the color translation tables, these systems could produce a variety of useful effects — zooming, double buffering, simple animation, and moving overlays on a fixed background. Their prevalence has helped merge graphics and image processing. Combining flexible use of raster memory with built-in general processors, these frame buffer systems have steadily taken over areas previously dominated by the random-scan vector systems. A good example of such a flexible raster system is described in England (1981); its successor is currently available as the Adage RDS 3000. The popularity of these raster systems has been hindered by slow image generation and their poor quality, "jaggy" vectors, especially when compared with random-scan "vector" systems. Overcoming these limitations has been the driving goal of many of the designs described in the following sections.

A confluence of several people's varying interests in graphics, text processing, and flexible personal computers led to the development in 1973 at the Xerox Palo Alto Research Center of the ALTO personal computer (Kay, 1977; Thacker, McCreight, Lampson, Sproull, and Boggs, 1979). Its basic design was a high-resolution B/W display (1 bit/pixel), dedicated processor, mouse for x,y input, and a high-speed interface to a local area network. It has been copied widely into successively newer technologies (Bechtolsheim and Baskett, 1980) and has spawned the new industry of personal, professional workstations. Although these systems are often used simply for text display with multiple fonts styles, their flexibility allows the rapid drawing of lines and some simple 2D textures. The ALTO also pioneered the use of multiple, overlapping windows of arbitrary size, inside each of which could be any combination of text, graphics, and images. To facilitate rapid movement of text about the screen and between screen and off-screen memory, the machine included a BitBlt (for Bit-Aligned Block Transfer) instruction. Variations and generalizations of this instruction have been widely adopted, often under different names (Raster-Ops, Pix-BLT).

Each of these three kinds of graphics systems has been optimized for a different mix of tasks: a) the high-performance 3-D interactive system, b) the color frame-buffer display, and c) the personal workstation. While these three kinds of systems have blended together over the years, each has certain strengths that its designers have tried to bolster:

a) for the high-performance 3-D system:
- fast geometric transformation and clipping of primitives (typically, lines and polygons)
- fast lighting calculations
- fast rendering of primitives

b) for color frame buffers:
- many bits per pixel and flexible ways of using them for multiple frames, background, overlays, etc.
- fast, general processor for executing a variety of algorithms, the code often repeated for many or all pixels

c) for general workstation display:
- high spatial resolution to display as many windows with as much information as possible
- fast text generation of multiple fonts, sizes and styles
- fast movement of text about the screen and to/from off-screen memory.

As we shall see in the next section, with the increasing similarity between the three kinds of systems, manufacturers have started to offer systems that attempt to satisfy two and occasionally all three of these areas with a single system, or more often, a family of systems.

3. TAXONOMY

Although the field is still too vague to arrange a definitive taxonomy, it may be useful to arrange the components around the simplified functional organization sketched in Fig. 1. It should be noted that a number of the referenced systems implement particular functions, such as rendering, with a totally different method than that indicated in the figure. They are listed in the figure to show the *extent* of their function in comparison to other systems. The brackets indicate the extent of each defined component; it is

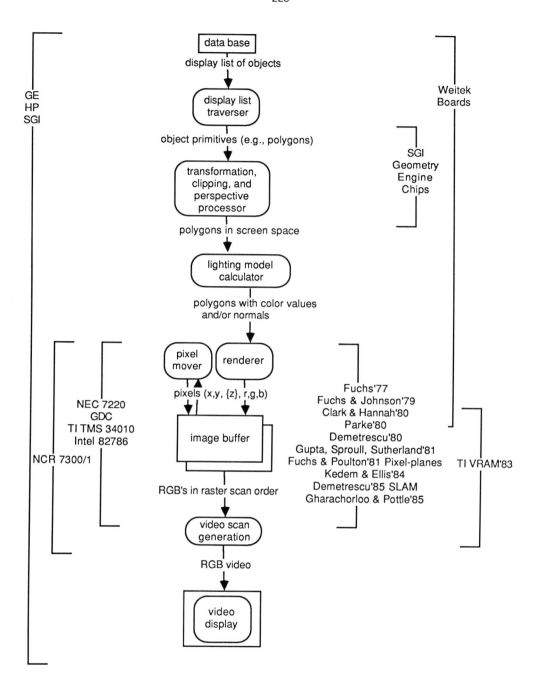

Fig. 1. Generic organization of 2D/3D raster systems.

understood that a working system constructed from any of the referenced machines would include all the functional units from data base to video monitor.

## 4. A VLSI SAMPLER

The systems reviewed here are roughly in chronological order, but the order is occasionally modified for clarity of presentation.

### a. *Graphics display controllers*

Among the earliest custom chips for graphics were the various video timing chips that generated the complex synchronization and blanking signals needed to drive a video display. These also controlled the timing of the access to the frame buffer memories for refreshing the screen. As silicon structures shrank in size and more transistors could be integrated into the same chip, the video chips increasingly took on tasks involved with the access to the frame-buffer memories for image generation. The first such chip that was generally available and could also handle low-level image generation, with help from some processor, was the NEC PD7220 GDC. It made possible the construction of frame buffers with much lower parts count and thus much lower cost. For example, an early system to use this chip, the Vectrix Corp. (Greensboro, NC) VX 128, contained a 480 × 672 pixel by 3 bits/pixel frame buffer, an internal 16-bit processor (Intel 8088), serial and parallel ports, separate package and power supply, and cost $2,000 in early 1983. The chip launched a flurry of under -$10,000 frame buffers and stimulated numerous "improved" versions from other chip vendors, some of which are described further below.

### b. *SGI's geometry engine*

This chip, which performs arithmetic operations on 4-element vectors, was introduced in Clark (1980) and more fully described in Clark (1982). It may have been the first custom VLSI chip for the 3D graphics market. This chip is the building block of a geometric transformation and clipping engine that is composed of a pipeline of a dozen of these chips. Each chip is configured to execute one part of the classic 3D processing pipeline as outlined in Sutherland (1968), Sproull and Sutherland (1968), and Sutherland and Hodgman (1974). The chip is basically a processor with four identical arithmetic units controlled by hardwired microcode that implements the various steps of matrix operations, clipping, and perspective divide. Each of the four units handles one of the elements of the 4-element vector in the commonly used homogeneous coordinate representation for handling 3D projective transformations. The arithmetic units handle 32-bit floating point representation, but due to silicon space limitations, many operations, such as multiply, take many clock cycles to perform. Nevertheless, with four arithmetic units per chip, and with a dozen chips in the pipeline, a 3D point is transformed in 15 microseconds. Improvements, made since 1982 in the commercial version of this chip from Silicon Graphics Inc., have enhanced its throughput considerably.

*c. Transformations and rendering with general purpose arithmetic chips*

The classic alternative to implementing a geometric transformation unit with custom chips is to implement it with general-purpose standard arithmetic units. Weitek (1986) describes the recommendations of one of the most popular suppliers of these fast artihmetic chips. Users of these general chips expect that their faster inherent speed will allow them to use fewer units than those of the slower custom variety from Silicon Graphics. In addition, the general-purpose chips allow coding of a wider variety of algorithms into the same unit. For example, the board-level products from Weitek not only perform geometric transformation, but also do rendering of planar and curved primitives. However, since the units have to transfer the pixels into an external frame buffer for screen refresh, their application to dynamically interactive systems has so far been limited. They do best on objects and scenes that take several seconds to transform and render.

*d. Dividing frame buffer among multiple processors*

Several designs have aimed to speed up the rendering of primitives into the frame buffer by dividing the frame buffer among multiple, usually identical processors. The frame buffer can be divided in an interleaved fashion so that every eighth pixel in a row and every eighth pixel in a column is assigned to the same memory. In this way a geometric primitive such as a polygon of line segment is sure to fall into many of the frame-buffer sections and thus sure to be processed by many of the processing elements (Fuchs, 1977). Fuchs and Johnson (1979) describe an implementation of this scheme using 8-bit microprocessors for the processing elements. Clark and Hannah (1980) describe a variation of this scheme with an additional layer of eight processors, each responsible for all the processors in every eighth row. These higher-level processors could perform the set-up calculations so that start-up costs for rendering a primitive are reduced.

Gupta, Sproull, and Sutherland (1981) describe a system with a similarly interleaved 64-piece frame buffer with processors optimized for data moving for workstation displays. In this system, each processor is connected to its 8 neighbors (up, down, left, right, and four diagonals) plus 3 others far away. Movement of a rectangular region can be accomplished by reading a footprint of 64 pixels into the processors, transferring the pixels among the processors and having each processor write its newly acquired pixel back into its memory. A similar system is described in Sproull, Sutherland, Thompson, and Minter (1983). An alternative subdivision of the screen is suggested in Parke (1980). Here the screen is successively subdivided in half. A memory portion and a rendering processor are stored at each leaf of a binary tree whose internal nodes each contain a splitting processor. A primitive travels down the tree, being steered to the region in which it is located and split if it straddles more than one region. The hope is that only a few of the primitives will end up at each root, and thus the total rendering load will be shared evenly among all the rendering processors. Various interleaved and successive subdivision schemes are analyzed in Parks (1982).

*e. Processor-per-polygon pipeline*

Demetrescu (1980) describes a novel architecture for rapid rendering of many polygons using a distributed Z buffer algorithm. The system consists of a pipeline of processing chips, each responsible for a single polygon. Each chip has one input and one output port. It receives a packet of color and Z values through the input port and sends either the same packet or a new packet to the output port. The packets move through the pipeline in raster-scan order so that each chip knows the X,Y address of the current pixel. To determine which packet a processor sends out, it first determines whether its polygon covers the current pixel at all — most of the time it will not. If it does, the processor compares its polygon's Z value with the Z value it received from its input and outputs the color and Z value of whichever has the smaller Z. Thus, the polygon that is output is the nearest one among all the ones in the pipeline. If the set-up values can be loaded into all the polygons fast enough, and if each processor can complete its task relating to a pixel in one pixel's video refresh time, then the system might run in real-time. The chip was designed and built, but not debugged (Demetrescu, 1986).

An extension to handle anti-aliasing is proposed in Weinberg (1981). In this scheme, the pipeline would not only pass color and Z values for a pixel, it would pass a sequence of polygon parts that would be visible within the pixel area. If a polygon processor found its polygon to be partially visible in the current pixel, it would add the portion of its polygon lying within the pixel to the sequence of inputs it received for the current pixel. Also, it would cull from the list any polygon parts that became obscured by its own polygon. The processor would then output all polygon parts still visible at that pixel. This enhancement significantly increases the amount of data passing through the pipeline and the amount of work required from each polygon processor. However, it allows a filter processor at the end of the pipeline to combine appropriately all visible portions of polygons for each pixel and, therefore, calculate a reasonable anti-aliased image.

*f. Processor-per-CSG-primitive tree machine*

Kedem and Ellis (1984) introduce a design for a machine that could render solid objects directly from their CSG (Constructive Solid Geometry) descriptions. A CSG tree consists of a primitive (such as cylinders and rectangular solids) at each leaf node and a set operator such as union and intersection at each internal node. The machine consists of a reconfigurable collection of processors onto which the CSG tree of the object is directly mapped. (It is assumed that the number of processors in the machine is typically larger than the nodes in the tree, although larger trees could be processed with multiple passes.) There are two kinds of processors, one to render a CSG primitive, such as a cylinder or a rectangular solid,

and another to perform the set operation on the output stream from two
renderers. Each rendering processor calculates the Z values and perhaps the
shade, for its primitive in each pixel, in raster scan order. This stream
of data is passed up the tree to a processor that performs the required set
operation on this and another stream. The output at the root of the tree
consists of a stream of packets, one for each pixel. The first color of
each pixel packet is the color of the visible surface. These values are
stored in a conventional frame buffer for screen refresh. The machine is
currently under construction.

*g. Rectangle-filling memory chip*

A memory chip that rapidly fills axially-oriented rectangles of constant
color is proposed in Whelan (1982). In this design, the addressing structures of both the row and column within a memory grid decode a minimum and
a maximum address and propagate the enable signal to all cells rows (or
columns) in between. Cells, whose row and column are both enabled, are
changed to the new value.

*h. Video RAM*

Introduced by Texas Instruments in 1983, the video RAM TMS4161 elegantly
solved the problem of accessing a high-resolution frame buffer memory for
screen refresh (Pinkham, Novak, and Guttag, 1983). The 64 kb × 1 DRAM includes an internal 256 × 1 bit shift register which can be accessed independently from the rest of the chip. In one memory cycle, an entire row is
transferred from the main memory to the shift register. Data can be shifted
out of this register at up to 25 MHz. During this read-out, the main memory is free to be accessed by the image generator. With standard DRAM's in
a high-resolution frame buffer, up to 50% or more of memory cycles may go
to screen refresh. This may significantly slow down the image generation
system. The use of the VRAM can reduce this access rate to less than 2%
(Guttag, Van Aken, and Asal, 1986). To achieve a similarly low rate with
standard RAM's, an elaborate memory organization with numerous extra parts
may be needed. However, if dual frame buffers can be implemented, then one
entire buffer can be devoted to screen refresh while the image generation
system is creating the next image in the other one. This effectively elimi-

nates the interference between the two systems, but at a substantial parts cost. (See Whitton, 1984 for a good discussion of this topic.)

*i. Scan-line access memory*

Demetrescu (1985) describes a memory chip enhanced to allow reading or writing of an entire row of the memory in a single cycle. The memory grid in the chip is treated as a rectangular part of one bit-plane of the frame buffer. Access to the chip is via op-codes and values that allow setting of the Y row, the X-Left and the X-Right edges for writing a specific span of the row (via read-modify-write), and setting of a 16-bit fill pattern. When writing a sequence of spans, the chip assumes reasonable values for the parameters in order to minimize the number of cycles taken up specifying them: the Y register increments, the X registers remain constant. Thus, filling a rectangular area can be accomplished with a single read-modify-write cycle per scan line. For instance, in order to fill a polygon, it is sufficient to specify the starting row and thereafter only the left and right boundaries for each successive row. Characters are written one scan-segment at a time. As each successive horizontal segment is loaded and the write command invoked, the proper addresses will be already in place (Y incremented, X's remain the same). A small system consisting of some dozen chips has been built and demonstrated (Demetrescu, 1986).

*j. Processor-per-pixel on a scan-line*

Gharachorloo and Pottle (1985) describe a rendering system that is a combination of the classic Watkins scan-line processor (Watkins, 1970) and a variation on the processor-per-polygon design of Demetrescu (1980). It sorts polygons by their top-most scan line and maintains an active queue of the polygons crossing the current scan line. For each of these, it calculates the left and right boundaries, and the starting and incremental Z values and colors. These are passed to a string of processors, one for each pixel on the scan-line. Each such processor performs a Z-buffer algorithm for its pixel on each data packet and passes the incremented values to its neighbor. The video data is scanned out of these pixel processors in the opposite direction of the packets: the packets travel right, the video data travels left.

### k. *TI's programmable graphics processor*

Texas Instruments introduced early 1986 a graphics controller chip, the TMS34010, that is a fully programmable single-chip 32-bit CPU. Although it still handles video control and timing, the chip is expected to be programmed in C to perform a variety of image-generating tasks more quickly than a general purpose 32-bit microprocessor. The processor contains 30 general purpose registers, stack instructions, a barrel shifter, field selection and control logic, and an instruction cache. It executes most instructions at 6 MHz. Its instruction set has been enhanced with graphics-specific codes such as block-move-and-modify RasterOps, and such variants of arithmetic operations as Maximum and Add-with-Saturate for combining multiple image patterns. The chip's designers appear to emphasize text generation and movement. It is not yet clear how suitable the chip will prove to be for image-generation tasks such as vector generation, polygon fill, and shading and lighting calculations. The approach is certainly reminiscent of the "wheel of reincarnation" effect desribed in Myer and Sutherland (1968). They note that designers tend to add more and more registers and functions to a display processor until it becomes essentially a CPU again, at which time it has become sufficiently slow and inefficient so that the designers put on it a small, fast display processor, and thus the cycle starts again. It should be no surprise that there are so many functions implemented within this chip; with some 200,000 transistors, it is three times the size of the popular Motorola 68000 32-bit CPU that is the main processor in many professional workstations (Guttag and Asal, 1987).

### l. *NCR low-cost integrated controller*

NCR recently announced its 7300 and 7301 chips set for the low-to-medium priced desk-top personal computers (Electronics, 1986b). The system facilitates low parts count by including within the 7300 character generation for two complete fonts as well as a look-up table and four 4-bit DAC's. The chip set can also control up to 8 windows.

### m. *Intel 82786 integrated controller*

Intel recently announced an integrated controller that is really three nearly-separate processors on the same chip (Electronics, 1986a). In addition to the usual frame buffer refresh control, there is novel high-level control of virtually any number of windows and a separate graphics processor for drawing into those windows. The graphics processor can generate text, lines, circles, and other geometric primitives. To control multiple windows, the CPU supplies the display processor with a map of the parts of various windows it wants displayed on the screen. The display processor fetches the appropriate pixels from main graphics memory.

*n. Other systems*

An early disclosure of a competitive graphics controller chip from Advanced Micro Devices appeared in the June 27, 1985 issue of *Electronic Design*. General Electric's Silicon Systems Technology Department has recently introduced the Graphicon 700 Graphics Processor which does high-speed rendering using several new custom chips of their own design (General Electric). Another new high-speed rendering system, this one from Hewlett-Packard, is described in Swanson and Thayer (1986). Lineback (1986) claims that nearly a dozen new designs are being developed currently in various chip houses in England, the USA and Japan. A description of several new systems can be found in Dill and Grimes (1986).

5. PIXEL-PLANES

*a. Overview*

The rest of this paper is devoted to a more detailed description of one particular VLSI-intensive graphics system, Pixel-planes. This high-speed rendering system features a 'frame buffer' composed of custom logic-enhanced memory chips that can be programmed to perform most of the time-consuming pixel-oriented tasks in parallel at each pixel. The novel feature of this approach is a unified mathematical formulation for these tasks and an efficient tree-structured computation unit that calculates inside each chip the proper values for every pixel in parallel. The current system contains 512 × 512 pixels × 72 bits per pixel implemented with 2,048 custom 3micron nMOS chips (63,000 transistors in each, operating at 10 million micro-instructions per second). New algorithms for rendering spheres (for molecular modeling), for adding shadows, and for enhancing medical images have been devised by various individuals within and also outside the group. The current system will shortly be placed in our graphics laboratory for use in molecular modeling, medical imaging, and architecture. We are building the next version of the system which should be available in 1988. This upgraded system will have chips that are much faster and that allow direct rendering of certain curved surfaces.

*b. Concept*

We are exploring the utility of a radical approach to raster graphics, having the front part of the system *specify* the objects on the screen in pixel-independent terms and having the memory chips themselves work directly from this description to generate the final image. The image primitives such as lines, polygons and spheres are each described by expressions (and operations) that are "linear in screen space", that is, coefficients A,B,C such that the value desired at each pixel is $Ax+By+C$, where $x,y$ is the pixel's location on the screen. Thus the information going to the 'frame buffer' is not address, data pairs (x,y addresses, RGB data), but ABC's and operation codes. In contrast to other raster systems, the most time-consuming calculations are not done by (1) general purpose computers or (2) special hardware that executes only a particular set of graphics functions. Instead, *Pixel-planes* is a rather general-purpose raster engine, especially powerful when most of the *pixel* operations can be described in linear (or *planar*) expressions. The system was introduced in Fuchs and Poulton (1981) and further developed in Fuchs, Poulton, Paeth, and Bell (1982). Basic algorithm descriptions to perform polygon rendering, Z-buffer tests, Gouraud-shading, as well as more elaborate algorithms such as spherical display and shading, and shadow casting can be found in Fuchs *et al.* (1985). Detailed description of a working prototype can be found in Poulton *et al.* (1985). Extensions of the linear tree to evaluate full six-coefficient quadratic expressions is described in Goldfeather and Fuchs (1986); algorithms to use the quadratic extensions to render solid models defined by Constructive Solid Geometry are described in Goldfeather, Hultquist, and Fuchs (1986).

*c. How it works*

The overall Pixel-planes system contains a fairly conventional graphics pipeline that traverses a hierarchical display list, computes viewing transformations, performs lighting calculations, clips polygons (or other primitives) that are not visible, and performs perspective division. The resulting colored polygon vertices in screen coordinates are then 'translated' into the form of data (A,B,C) for linear expressions together with instructions for the 'smart' frame buffer. An Image Generation Controller converts word-parallel A,B,C + Instructions to the bit-serial form required by the enhanced memory chips. A video controller scans out video data and refreshes a standard raster display (see Fig. 2). In our prototype system, the graphics pipeline and the translator have been implemented on three Mercury Systems ZIP 3232 numerical array processors, connected in a pipeline by special hardware of our design. The pipeline provides an honest 12 MFlops performance and can process 25,000 smooth-shaded, Z-buffered triangles per second. (Quadrilaterals are about 20% slower.) Shadows are cast at about 11,000 triangles per second, using shadow volumes. About 13,000 smooth-shaded, Z-buffered, interpenetrating spheres can be rendered per second. Although it is quite fast, the pipeline represents the speed-limiting factor in the current system.

The heart of the system is the 'smart' frame buffer, an array of custom, VLSI, processor-enhanced memory chips. These chips have three main parts: a conventional dynamic memory array that stores all pixel data for a 64-pixel column on the screen, an array of 64 tiny one-bit ALU's, and a multiplier/accumulator (M/A) that generates linear expressions simultaneously for all pixels. All ALU's in the system execute the same micro-instruction at the same time (Single-Instruction-Multiple-Data fashion), and all memories receive the same address (each pixel ALU operates on its corresponding bit of data) at the same time. The multiplier provides the power of two 10-bit M/A's at every pixel, but at much less expense in silicon area on the chip. The part of the M/A that is common to the pixels in a single column is factored out (10 stages of X-multiplier and the first 4 stages of the Y-multiplier). The six stages of the Y-multiplier can be built as a binary tree, since Y-products for a column are closely related, thereby reducing the cost in silicon area to about 1.2 bit-serial M/A stages per pixel for the entire linear expression evaluator. The current prototype chips have 70% of their area devoted to memory, 30% to processing circuits; each chip contains two identical 64-pixel modules like the one shown in the figure. (A U.S. patent [No. 4,590,465] has recently been granted for the basic design of the system; another is pending.)

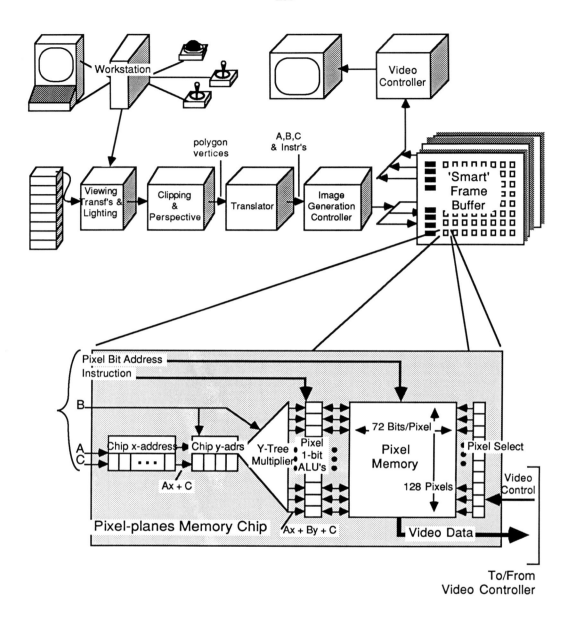

Fig. 2. Pixel-planes functional organization

## 6. SUMMARY

The continuing decline of DRAM prices and the steady increase in capabilities that can be squeezed into a single chip are fueling an explosion of interest in graphics-specific chips among the chip design houses around the world. Most suppliers are aiming for general capabilities and low chip count in order to reduce parts cost to increase their market size. A small, but increasing number of suppliers (Silicon Graphics, Weitek, General Electric, HP) are focusing on high-performance 3D graphics workstations. Encouraging results continue to be reported from the research community, where working prototypes using custom chips are starting to appear.

*Acknowledgements*

*An earlier version of the survey section of this paper was presented at the 1986 International Summer Institute, "State of the Art in Computer Graphics", at the University of Stirling, Scotland. The author is grateful for assistance with this section from Dr. Rae Earnshaw of Leeds University, Mr. Julian Ball of Raven Computers, Bradford, Yorkshire, England, and Dr. Melanie Mintzer. The author also thanks long-time colleague Professor John Poulton for his diagram on which Fig. 2 was based.*

*This research was supported in part by the (U.S.) Defense Advanced Research Projects Agency (monitored by the U.S. Army Research Office, Research Triangle Park, North Carolina) under Contract DAAG29-83-K-0148, by the National Institutes of Health under Grant R01-CA39060, and by the National Science Foundation under Grants ECS-8300970 and DCI-8601552.*

## REFERENCES

Abram, G.D. and Fuchs, H. (1984). VLSI architectures for computer graphics. In: *Proc. NATO ASI on Microarchitecture for VLSI Computers*, P. Antognetti, F. Anceau, and J. Vuillemin (eds.), Springer-Verlag, Berlin, pp. 189-205.

Bechtolsheim, A. and Baskett, F. (1980). High-performance raster graphics for microcomputer systems, *Computer Graphics* 14, pp. 43-47.

Clark, J. (1980). A VLSI geometry processor for graphics, *IEEE Computer* 13, pp. 59-68.

Clark, J. (1982). The geometry engine: A VLSI geometry system for graphics, *Computer Graphics* 16, pp. 127-133.

Clark, J. and Hannah, M. (1980). Distributed processing in a high-performance smart image memory, *Lambda* (since 1981, called *VLSI Design*), 1, pp. 40-45.

Demetrescu, S. (1980), A VLSI-based real-time hidden-surface elimination display system, M.Sc. Thesis, Department of Computer Science, California Institute of Technology.

Demetrescu, S. (1985). High speed image rasterization using scan line access memories. In: *Proc. 1985 Chapel Hill Conf. on VLSI*, H. Fuchs (ed.), Computer Science Press, Rockville, pp. 221-244.

Demetrescu, S. (1986). Personal communication.

Dill, J. and Grimes, J. (eds.) (1986). *IEEE Computer Graphics and Applications*, Vol. 6, No. 10.

Electronics (1986a). Intel designs a graphics chip for both CAD and business use, *Electronics*, May 19, 1986, pp. 57-60.

Electronics (1986b). NCR aims its graphics chips at PC instead of work stations, *Electronics*, May 19, 1986, pp. 61-63.

England, N. (1981). Advanced architectures for graphics and image processing, *Proc. IEEE* 301, pp. 54-57.

Evans and Sutherland Computer Corporation (1971). Line drawing system model I system reference manual, Evans and Sutherland Computer Corporation, Salt Lake City.

Everett, R.R. (1952). The whirlwind I computer, *Review of Electronic Digital Computers*, February 1952, p. 70 (as cited in Newman and Sproull, *Principles of Interactive Computer Graphics*, New York, McGraw-Hill, 1973 [first edition], pp. xxi-xxii).

Fuchs, H. (1977). Distributing a visible surface algorithm over multiple processors. In: *Proc. 1977 ACM Annual Conf.*, pp. 449-451.

Fuchs, H., Goldfeather, J., Hultquist, J., Spach, S., Austin, J., Brooks, F., Eyles, J., and Poulton, J. (1985). Fast spheres, shadows, textures, transparencies, and image enhancements in pixel-planes, *Computer Graphics* 19, pp. 111-120.

Fuchs, H. and Johnson, B. (1979). An expandable multiprocessor architecture for video graphics. In: *Proc. 6th Annual Symp. on Computer Architecture*, ACM-IEEE, New York, pp. 58-67.

Fuchs, H. and Poulton, J. (1981). PIXEL-PLANES: A VLSI-oriented design for a raster graphics engine, *VLSI Design* 2, pp. 20-28.

Fuchs, H., Poulton, J., Paeth, A., and Bell, A. (1982). Developing pixel-planes, a smart memory-based raster graphics system. In: *Proc. Conf. on Advanced Research in VLSI*, P. Penfield, Jr. (ed), Artech House, Dedham, pp. 137-146.

General Electric Graphicon 700 Product Literature, General Electric Silicon Systems Technologies Department, Research Triangle Park.

Gharachorloo, N. and Pottle, C. (1985). Super buffer: A systolic VLSI graphics engine for real time raster image generation. In: *Proc. 1985 Chapel Hill Conf. on VLSI*, H. Fuchs (ed.), Computer Science Press, Rockville, pp. 285-306.

Glassner, A. and Fuchs, H. (1985). Hardware enhancements for raster graphics. In: *Proc. 1985 ASI on Fundamental Algorithms in Computer Graphics*, R.A. Earnshaw (ed.), Springer-Verlag, Berlin, pp. 631-658.

Goldfeather, J. and Fuchs, H. (1986). Quadratic surface rendering on a logic-enhanced frame-buffer memory system, *IEEE Comp. Graphics Appl.* 6, pp. 48-59.

Goldfeather, J., Hultquist, J., and Fuchs, H. (1986). Fast constructive solid geometry display in the pixel-powers graphics system, *Computer Graphics* 20, pp. 107-116.

Gupta, S., Sproull, R., and Sutherland, I.E. (1981). A VLSI architecture for updating raster-scan displays, *Computer Graphics* 15, pp. 71-78.

Guttag, K. and Asal, M. (1987). A VLSI 32-bit graphics system processor, submitted for publication.

Guttag, K., Van Aken, J., and Asal, M. (1986). Requirements for a VLSI graphics processor, *IEEE Comp. Graph. Appl.* 6, pp. 32-47.

Ikedo, T. (1984). High-speed techniques for a 3-D color graphics terminal, *IEEE Comp. Graph. Appl.* 4, pp. 46-58.

Kajiya, J.T., Sutherland, I.E., and Cheadle, E.C. (1975). A random-access video frame buffer. In: *Proc. IEEE Conf. on Computer Graphics, Pattern Recognition, and Data Structure*, IEEE, New York, pp. 1-6.

Kay, A. (1977). Microelectronics and the personal computer, *Scient. Amer.* 237 (3), pp. 230-244.

Kedem, G. and Ellis, J. (1984). Computer structures for curve-solid classification in geometric modelling, Technical Report TR137, Department of Computer Science, University of Rochester.

Lineback, J.R. (1986). The scramble to win in graphics chips, *Electronics*, May 19, 1986, pp. 64-65.

Myer, T.H. and Sutherland, I.E. (1968). On the design of display processors, *Comm. ACM* 11, pp. 410-414.

Parke, F. (1980). Simulation and expected performance analysis of multiple processor Z-buffer systems, *Computer Graphics* 14, pp. 48-56.

Parks, J.K. (1982). A comparison of two graphics computer designs, M.Sc. Thesis, Computer Science Department, University of North Carolina at Chapel Hill, C.S. Technical Report TR82-001.

Pinkham, R., Novak, M., and Guttag, K. (1983). Video RAM excels at fast graphics, *Electronic Design* 31, pp. 161-172.

Poulton, J., Fuchs, H., Austin, J., Eyles, J., Heinecke, J., Hsieh, C.-H., Goldfeather, J., Hultquist, J., and Spach, S. (1985). Implementing a full-scale pixel-planes system. In: *Proc. 1985 Chapel Hill Conf. on VLSI*, Henry Fuchs (ed.), Computer Science Press, Rockville, pp. 35-60.

Schachter, B. (1981). Computer image generation for flight simulation, *IEEE Comp. Graph. Appl.* 1, pp. 29-68.

Schumacker, R., Brand, B., Gilliland, M., and Sharp, W. (1969). A study for applying computer generated images to simulation, AFHRL-TR-69-14, Air Force Human Resources Lab, Wright-Patterson AFB, Ohio.

Sproull, R.F. and Sutherland, I.E. (1968). A clipping divider. In: *Proc. Fall Joint Computer Conf.*, Thompson Books, Washington, D.C., pp. 765-775.

Sproull, R.F., Sutherland, I.E., Thompson, A., and Minter, C. (1983). The 8 by 8 graphics display, *ACM Trans. Graphics* 2, pp. 32-56.

Sutherland, I.E. (1966). The ultimate display. In: *Proc. IFIP Congress 65: Information Processing*, Vol. 2, W.A. Kalenick (ed.), Spartan Books, 1966.

Sutherland, I.E. (1968), A head-mounted display. In: *Proc. Fall Joint Computer Conf.*, Thompson Books, Washington, D.C., pp. 757-764.

Sutherland, I.E. and Hodgman, G.W. (1974). Reentrant polygon clipping, *Comm. ACM* 17, pp. 32-42.

Swanson, R.W. and Thayer, L.J. (1986). A fast shaded-polygon renderer, *Computer Graphics* 20, pp. 95-101.

Thacker, C.P., McCreight, E.M., Lampson, B.W., Sproull, R.F., and Boggs, D.R. (1982). ALTO: A personal computer. In: *Computer Structures: Principles and Examples*, D.P. Siewiorek, C.G. Bell, and A. Newell (eds.), McGraw-Hill, New York, pp. 549-572.

Watkins, G.S. (1970). A real-time visible surface algorithm, Ph.D. Dissertation, University of Utah, UTEC-CSc-70-101.

Weinberg, R. (1981). Parallel processing image synthesis and anti-aliasing, *Computer Graphics* 15, pp. 55-61.

Weitek Corporation (1986). Preliminary data documentation, board level graphics processors, and scientific processors, Weitek Corporation, Sunnyvale.

Whelan, D. (1982). A rectangular area filling display system architecture, *Computer Graphics* 16, pp. 147-153.

Whitton, M.C. (1984). Memory designs for raster graphics displays, *IEEE Comp. Graph. Appl.* 4, pp. 48-65.

Wientjes, B., Guttag, K., and Roskell, D. (1986). First graphics processor takes complex orders to run bit-mapped displays, *Electronic Design* 34, pp. 73-81.

Williamson, R. and Rickert, P. (1983). Dedicated processor shrinks graphics systems to three chips, *Electronic Design* 31, pp. 143-148.

KNOWLEDGE BASED INTERPRETATION OF MEDICAL IMAGES

John Fox and Nicholas Walker

*Imperial Cancer Research Fund Laboratories
London, U.K.*

ABSTRACT

*Classical medical imaging research has concentrated on new imaging technologies, on methods for improving image quality, and on techniques for extracting clinically useful parameters from images. Relatively little attention has been given to combining imaging systems with methods for interpreting clinical data. The emergence of expert systems has raised the possibility that imaging techniques can be integrated with tools for clinical decision making and problem solving. Five general schemes for combining these technologies are discussed. The interpretation of biomedical images is particularly problematic because of statistical, structural and temporal variation in morphology. Particular attention is paid to the processes which are needed to transform a pixel array into a symbolic form suitable for interpretation of the morphology. Some ways in which knowledge of shape, structure, and object taxonomy may contribute to the interpretation are discussed.*

1. INTRODUCTION

Medical data are frequently collected in the form of images. These may be single two-dimensional images like radiographs, ultrasound images and gamma-camera scans. Or they may be collections of images representing volumetric structures (e.g. tomographic images) or temporal processes (e.g. cine-angiograms). The computer's role may be in the production of the image, as in tomography, or in the processing of the images to yield clinically useful information. Several distinct traditions have emerged in computer based imaging.

The "image processing" tradition was one of the first to be established. It is primarily concerned with improving image quality to facilitate assessment by human observers. Some imaging devices produce poor quality results because of intrinsically low signal to noise ratios or low spatial resolution (e.g. gamma-cameras). Techniques for selectively filtering unwanted noise, such as low pass filtering in the spatial frequency domain, or selectively enhancing statistically abnormal features are well-established. Histogram equalization techniques and high spatial frequency enhancement have proved particularly useful as with X-Ray Xerography (Hall *et al.*, 1971).

"Pattern recognition" techniques extend the role of the computer into the realm of interpretation. Quantitative assessments or comparisons of image parameters can provide useful information which may not be immediately apparent to the clinician in a raw or enhanced image. Additional statistical techniques can be used to exploit these quantitative parameters in order to classify images into clinically important categories. An example is Stoecker *et al.* (1986) who use a variety of measures derived from the image to classify skin lesions.

A common feature of imaging work is its focus solely on the image - on improving its quality, extracting useful data from it and so on. Relatively little attention has been given to bringing non-image information, such as the patient's medical history, explicitly into the interpretation process. In recent years two potentially relevant lines of work have been developed within Artificial Intelligence (AI); image understanding systems and expert systems.

## 2. IMAGE UNDERSTANDING SYSTEMS

Image understanding systems differ from traditional imaging systems in their attempt to describe the world which is viewed through the imaging device. By "describe" we mean to cover a range of capabilities including the separation and identification of structures, objects or events; the ability to reason symbolically about how they are organized in space and time; how they are causally or functionally related, and so on. Perhaps the most developed of these is ACRONYM (see Brooks, Geiner, and Binford, 1979; Brooks, 1981).

Most image understanding work is linked to robotic manufacturing, satellite images or military recognition tasks. While the aims are exciting the systems are currently quite limited, notably in the range of domains the systems can function in, and the relative poverty of the symbolic image representations. Certainly no system yet built would qualify as a general purpose vision system. When good performance is achieved on a task it is generally dependent on "top down" interpretation from knowledge of the task domain. This works so long as the domain is limited and contains small numbers of objects. Binford (1982) provides a critical review.


## 3. KNOWLEDGE BASED EXPERT SYSTEMS

Knowledge based expert systems have been more concerned with interpretation and decision making than signal or image processing. However they may offer important techniques for combining disparate kinds of information used in medical problem solving, including images though little work has been done to date.

Knowledge based systems are computer programs which explicitly make use of knowledge about a subject ("domain"), a task, the people who use the computer system, or even knowledge about their own performance. Their task may be to interpret typed language, spoken commands or visual images; to control an instrument or a robot; to advise on decisions or solve problems. We speak of knowledge, rather than information or data, in order to emphasize that the aim is for knowledge to be represented explicitly in the computer rather than implicitly in abstract representations like numbers or computer algorithms (e.g. Levesque, 1987). How and why we aim to do this will be discussed in a moment.

Expert systems are knowledge based systems which have usually had two specific features. First they have been primarily concerned with making decisions — as opposed to interpreting sounds, images etcetera; second they have usually been designed to interact with human users — they are not autonomous robots or laboratory instruments. Medical expert systems, for example, are typically programs which can be consulted by a physician wishing to answer specific questions about diagnoses, investigations or treatments. The expert system may consider specific data about the patient; it may remind the enquirer of possible interpretations of data, or important data that need to be collected; it may offer pointers to the relevant literature, or even given "opinions" on possible diagnoses or management plans.

Expert systems have received great attention in the last few years and the medical imaging community has begun to ask whether, and in what ways, the techniques might be helpful to them. After summarizing the technical ideas introduced in expert systems we consider a number of different ways in which they might be used to assist in the interpretation of patient data presented in the form of images. For our purposes the particularly distinctive features of knowledge based systems which seem pertinent to medicine and imaging are:

*a. Qualitative representation of knowledge*

A great deal of research in AI has been concerned with logical reasoning, theorem proving and the machine representation of human knowledge. These varied interests have converged in the development of techniques for symbolic, qualitative reasoning. In vision the emphasis may be placed on the qualitative topology of the objects in a visual scene, rather than their precise geometry for example. Similarly we may be more concerned with the temporal topology (what happens before, during or after what) than the accurate chronometry of a sequence of photographic frames, and more interested in the reasons for believing hypotheses than the mathematical probability that they are correct. If ... then ... production rules are frequently used to implement reasoning rather than quantitative calculation. Although numerical data may be included in an expert system knowledge base they are invariably used within a qualitative framework.

Although decisions about diagnoses, desirable investigations, treatments and so on can be viewed with well-established quantitative models (e.g. those which view a decision making process as an attempt to maximize the expected value of a decision outcome) human judgements exploit many qualitatively different sources of information. Among the more obvious of these in medicine are models of causal mechanisms underlying diseases; information about structure and anatomy; information about the function of organs and physiological or biochemical systems; information about the taxonomic relationships among diseases, symptoms, and so on. Although many argue for the use of both qualitative and quantitative methods (such as the combination of statistical and logic-based inference techniques; see Kanal and Lemmer, 1986) the principle theoretical questions that we are concerned with are the new capabilities that qualitative techniques may offer.

## b. Explicit representation of knowledge

Conventional software encodes knowledge implicitly in algorithms or abstractly as numbers. Knowledge based systems attempt to make both facts and programs explicit in symbol structures that the computer can search and modify as well as execute. Facts like "region 1 is inside region 2" are encoded symbolically in such a way that the machine has access to information which defines what a "region" is, and what it means for one thing to be "inside" another.

"Procedural knowledge" (knowledge about how to do things) is also explicitly encoded in the sense that the system can examine the procedures, decide when to use them and even modify or adapt them for particular circumstances. In a pure expert system (a pure system has probably never been implemented but no matter) all domain knowledge is explicit.

---

**GIVEN THE FACTS:**

    complaint of John Smith is dyspepsia

    peptic ulcer is a cause of dyspepsia

    duodenal ulcer is a kind of peptic ulcer
    gastric ulcer is a kind of peptic ulcer

**AND THE HYPOTHESIS GENERATION RULE:**

    if complaint of Patient is Complaint
       and Disease-class is a cause of Complaint
       and Specific-disease is a kind of Disease-class
    then Specific-Disease is a possible cause of Complaint of Patient

**IMPLY:**

    duodenal ulcer is a possible cause of peptic ulcer of John Smith
    gastric ulcer is a possible cause of peptic ulcer of John Smith.

---

*Fig. 1. Given initial facts about a patient, text book knowledge about conditions which cause dyspepsia, and taxonomic knowledge of different kinds of those conditions, the qualitative logical rule generates and infers specific possible causes of the patient's condition. This and later examples have been implemented in the PROPS2 knowledge engineering package (Frost et al., 1986; Duncan, 1986; Fox et al., 1986).*

Figure 1 illustrates an explicit, qualitative representation of medical facts and a rule for formulating diagnostic hypotheses.

*c. The use of strategic or meta-level knowledge*

The first wave of research into medical expert systems was primarily concerned with learning how to represent knowledge in an explicit logical form that could be exploited by a computer in arriving at decisions or solving problems, using methods for making decisions and solving problems that were usually built into the code of the expert system software. More recently interest has turned to explicitly representing the decision making and problem solving strategies which underly good clinical judgements (see Davis and Buchanan, 1984). When should we try to eliminate possible diagnoses rather than confirm specific hypotheses? How much diagnostic detail should be collected before committing to treatment? How should the clinician cover for several possible but uncertain conditions? Figure 2 illustrates how a simple diagnostic strategy can be represented explicitly as a set of rules. The ability of a machine to reason at the "meta-level" about its own knowledge and methods offers both an important source of flexibility and a way of formalizing knowledge based theories of decision making (Fox, 1984; Fox et al., 1987).

```
if complaints of Patient include Condition
   and text book causes of Condition include Cause
   and Cause is not eliminated
then possible diagnoses of Condition include Cause

if number of possible diagnoses of Condition > 4
   and possible diagnoses of Condition include Cause
then check Cause is eliminated

if eliminating signs of Cause include Sign
   and Sign is confirmed
then Cause is eliminated.
```

*Fig. 2. This group of rules implements a simple strategy for eliminating diagnostic possibilities when the set of possibilities becomes too large to manage conveniently. Rule 1 generates hypotheses on the basis of the patient's complaint and the text book causes of the complaint. Rule 2 monitors the number of possible diagnostic hypotheses; if this rises above 4 then it creates "goals" whereby the system will try to eliminate alternatives until the total falls to the manageable number 4. Rule 3 specifies the general conditions in which a diagnostic hypothesis can be eliminated.*

*d. The user interface*

A feature often considered important in designing medical expert systems, and which is in large part a consequence of making data, facts and meta-level knowledge explicit, is the ability for the user to engage in a question-answer dialogue with the system about what it knows, how it has arrived at its conclusions, whether such-and-such a hypothesis is supported, and so forth. Questions from the user may sometimes be answered simply by retrieving information stored in the knowledge base — an expert system knowledge representation can be highly intelligible (see Figs. 1 and 2). In other cases a user's questions may cause the expert system to reason out the answer, or to set itself a task to collect additional data. These features encourage the view of an expert system as an active partner in clinical decision making; a significant change from the imaging community's traditional view of the computer as a subservient instrument.

4. EXPERT SYSTEMS AND MEDICAL IMAGING

Since medical images are just one form of patient data the question arises whether expert systems could help to interpret them. Because expert systems offer the possibility of integrating different types of knowledge, and use that knowledge flexibly, they offer some attractive possibilities. For example, based on general medical knowledge of the clinical presentation of different kinds of lesions and their typical anatomical locations the knowledge base could be used to generate hypotheses about what to expect in an image (cf. Fig. 1). Given the fact that:

    symptoms of cerebellar tumour include dizziness

and the patient data:

    history of John Smith includes dizziness

this rule:

    if history of Patient includes Symptom
        and symptoms of Lesion include Symptom

>     and Lesion is not eliminated
> then possible causes of Symptom of Patient include Lesion

will generate the following conclusion:

>     possible causes of dizziness of John Smith include cerebellar tumour.

Furthermore, by analogy with the rule set in Fig. 2, control rules could direct the strategy for processing, encoding or searching images in different ways depending on the hypothesis which is being pursued. Given the above conclusion and these facts:

>     common locations of cerebellar tumour include posterior fossa
>     morphologies of cerebellar tumour include dense sphere

and this rule:

>     if possible causes of Symptom of Patient include Lesion
>         and common sites of Lesion include Site
>         and morphologies of Lesion include Morphology
>     then check presence of Morphology at Site

the following goal is generated:

>     check presence of dense sphere at posterior fossa.

These facts illustrate the sort of knowledge that might be used to generate hypotheses which direct a process of interpreting a brain scan. However the rules are completely general; they are not specific to particular types of scan or imaging device. For example, the following facts are relevant to the interpretation of a chest X-ray:

>     history of John Smith includes erythema nodosum
>     symptoms of sarcoidosis include erythema nodosum
>     common locations of sarcoidosis include hilar regions
>     morphologies of sarcoidosis include lymphadenopathy

and will cause the rules above to generate the goal:

check presence of lymphadenopathy at hilar regions.

These rules illustrate how we might begin to explore the use of qualitative clinical knowledge in machine understanding of biomedical images, though little work of this kind has yet been carried out. Our examples are realistic in the sense that they have been implemented and are executable in an existing package (PROPS2: Frost et al., 1986; Duncan, 1986; Fox et al., 1986). However they gloss over substantial complexities. For example, what does it mean to "check" an image for the presence of a visual structure? This raises the general question of how image processing systems might be coupled to knowledge based expert systems.

5. COUPLING IMAGE PROCESSING AND IMAGE INTERPRETATION

Figure 3 illustrates five different ways in which we might couple an imaging system to the interpretation mechanisms of an expert system. All the schemes assume two common components which we shall assume but not try to describe in detail here as there are so many different possibilities.

First there will be some sort of signal processing subsystem which delivers a quantised representation of the input signal as an array of pixels. We shall consider just two picture dimensions here, but the array might have more dimensions if, say, volumetric or temporal information are required. Second we assume the presence of a knowledge based system which is capable of making logical inferences of clinical significance on the basis of image features or structures once these have been extracted from the pixel array. This system might incorporate knowledge about parameter values; the appearance of particular types of lesion; the clinical symptoms and signs that are associated with different lesions, and so forth.

Multi-pixel features (e.g. contours, patches, textures) or structures (e.g. a dark patch adjacent to a curvilinear contour; an increase in patch diameter over time) are only implicit in the pixel array and must be extracted before the expert system can do any useful work. The pivotal question for coupling imaging and expert systems appears to be this. What general schemes are there for extracting features or structures from the pixel array and presenting them to the expert system?

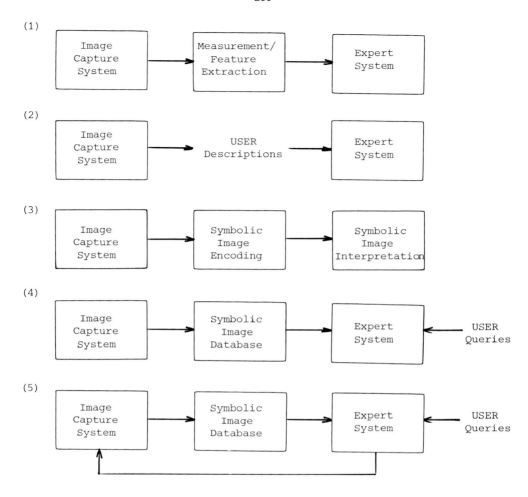

*Fig. 3. Five general schemes for coupling imaging devices and knowledge based expert systems. Described in text.*

Scheme (1) represents an apparently simple coupling of the two subsystems using conventional pattern recognition procedures. Scheme (2) requires a human participant to encode image features or structures while Schemes (3), (4) and (5) demand additional programs — currently beyond the state of the art — for creating and managing a symbolic description of image contents. Let us consider some of the characteristics of these schemes in more detail.

*Scheme 1: Quantitative coupling*

In this simple scheme conventional algorithms are applied to the pixel array to yield (i) singleton parameters (e.g. average pixel value; DC level in the frequency domain), (ii) parameter groups (e.g. cross-correlation values for images in a library; the power spectrum of the image), (iii) derived boolean descriptors (e.g. normal/abnormal). Such data could be passed for interpretation by an expert system prepared using one of the many commercial expert system shells which are now available. Given the output from an initial set of general algorithms, the expert system could use its internal knowledge to select further ones to further restrict the range of possible image interpretations.

This may be a practical scheme for images where the clinical significance of particular parameters is known. It should be remarked, however, that it is a conservative application of expert system technology. Indeed the approach may not warrant the use of an expert system at all since the representation of the image is so impoverished. Since it does not describe the image in natural terms — objects and their parts; organs, tissues, lesions and their spatial topology etcetera — the opportunity to use the spatial, temporal or causal forms of reasoning that knowledge based systems offer is very limited. Such a system could not be said to "understand" the image; it could not offer medically intelligible explanations of its conclusions for example. It is acting as a sophisticated look-up table.

This is not to dismiss the quantitative coupling approach. It may well be a clinically valuable extension to existing pattern processing techniques, particularly as some of the more exotic possibilities of image understanding are still in the realm of AI research rather than practical development.

*Scheme 2: User encoding of image data*

While people cannot perceive the power spectrum of an image, or reliably report its statistical properties, they can identify structures and shapes, areas of interest, possible abnormal formations, and so on. An expert system might therefore be developed to operate interactively with a human observer viewing the output of the imaging system. The expert system might be able to ask questions like "do you see any areas of high activity in the posterior portion of the scan?" and, if the answer is positive, "is the area circular or elongated?", "is it close to the cranial wall?" and so on. **Ellam and Maisey (1986) give a promising example of this approach applied to the interpretation of thyroid scans.**

Some of the advantages and possible disadvantages of the approach are obvious. It would be technically straightforward to implement and an experienced observer will be able to make some judgements more effectively than a computer for some time to come. Against this are problems: the procedure may be time consuming; even skilled observers have lapses and make errors; the features being described are not quantified and, indeed, the computer does not have access to the quantitative pixel data at all.

*Scheme 3: Automatic symbolic encoding of the image*

Scheme 1 is practical but depends upon pixel- and array-oriented computations on the image rather than on object-oriented computations which deliver an encoding in terms of organs and tissues and the relationships between them. Scheme 2 is also practical but has the opposite problem — an observer can describe objects and their spatial relationships but the underlying pixel data are not available for quantitative analysis. Clearly we would prefer to have programs which can accept pixel arrays and output explicit symbolic descriptions.

AI work in image understanding has developed some way towards providing general techniques for describing images. A lot of time has been spent in developing edge finders which perform well in the presence of noise, for example. A brief introduction to two of these is given later. However all these operators, whether region or edge based, only produce another image — that is, a set of numbers arranged in an array. They are more closely related to real aspects of the world, (surface discontinuities; the boundaries of organs) than the original pixel array, but no interpretation can be put on them directly.

There are two general approaches to beginning low level interpretation. In many systems the strategy is to use a top-down model that works over the primitives looking for possible instances of the (small number of) objects described in its knowledge base. This is not always a natural strategy, and slightly absurd recognition rules may be needed; for example in Ohta (1980) a region has to touch the bottom of the image to be called a road.

Alternatively programs look for suggestive structures. ACRONYM (Brooks, 1981) looks for instances of a 'ribbon' — two matching edge segments lying on either side of a hypothetical axis — which is taken to be the two-dimensional projection of a cylinder. ACRONYM uses an intervening process which marks the position of possible ribbons given the edges.

The insertion of intervening processes, to find relationships between primitives which are unlikely to have arisen by chance, is termed perceptual grouping. Grouping seems to be an important step from images to symbolic representations. Continuity relationships — having a symmetry axis, being parallel, lying on a single curvilinear contour, or simply being near to each other in the image — simultaneously dissect the image into unrelated parts which can be interpreted separately, and also organize it into structures, which can then form elements of the shape description. These continuity relationships are exactly those that remain invariant over both large ranges in the viewing angle of a particular object, and often over different instances of objects within a class.

Many of the available techniques for representing objects, their shapes, and spatial distribution, are similar to, and often derived from, those developed in computer graphics. Typically a graph representing the edges, faces, and vertices of block-like structures or a tree structure detailing a spatial arrangement of a small number of volumetric primitives is stored and manipulated using geometrical transformations. Descriptions of complex shapes that allow recognition and subsequent reasoning about what that shape might mean, in terms of function (or malfunction) is only just beginning (Connell and Brady, 1987; Lowe, 1987). Unfortunately little of this work has been carried out on the kinds of image routinely produced in biomedical laboratories.

There are a number of reasons why automatic symbolic description of medical images is particularly difficult. These include:

- Current feature extraction techniques are not very robust while medical images are frequently of low resolution and very noisy. This situation is currently improving. However it seems clear that it is not possible to extract features effectively without a heavy computational overhead. Happily this can be effected in a highly parallel way and specialized chips are becoming available which can carry out parallel extraction of simple features like edges, across an image in near real-time.

- High statistical variability in image parameters and object geometry (lesion morphology and "normal" anatomical features) implies complications in (i) deriving the "true" topologies represented by feature groupings, (ii) identifying and characterizing the underlying topologies of objects and structures in a way that is insensitive to biological variation, and (iii) representing these topologies in a form that is amenable to symbolic manipulation.

- Most object description techniques like computer graphic techniques make simplifying assumptions, which are unrealistic for the biological case. For example they commonly assume that objects are rigid (whereas biological structures are flexible and easily distorted) and do not change over time (whereas normal ageing and pathological processes may produce continuous changes of morphology).

We are some way from designing a system that can automatically produce a good symbolic representation of the shape of a nonrigid, poorly defined structure in the presence of other structures. This is of course the kind of image which is routine in, say, medical radiology. Later sections cover in more detail the approach to symbolic encoding of biomedical images which is being pursued in our laboratory.

*Scheme 4: The image as a database*

An assumption we might easily make about a symbolic image description is that it could explicitly contain all possible descriptions of the image, just as the pixel array contains all the information in the original two-dimensional signal (assuming the spatial sampling frequency is high enough). Although in some simple image understanding problems it may be possible to precompute all possible descriptions this is not generally true.

Firstly it is well-known that images are often underspecified, in the sense that (however high the fidelity of the pixel array) there will frequently be more than one consistent interpretation of the image; the most obvious example of this is that an infinite number of three-dimensional objects can project the same shape onto a two-dimensional image plane. Some ambiguities in the image can only be resolved by means of external information — such as clinical knowledge about the kinds of lesion that are possible. The second reason that we cannot expect to have a complete symbolic description of the image is that descriptions are not absolute but relative to a purpose or question — what a radiologist wants to extract from an image may depend upon the diagnosis that is suspected, or the treatment that is being evaluated.

The image description should not therefore be viewed as a closed structure but as a database of descriptions which the expert system can search, and add to, as new descriptions are derived which are specific to the clinical questions at hand. A fundamental research question is now obvious. Namely, what is the optimal set of image measurements and descriptions that the symbolic image processor should record in the primary database, which is complete enough to support a wide range of queries across domains?

*Scheme 5: Active control of imaging*

The fourth scheme emphasized the ability to update the symbolic image database with information or inferred interpretations which reflect the context of a user's queries. The next obvious question is whether future expert systems will need to control the image capture process and very low level processing. That is, should it be possible for the expert system to adjust the parameters of the imaging device, the quantisation of the image, or modify the symbolic encoding of the image?

The requirement seems plausible for two reasons. First, imaging research has produced a wide range of operators for computing edges, regions, textures, and so on. So far however there is no universal operator; each has particular strengths and weaknesses. Unless one computes all possible operations, which may be impractical, there is a risk of losing potentially significant information because the wrong operator is used. For example, different features have different sizes and operators with different resolutions are needed to extract them (see below). Human vision seems to solve this problem by actively modifying the image capture process; the human retina only provides high resolution in the foveal region, with resolution decreasing dramatically towards the periphery. This area of fine resolution is moved about to look at the interesting areas of the image. A radiologist scans a radiograph both under the guidance of clinical knowledge (knowing what he/she is looking for) and low resolution information coming in from the periphery. There may be benefits in other kinds of active control in machine imaging.

The second argument is related but is concerned with image enhancement. When specific clinical questions are raised it may be useful to collect image data in a way that reflects these questions. For example high frequency information may mask structures recorded in the low spatial frequencies. If we are looking for structures which are expected in the low frequencies, such as a large, fuzzy patch, then reprocessing the image with high spatial frequencies removed might be helpful.

This question of whether the image capture subsystem should operate in a fixed way, whether or not it should be independent of the clinical interpretation task which is being carried out, seems to be a variant of the earlier argument that the database of symbolic descriptions is not closed. The set of measurements we make on the image should not necessarily be viewed as closed either, because the choice of measurements makes implicit assumptions about what we are making the measurements for. If a "neutral" set of operators cannot be found, and the image capture process has to reflect idiosyncrasies of the clinical situation, then the expert system may require two very different kinds of knowledge: (i) "domain" knowledge about the clinical significance of particular structures, and (ii) a second "imaging" knowledge base which embodies know-how about imaging itself (e.g., see Nazif and Levine (1984) for a system where the very low level operations are made explicit as production rules and thus represent a body of knowledge about imaging).

## 6. LOW LEVEL OPERATORS FOR SYMBOLIC ENCODING OF IMAGES

A vast amount of research has gone into the very earliest stages of vision; extracting discontinuities from the pixel array. When the images are entirely noise free this is simple, but traditional edge finders are very sensitive to additive noise. Whole books have been written on edge and line finders but an overview of two very popular ones may be useful.

Traditionally an "edge" is the step discontinuity that lies between two regions of uniform but different light intensity. In reality natural image regions are rarely of uniform intensity and the step function must be smoothed into a ramp or similar function. There is a peak in the gradient of the image intensities at this type of edge but obtaining the slope by simply differentiating the image will amplify the noise as well as the discontinuity, so the signal to noise ratio has to be improved by smoothing. This can work well because the signal is present over an area of the image while the noise should be uncorrelated over this same area.

It is possible to demonstrate the optimality of a particular operator by showing that it minimizes some error function. This function should measure the performance in the presence of noise (generally modelled with gaussian white noise) in terms of the probability of missing edges that are present and of finding edges that are not. It should also measure how precisely the operator places the edge compared to its real position. The effect of minimizing this measure depends strongly on the particular function that is used to model an edge, usually either a step or a ramp.

*a. The Marr-Hildreth operator*

This operator was put forward more on a heuristic than an analytical basis, but it has proved to be generally useful and at least provides a baseline measure against which to compare others. The Marr-Hildreth edge finder (Marr and Hildreth, 1980) filters the image with a gaussian of a particular size (variance) and places edges where the second differential of this new image is zero. This appears to work well because there is no arbitrary thresholding. Unfortunately, in areas where there is no strong signal, as in flat expanses of uniform intensity, many zero-crossings are produced from the values randomly wandering around zero. In practice thresholding on the gradient at the zero-crossing, or some corroboration between zero-crossings produced by different sized operators, can be used to filter out the unwanted noise.

Choosing the size of the gaussian is arbitrary (a feature of all the operators mentioned here). There is a tradeoff between the precision with which an edge is located and the signal to noise ratio - the larger the gaussian the poorer the edge location but the better the signal to noise ratio. Convolving with a gaussian is a low pass filtering operation so details in the image whose spatial frequency is higher than the cut off frequency of the particular gaussian are not found. The laplacian $\nabla^2$ can then be applied to detect intensity changes in the image in an orientation independent way (Marr and Hildreth, 1980):

$$\nabla^2[g(x,y)*I(x,y)] = \nabla^2 g(x,y)*I(x,y) \tag{1}$$

The loci of zero-crossings at a given scale can be found by searching in:

$$\nabla^2 g(x,y) = -\frac{1-(x^2+y^2)}{2\sigma^2} \frac{\exp\frac{-(x^2+y^2)}{2\sigma^2}}{\pi\sigma^4}. \tag{2}$$

There are several problems with this filter. Firstly not all zero-crossings correspond to step edges. Zero-crossings represent places where the first differential of the image is zero. This occurs at peaks in the slope, but also at saddle points, as occur in a staircase function; the operator is symmetrical and thus locates the edge equally well both across the edge (which we want) but also along the edge. This spuriously good location along an edge is at a cost of a worse signal to noise ratio. An important point in its favour is that it can be closely apprximated by a difference of gaussians (DOG) filter.

b. *The Canny operator*

We really want an edge finder which uses the fact that edges are symmetrical structures; by integrating along the edge we can improve the performance. We require an asymmetrically oriented operator, though this entails extra computation because it must be applied at a number of different orientations.

Canny (1983) used variational principles to find an optimal one-dimensional step edge finder but decided that the resulting function so closely matched the first differential of a gaussian that this was the best choice. In effect the Canny operator is rather similar to the Marr-Hildreth but it finds the zero-crossings in the single dimension across the edge (and not along it) by finding peaks, in the direction across the edge, in the first differential of the image:

$$\frac{\partial^2}{\partial n^2} g(x,y) * I(x,y) = 0, \text{ where} \qquad (3)$$

$$n = \frac{\nabla g(x,y) * I(x,y)}{|\nabla g(x,y) * I(x,y)|} . \qquad (4)$$

Again a gaussian of arbitrary variance is applied to the image before finding these points. In practice the operator works as follows.

A gaussian is applied to the image and the result is differentiated in the X and Y directions by taking pixel differences. Two arrays are produced from these values; one contains the amplitude of the sum of squares of these values, the other contains the direction across the possible egde found from the direction of maximal gradient. Those pixels which represent peaks in the amplitude in the direction across the edge are found and the position of the peak is estimated to subpixel accuracy using the neighbouring pixels in the edge direction. The amplitude of these peak pixels is thresholded with a non-linear threshold; all those peaks with an amplitude greater than some upper threshold are found and represent edges. Also found are those peaks above some lower threshold which are "linked" (i.e. are in the 8-neighbourhood) either directly to the first type of edge or to them through a peak above the lower threshold. Thus every edge is either above the higher threshold, or it is above the lower threshold and linked by a chain of other edges to one of the edges above the higher threshold.

Analogous operators can be found for ridges in the images which would correspond to line or region detectors depending on the scale.

These operators suggest that a range of gaussians of different variance could be applied to the image in order to obtain sets of features at different scales. This would seem to imply a vast increase in computation but in fact this is not necessarily the case. Because the smoothing filter is a gaussian the central limit theorem shows that it can be approximated by any simple function if applied repeatedly. Since it is a low pass filter and we are removing the highest spatial frequencies Shannon's sampling law shows how we can resample the resulting smoothed image into an array smaller than the original image without any loss of information (provided we correctly match the smoothing and resampling). Combining these two facts means that we can generate a range of images, with different gaussians applied to them, by the application of a simple function (e.g. something that roughly resembles a gaussian) followed by resampling the resulting image into a smaller array. The process is then repeated. Ultimately, we end up with a single value which is the average grey level of the original image.

If we smooth and resample into arrays of half the dimension of the previous array (and thus a quarter of the area if the image is 2-dimensional) the memory and processing requirements only increase by about one third compared with using a single gaussian on the original image. In fact, to obtain all the features the smoothing and resampling probably ought to be finer than this and thus the computational requirements will be greater. On a highly parallel computer the whole range of scales can be obtained in a time proportional to the logarithm of the original image dimension. A further advantage is if the Marr-Hildreth scheme is being used the DOG filter can be obtained by taking the difference between neighbouring arrays.

The usefulness of obtaining descriptions of the image in which the scale of the features is explicitly represented is discussed in the next section.


7. SYMBOLIC ENCODING AND REPRESENTATIONAL HIERARCHIES

A fundamental principle which has emerged from work done in image understanding is the importance of representational hierarchies at all points in the signal-to-symbol transformation process. The following description of the use of such hierarchies is not comprehensive but should illustrate the concepts involved.

*a. Scale hierarchies*

Images contain information at a broad range of spatial resolutions or scales. The lungs in a chest X-ray can be described as containing definite structures, or as a textured field, or as a grey area of a particular shape. An edge can be wavy at one scale but straight at higher scale (lower resolution). To recover the information at different scales image processing operators must have a range of sizes - or equivalently a range of pixel neighbourhoods over which their output is calculated. As already discussed operators with a large neighbourhood miss fine detail but are relatively insensitive to noise, while the converse is true for small neighbourhoods. The result of using a range of operator sizes is an explicit representation of features at different scales - a multiresolution representation. An image and its representation at two different scales is illustrated in Fig. 4. The operator used is the Marr-Hildreth, with a small threshold on the gradient of the zero-crossing. All those zero-crossings whose change in position between neighbouring scales is greater than another threshold have been removed thus leaving only those features which remain constant over small changes in scale.

Many low level visual processing algoritms are inherently multiresolutional; the algorithm is applied to a coarse representation of the image, and the results are then used to constrain the processing of higher resolution features.

Two major assumptions in these algorithms are that the world consists of smooth and rigid things. At low spatial resolution irregularities of surfaces are smoothed out. Analogously, if examined over a short enough time interval (high temporal resolution) even highly plastic objects will be locally rigid. If multiresolution processing of a sequence of images links low spatial resolutions to high temporal resolutions, and vice versa (as appears to be the case in human vision), then a coarse to fine progression will start with smooth, rigid surfaces, and proceed to rough, plastic ones.

The output of these algorithms is also multiresolution and the overall system can be made more robust if the multiresolution approach carries through into the symbolic levels, as in shape hierarchies, hierarchies of objects and their parts, and taxonomic or "is a" hierarchies.

## b. *Shape hierarchies*

Recognition is the successful match of an object description with a model description. However, in a multiresolution representation the object and the model will both have a set of increasingly detailed descriptions. Recognition then becomes a succesful match at some resolution. If, for example, we represent 3-D objects with cylinder-like primitives (Marr and Nishihara, 1978; Brooks *et al.*, 1979) then a human figure consists of a round blob (head) on an elongated blob (trunk), with elongated blobs representing arms and legs. As we pursue more and more detailed descriptions we can match object and model more closely, progressively achieving a more precise classification. At some point in our multiresolution hierarchy of shape descriptions we recognise a human being, at another a female and at another Joan Smith.

The hierarchical representation of shape contributes to the robustness of the recognition process. It is not very useful if our recognition system simply fails when it tries to match an object instance with a stored description, as might happen with a single resolution representation. It would be rather more flexible if it returned some information about where in the shape hierarchy the failure occurs, e.g. the object is a person, though the sex is undetermined, or the object appears like a round, dense tumour though its detailed morphology cannot be seen. With increasing noise, obscuring structures, and so on the shape hierarchy can be used to limit the possible mappings between the external object representation and the internal descriptions. Once this has been done and we have extracted the maximal information we can from the image, we can then apply top down, knowledge driven processes which provide further interpretation constraints.

## c. *"Part" hierarchies*

Parts of objects are objects in their own right. Sometimes we can only recognise the overall object when we have recognised some of its parts. Different objects may contain the same parts. An object representation must therefore represent the relationship between the parts and the whole, and similarly the subparts and the parts, and so on. The parts of an object tend to be smaller than the parent object, so the shape hierarchy will not

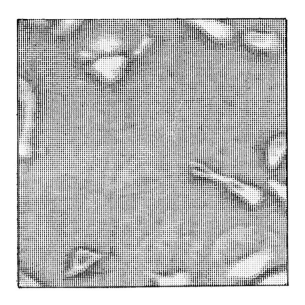

Fig. 4. A multiresolution view of a complex aggregate of cells.

represent the presence of small sub-objects at the lower resolutions. Even partial recognition of an object constrains the possible labellings assigned to its parts, and vice versa.

*d. "Is a" hierarchies*

The taxonomy of shapes - the class or category structure of objects which the knowledge based system deals with - is related to the hierarchy of shapes along the resolution dimension. An "is a" hierarchy can arise from a whole range of abstractions, not just those based on shape. However, if the expert system is to reason about unknown images the shape classification will presumably be the most significant.

*e. Parameter quantisation*

A common way of converting between the parametric output of visual descriptions and the qualitative reasoning of an expert system, is to represent values as overlapping ranges. The length of an object may be described as 'very short', 'short', 'long' or 'very long' with there being some overlap between these. The discrimination of one object from another may require a finer quantisation for these ranges than another discrimination and it is not obvious that any particular number of these qualitative descriptions or any particular length over which they should extend is the correct one. How short is 'short'? Within a multiresolution description the number of descriptions increases with resolution and the individual range of each of the descriptions decreases. This can help to avoid the arbitrary nature of parameter quantisation.

## 8. SUMMARY AND CONCLUSIONS

Medical image processing has made considerable progress in the development of new imaging techniques and computer based image enhancement. Work in artificial intelligence, notably in knowledge based expert systems has aroused interest in the possibilities of combining image processing technology with clinical knowledge for interpreting image data. Some of the basic concepts of expert systems have been outlined, notably the use of explicit and qualitative representations of knowledge. Some ways in which these concepts might be exploited in image interpretation have been illustrated. Five general schemes for coupling knowledge based expert systems to image processors have been outlined. Two of these are currently practical, the remainder are restricted by our inability to create symbolic representations of the structures typically found in biomedical images, notably because of their variability and non-rigidity. Some ways in which knowledge of shape, structure, and object taxonomy may contribute to the interpretation of images have been discussed.

Some clear areas for further research include (i) exploring in detail the limits of existing symbolic encoding techniques for biomedical images, (ii) identifying the optimal set of information types which should be encoded in a symbolic image description database, (iii) developing ways of symbolically representing the characteristic spatial and temporal topologies of biological objects and events, (iv) identifying ways in which non-image domain knowledge can be used to constrain the interpretation of image elements, and (v) establishing the degree to which the low level image capture mechanisms need to be controlled by knowledge-intensive interpretation processes.

*Acknowledgements*

*We would like to thank Andrew Todd-Pokropek (University College, London), Bernard Buxton (GEC Hirst Research Laboratories) and Sue Ellam (Guy's Hospital, London) for their very helpful comments on earlier drafts of this paper.*

## REFERENCES

Binford, T. (1982). Survey of model-based image analysis systems, *Int. J. Robotics Res.* 1, pp. 18-64.
Brooks, R., Greiner, R., and Binford, T. (1979). The ACRONYM model-based vision system, *Proc. IJCAI 6*, Tokyo, pp. 105-113.
Brooks, R. (1981). Symbolic reasoning among 3-D and 2-D images, *Artificial Intelligence* 17, pp. 285-348.
Canny, J. (1983). A variational approach to edge detection, *Proc. Nat. Conf. on Artificial Intelligence AAAI-83*.
Connell, J.J. and Brady. M. (1987). Generating and generalising models of visual objects, *Artificial Intelligence* 31 (2), pp. 159-184.
Davis, R. and Buchanan, B.G. (1984). Meta-level knowledge. In: *Rule based*

*expert systems*, B.G. Buchanan and E.H. Shortliffe (eds.), Addison-Wesley, Reading.
Duncan, T. (1986). PROPS 2 reference manual, Imperial Cancer Research Fund Laboratories, Biomedical Computing Unit Technical Report.
Ellam, S.V. and Maisey, M.N. (1986). A knowledge based system to assist in medical image interpretation: design and evaluation methodology. In: *Research and development in expert systems III*, M. Bramer (ed.), Cambridge U.P.
Fox, J. (1984). Formal and knowledge-based methods in decision technology, *Acta Psychologica* 56, pp. 303-331.
Fox, J., Frost, D., Duncan. T., and Preston, N. (1986). The PROPS 2 primer, Imperial Cancer Research Fund Laboratories, Biomedical Computing Unit Technical Report.
Fox, J., Glowinski, A., O'Neil, M., and Vekaria, A. (1987). ESPERANTO: a knowledge based framework for decision theory, Imperial Cancer Research Fund Laboratories, Biomedical Computing Unit Technical Report.
Frost, D., Fox, J., Duncan, T.D., Preston, N., and Hajnal, S. (1986). The PROPS 2 knowledge programming package, Imperial Cancer Research Fund Laboratories, Biomedical Computing Unit Technical Report.
Hall, E.L., Kruger, R.P., Dwyer, S.J., Hall, D.L., McLaren, R.W. and Lodwick, G.S. (1971). A survey of preprocessing and feature extraction techniques for radiographic images, *IEEE Trans. Computers* C-20, pp. 1032-1044.
Kanal, L.N. and Lemmer, J.F. (eds.)(1986). *Uncertainty in artificial intelligence*, North-Holland, Amsterdam.
Levesque, H. (1987). Chapter on knowledge representation, *Annual Review of Computer Science, Volume 1*. In preparation.
Lowe, D.G. (1987). Three-dimensional object recognition from single two-dimensional images, *Artificial Intelligence* 31, pp. 255-396.
Marr, D. and Hishihara, H.K. (1978). Representation and recognition of the spatial organisation of three dimensional shapes, *Proc. Roy. Soc. London B* 200, pp. 269-294.
Marr, D. and Hildreth, E.C. (1980). Theory of edge detection, *Proc. Roy. Soc. London B* 207, pp. 187-217.
McKeown, D. (1985). Rule-based interpretation of aerial imagery, *IEEE PAMI-7*, pp. 570-585.
Nazif, A. and Levine, M. (1984). Low level image segmentation: an expert system, *IEEE PAMI-6*, pp. 555-577.
Ohta, Y. (1980). A region-oriented image-analysis system by computer, Ph.D. thesis, Kyoto University, Dept. of Information Science.
Stoeker, W., Moss, R., Chu, K., Lin, S., Prasad, R., and Poneleit, K. (1986). Skin cancer recognition by computer vision. In: *MEDINFO* 1986, R. Salamon, B. Blum, M. Jurgensen (eds.), North-Holland, Amsterdam, pp. 642-648.

Part 2
# Selected Topics

2.1 Analytic Reconstruction Methods
2.2 Iterative Methods
2.3 Display and Evaluation
2.4 Applications

## Section 2.1
# Analytic Reconstruction Methods

# THE ATTENUATED RADON TRANSFORM

Frank Natterer

*Universität Münster*
*West Germany*

## ABSTRACT

*The attenuated Radon transform comes up in single particle emission computed tomography (SPECT). We describe some mathematical properties of the attenuated Radon transform and line out numerical procedures using the conjugate gradient algorithm. In the case of constant attenuation, a complete solution is possible, including the attenuation correction problem. We conclude with a suggestion for the attenuation correction problem in the general case.*

## 1. THE ATTENUATED RADON TRANSFORM AND SOME SPECIAL CASES

Let $f$ be a function in $\mathcal{R}^2$ (the activity distribution) which vanishes outside the unit disk, and let $\mu$ be the attenuation distribution. The attenuated Radon transform is defined to be

$$R_\mu f(\theta,s) = \int_{x \cdot \theta = s} f(x) \, e^{-\int_0^\infty \mu(x+t\theta^\perp) dt} \, dx, \qquad (1)$$

i.e. $R_\mu f$ is the weighted line integral of $f$ over the line $x \cdot \theta = s$, $\theta \in S^1$ a unit vector, $s \in \mathcal{R}^1$, the weight being defined by a line integral over $\mu$. The notation we use is explained to some detail in the tutorial article of the present author in this volume, Natterer (1987). In the following we shall make use of this article without explicit reference. For the application to SPECT see Budinger *et al.* (1979).

If $\mu$ is constant in a convex set containing the activity distribution, then $R_\mu$ can be replaced by the exponential Radon transform

$$T_\mu f(\theta,s) = \int_{x \cdot \theta = s} e^{\mu x \cdot \theta^\perp} f(x) \, dx = \int_{\mathcal{R}^1} e^{\mu t} f(s\theta + t\theta^\perp) dt \qquad (2)$$

where $\mu$ is a real parameter. In PET (positron emission tomography) the relevant integral transform is $e^{-R\mu}Rf$, with $R = R_0$ the (ordinary) Radon transform.

## 2. SOME MATHEMATICAL PROPERTIES OF $R_\mu$, $T_\mu$

The important central slice theorem for the Radon transform generalizes to $T_\mu$ but not to $R_\mu$. Defining the 1D respectively 2D Fourier transform by

$$\hat{g}(\sigma) = (2\pi)^{-1/2} \int_{\mathcal{R}^1} e^{-is\sigma} g(s) \, ds \tag{3}$$

$$\hat{f}(\xi) = (2\pi)^{-1} \int_{\mathcal{R}^2} e^{-ix\cdot\xi} f(x) \, dx \tag{4}$$

the central slice theorem reads

$$(T_\mu f)^\wedge (\Theta,\sigma) = (2\pi)^{1/2} \hat{f}(\sigma\Theta + i\mu\Theta^\perp) \tag{5}$$

with $i = \sqrt{-1}$. Note that on the left hand side we took the 1D Fourier transform with respect to the second variable. The proof of Eq. (5) is as follows. We have

$$(T_\mu f)^\wedge (\Theta,\sigma) = (2\pi)^{-1/2} \int_{\mathcal{R}^1} e^{-is\sigma} \int_{\mathcal{R}^1} e^{t\mu} f(s\Theta + t\Theta^\perp) \, dt \, ds. \tag{6}$$

Putting, for $\Theta$ fixed, $x = s\Theta + t\Theta^\perp$, the integral can be written as

$$(T_\mu f)^\wedge (\Theta,\sigma) = (2\pi)^{-1/2} \int_{\mathcal{R}^2} e^{-ix\cdot\Theta\sigma + \mu x \cdot \Theta^\perp} f(x) \, dx$$

$$= (2\pi)^{-1/2} \int_{\mathcal{R}^2} e^{-ix\cdot(\sigma\Theta + i\mu\Theta^\perp)} f(x) \, dx$$

$$= (2\pi)^{1/2} \hat{f}(\sigma\Theta + i\mu\Theta^\perp). \tag{7}$$

This generalized central slice theorem is not as useful as in the special case $\mu = 0$, since Eq. (5) gives $\hat{f}$ on a surface in $\mathbb{C}^2$, the two-dimensional space of two complex variables. So we have to use Cauchy's theorem in $\mathbb{C}^2$ to turn Eq. (5) into an inversion formula by means of the inverse Fourier transform. This has been done in Natterer (1979) and in Markoe (1986).

Backprojection operators $R_\mu^*$, $T_\mu^*$ can be defined by

$$R_\mu^* g(x) = \int_{S^1} e^{-\int_0^\infty \mu(x+t\theta^\perp)\,dt} g(\theta, x\cdot\theta)\,d\theta \tag{8}$$

$$T_\mu^* g(x) = \int_{S^1} e^{\mu x \cdot \theta^\perp} g(\theta, x\cdot\theta)\,d\theta \tag{9}$$

for a function g of $(\theta,s)$. As in the case of the ordinary Radon transform, they can be viewed as $L_2$-adjoints of $R_\mu$, $T_\mu$. The role of the operator $R^*R$ in the ordinary case is played by

$$R_{-\mu}^* R_\mu f(x) = 2 \int |x-y|^{-1} \cosh\left\{\int_x^y \mu\,dt\right\} f(y)\,dy \tag{10}$$

$$T_{-\mu}^* T_\mu f(x) = 2 \int |x-y|^{-1} \cosh\{\mu|x-y|\} f(y)\,dy. \tag{11}$$

Equation (10) tells us that the function $R_{-\mu}^* g$, $g = R_\mu f$, is a blurred version of f, the blurring function being non-isotropic. We see that summation methods (see Barrett (1987)) can be used for the attenuated Radon transform as well. Equation (11) can be written as

$$T_{-\mu}^* T_\mu f = k * f, \qquad k(x) = \frac{2}{|x|} \cosh(\mu|x|). \tag{12}$$

Thus we come to the inversion formula

$$f = h * T_{-\mu}^* T_\mu f, \qquad \hat{h} = 1/\hat{k} \tag{13}$$

for the exponential Radon transform. This formula requires a backprojection, prior to a 2D convolution. It corresponds to the $\rho$-filtered layergram method of CT.

A formula which corresponds to the filtered backprojection algorithm has been obtained by Tretiak and Metz (1980) in the following way. Start out from

$$(T_{-\mu}^* h) * f = T_{-\mu}(h * T_\mu f) \tag{14}$$

which holds for any function h and which is derived very much in the same way as the central slice theorem (5). Then try to determine h in such a way that

$$T^*_{-\mu} h = \delta \qquad (15)$$

with $\delta$ the Dirac $\delta$-function. It turns out that

$$\hat{h}(\sigma) = c \begin{cases} |\sigma|, & |\sigma| > \mu \\ 0, & |\sigma| \leq \mu \end{cases} \qquad (16)$$

with a suitable constant c does the job. With this h we get

$$f = T^*_{-\mu} (h * T_\mu f). \qquad (17)$$

This Tretiak-Metz formula does a 1D convolution prior to backprojection, hence it corresponds to the filtered backprojection algorithm.

For general $\mu$, an analytic inversion formula cannot possibly exist. Therefore, iterative methods (e.g., ART) have to be used to solve $R_\mu f = g$ for f. Heike (1986) obtained excellent results by applying the conjugate gradient algorithm (CG) to a properly discretized and regularized version of $R_\mu f = g$. Therefore it may be useful to explain that algorithm. For the mathematical background see Luenberger (1973).

CG is a method for minimizing $\|Ax - b\|$. It works as follows. Choose $x^0$ and put $d^0 = b - Qx^0$, $Q = A^*A$. Then compute $x^t$, $d^t$ recursively from

$$x^{t+1} = x^t + \alpha^t d^t, \quad \alpha^t = -\frac{(g^t, d^t)}{(d^t, Qd^t)}, \quad g^t = -A^*b + Qx^t \qquad (18)$$

and

$$d^{t+1} = -g^{t+1} + \beta^t d^t, \quad \beta^t = \frac{(g^{t+1}, Qd^t)}{(d^t, Qd^t)}. \qquad (19)$$

The CG algorithm has the following nice optimality property (Luenberger, 1973, chapt. 8.4). Let $x^0 = 0$. Then

$$x^t = P_t(Q) A^* b \qquad (20)$$

with $P_t$ a polynomial of degree $< t$. Among all polynomials $P$ of degree $< t$, $P_t$ minimizes

$$(x, Q(I - QP(Q))^2 x) \tag{21}$$

where $x = A^+ b$; $A^+$ is the Moore-Penrose generalized inverse of A.

In terms of the singular value decomposition of $A^*$, the lemma says that

$$x^t = P_t(Q) A^* b = \sum_{k=1}^{p} \sigma_k P_t(\sigma_k^2) (b, u_k) v_k \tag{22}$$

and $P_t$ minimizes

$$\sum_{k=1}^{p} (x, v_k)^2 \sigma_k^2 (1 - \sigma_k^2 P(\sigma_k^2))^2 \tag{23}$$

among all polynomials of degree $< t$. This means that $1 - \sigma_k^2 P_t(\sigma_k^2)$ is small for $(x, v_k)^2 \sigma_k^2$ large, i.e. $\sigma_k P_t(\sigma_k^2)$ is close to $\sigma_k^{-1}$ in some sense. Comparing Eq. (22) with

$$x = A^+ b = \sum_{k=1}^{p} \sigma_k^{-1} (b, u_k) v_k \tag{24}$$

we see that CG provides a good approximation in those subspaces $\text{sp}\{v_k\}$ for which $(x, v_k)^2 \sigma_k^2$ is large. This means that CG not only tends to pick up soon the contributions of the large singular values, it also takes into account the size of this contribution as measured by $(x, v_k)^2$. For instance, it does not work on the component $v_k$ if $v_k$ is not really contained in x, irrespective of the size of $\sigma_k$.

Applying CG to the minimization of

$$\| R_\mu f - g \|^2 + \gamma^2 \| f \|^2 \tag{25}$$

leads to an efficient and accurate inversion method.

## 3. THE ATTENUATION CORRECTION PROBLEM

In practice, one has to determine $\mu$, prior to solving $R_\mu f = g$ for $f$. In the constant attenuation case, an elegant mathematical solution has been given by Hertle (1986).

Let $g = T_\mu f$. Then, from Eq. (5),

$$\hat{g}(\Theta,\sigma) = (2\pi)^{1/2} \hat{f}(\sigma\Theta + i\mu\Theta^\perp). \tag{26}$$

Differentiating with respect to $\sigma$ we get

$$\frac{\partial}{\partial \sigma} \hat{g}(\Theta,0) = (2\pi)^{1/2} \Theta \cdot \nabla \hat{f}(i\mu\Theta^\perp). \tag{27}$$

Differentiation with respect to $\phi$, $\Theta = (\cos\phi, \sin\phi)$ yields

$$\frac{\partial}{\partial \phi} \hat{g}(\Theta,0) = (2\pi)^{1/2} (-i\mu\Theta) \cdot \nabla \hat{f}(i\mu\Theta^\perp). \tag{28}$$

Here, $\nabla = \left(\frac{\partial}{\partial \xi_1}, \frac{\partial}{\partial \xi_2}\right)$ denotes the gradient.

From Eqs. (27) and (28) we conclude that

$$\frac{\partial}{\partial \sigma} \hat{g}(\Theta,0) \Big/ \frac{\partial}{\partial \phi} \hat{g}(\Theta,0) = 1/(-i\mu) \tag{29}$$

provided the denominator does not vanish. Equation (29) gives $\mu$ in terms of the data $g = T_\mu f$. It therefore constitutes a complete solution to the problem of finding $\mu$, while the Tretiak-Metz inversion formula completely solves the problem of finding $f$ from $T_\mu f = g$ once $\mu$ has been found. Thus, in the constantly attenuated case, we have a complete and fully satisfactory solution to the problem of solving $T_\mu f = g$ both for $f$ and $\mu$.

In the general case, finding $\mu$ from $g = R_\mu f$ is not so easy. One possibility, which has been tried with some success in Natterer (1983), makes use of consistency conditions which are satisfied by $g = R_\mu f$. By this we mean the following. If one picks out an arbitrary function $g$ of two variables $\Theta$ and $s$, then this function is not likely to be the attenuated Radon transform of a nice function $f$. In order for this to be the case, $g$ has to satisfy the equations

$$\int_0^{2\pi}\int_{\mathcal{R}^1} e^{ik\phi + \frac{1}{2}(I+iH)R\mu} s^m g d\phi ds = 0, \qquad k > m, \tag{30}$$

see Natterer (1986).

Here, k and m are integer, H is the Hilbert transform (see e.g. Barrett (1987)), and $\Theta = (\cos\phi, \sin\phi)$. The idea is now to solve (30) for $\mu$. This is a non-linear system which can solved by linearization and regularization techniques. The success of this technique depends to a large extent on the actual activity distribution. No general statement about the usefulness of this approach is possible by now.

## REFERENCES

Barrett, H.H. (1987). Fundamentals of the Radon transform. *These proceedings*.
Budinger, T.F., Gullberg, G.T., and Huesman, R.H. (1979). Emission computed tomography. In: *Image Reconstruction from Projections*, G.T. Herman (ed.), Springer-Verlag, Berlin, pp. 147-246.
Heike, U. (1986). Single-photon emission computed tomography by inverting the attenuated Radon transform with least-squares collocation, *Inverse problems* 2, pp. 307-330.
Hertle, A. (1986). The identification problem for the constantly attenuated Radon transform. Preprint, Fachbereich Mathematik, Universität Mainz.
Luenberger, D.G. (1973). *Introduction to linear and nonlinear programming*, Addison-Wesley, Reading.
Markoe, A. (1987). Fourier inversion of the attenuated Radon transform, *SIAM J. Math. Anal.*, to appear.
Natterer, F. (1979). On the inversion of the attenuated Radon transform, *Numer. Math.* 32, pp. 431-438.
Natterer, F. (1983). Computerized tomography with unknown sources, *SIAM J. Appl. Math.* 43, pp. 1201-1212.
Natterer, F. (1986). *The mathematics of computerized tomography*, Wiley-Teubner, Stuttgart.
Natterer, F. (1987). Regularization techniques in medical imaging. *These proceedings*.
Tretiak, O. and Metz, C. (1980). The exponential Radon transform, *SIAM J. Appl. Math.* 39, pp. 341-354.

# INVERSE IMAGING WITH STRONG MULTIPLE SCATTERING

S. Leeman[1], V.C. Roberts[1], P.E. Chandler[2], L.A. Ferrari[2]

1: King's College School of Medicine and Dentistry, London, U.K.
2: University of California Irvine, U.S.A.


ABSTRACT

*A method for mapping the scattering interaction density of a certain class of objects from a measurement of the scattering amplitude is developed for the case of a longitudinal acoustic or scalar electromagnetic wave scattering from an inhomogeneous medium consisting of wave velocity fluctuations. The inversion procedure is exact, and explicitly takes into account all orders of multiple scattering. The interaction parameter map constitutes an image of the object. An interesting feature is that progressively higher resolution of the recovered image is obtained as higher order scattering is progressively incorporated into the inversion procedure.*


## 1. INTRODUCTION

Many of the new medical imaging methods that have been proposed recently may be regarded as implementations of the inverse scatter problem. In these approaches, a known input wave penetrates into and interacts with some object under investigation, and the emergent waves, representing the result of the interaction, are detected and measured outside the object. The aim is to reconstruct a mapping (image) of the interaction parameters throughout a region of interest inside that object.

Foremost amongst these techniques in medical applications, is the well-known reconstruction-from-projections imaging method, implicit in which is the notion that the probing radiation propagates in a ray-like ("geometric optics") fashion, so that the interaction density may be summed along well-defined pathways (line integrals) in order to form the projection. In this context, the wave nature of the radiation may be disregarded to good approximation. For certain radiations, however, significant scattering results, and, under these conditions, the ray approximation is no longer valid. This is a particular problem when attempting to do computer based medical imaging with ultrasound. However, methods may be devised which reconstruct maps of the scattering interaction density from measurements of the diffracted and scattered field. Such approaches may be referred to as inverse scatter imaging techniques.

These imaging methods may be regarded as consisting of three basic components: (a) an underlying physical model which specifies the propagation and scattering of the wave in the object; (b) a data acquisition configuration, which may be chosen in order to minimise eventual artefacts in the final image; and (c) a computational model, which is essentially the algorithm whereby the desired mapping is recovered from the measured data set. In practice, most effort has been devoted to the last aspect of the problem.

The ultrasound methods, in particular, may be further classified into three major groupings, according to their data acquisition configurations: (a) through-transmission methods, which may be regarded as utilising the forward-scattered field but are actually implemented as conventional reconstruction-from-projections methods; (b) reflectivity tomography methods, which utilise the backscattered field only, and whose reconstruction algorithms are very close to those based on ray (but not necessarily straight-line path) considerations; (c) diffraction tomography techniques, which measure the angle-scattered fields, and for which a host of new, unfamiliar, reconstruction algorithms have arisen.

In this communication, only diffraction tomography is considered. The computational methods employed in this context fall into essentially two main groups. (i) Inverse Fourier techniques are analogues of the filtered-backprojection and central-slice-theorem reconstruction algorithms of conventional straight-line transmission computerised tomography (CT) methods. At present, it is essential to assume that multiple scattering effects are negligible. (ii) Successive approximation schemes, which are the analogue of the ART CT reconstruction algorithm, have also been proposed. These are much more tolerant of multiple scattering effects, but are computationally extremely demanding. Another problem is that the convergence of these methods to a realistic endpoint has been shown only with simulated data for very simple, small, two-dimensional objects.

Methods which depend upon discretising the forward problem, and then recovering the scattering function by the inversion of the resulting (very large) matrix equation, may also be regarded as belonging to this group.

The theoretical basis of a different reconstruction method is presented here: it is not based on any CT analogue, and it is applicable to scattering from three dimensional objects, with multiple scattering allowed, for a variety of radiations. The inversion procedure is exact, and explicitly takes into account *all* orders of multiple scattering. An interesting feature is that progressively higher resolution of the recovered image is obtained as higher order scattering is progressively incorporated into the inversion procedure. There is no limitation on the strength of the interaction, but the method has been developed, up to now, to be applicable only to a restricted class of objects, viz., those for which the scattering density function can be written as a truncated Laplace transform. The technique is based on earlier approaches (as, indeed, are all the ultrasound diffraction tomography methods) developed in a rather different context, for which full details and references are given by Leeman (1970). However, it has not previously been reformulated for the acoustical imaging case.

## 2. THE PHYSICAL MODEL

Consider the case that the radiation propagating in the object may be described by the Helmholtz wave equation:

$$\nabla^2 \chi(\underline{r}) + k^2 \chi(\underline{r}) = \rho(\underline{r}) \chi(\underline{r}) \tag{1}$$

with $\chi(\underline{r})$ denoting the wave at spatial location $\underline{r}$, $\rho(\underline{r})$ representing the scattering interaction density, and with $k=\omega/c$. Here, $\omega$ is the (circular) frequency of the field, and $c$ is a mean value for the wave velocity over the scattering region. The Helmholtz wave equation is applicable to longitudinal ultrasound waves or scalar electromagnetic waves propagating in velocity-inhomogeneous media. For simplicity, the medium will be regarded as lossless (k real).

It is possible to express the Helmholtz equation in its integral formulation;

$$\chi(\underline{r}) = \chi_0(\underline{r}) + \int_V d\underline{r}' G(\underline{r},\underline{r}') \rho(\underline{r}') \chi(\underline{r}') \qquad (2)$$

where $\chi_0$ denotes the incident field (i.e. the wave that would exist in the absence of the velocity fluctuations), V is the bounded volume in which the scattering occurs, and G denotes the Green's function appropriate for this scattering configuration,

$$G(\underline{r},\underline{r}') = -\exp\{ik|\underline{r}-\underline{r}'|\}/4\pi|\underline{r}-\underline{r}'|. \qquad (3)$$

In an entirely symbolic way, the integral equation can be succinctly written as

$$\chi = \chi_0 + G\rho\chi \qquad (4)$$

with G denoting an (integral) operator representing the Green's function.

The underlying structure of the basic integral equation is now apparent, and it may be solved by iteration to yield the so-called Born-Neumann expansion for the field:

$$\chi = \chi_0 + G\rho\chi_0 + [G\rho]^2 \chi_0 + \ldots \qquad (5)$$

This gives a valid solution, provided that the expansion converges. Terms in this expansion involving $[G\rho]^N$, with N > 1, may be interpreted as describing multiple scattering events of order N.

The full series solution for the field $\chi$ is extremely cumbersome to evaluate in any realistic case. In practice, therefore, the series is terminated, in order to give an approximate solution. In particular, the first Born approximation is given by

$$\chi_B \equiv \chi_0 + G\rho\chi_0. \qquad (6)$$

This approximation is of some considerable interest, since the inverse problem can be solved exactly when it is valid [e.g. Leeman, 1980].

Clearly, the validity of the (first) Born approximation implies that multiple scattering effects are negligible.

In general, $\chi_0$ may be considered known or accurately measured, and the form of the integral equation, (4), suggests that the field may everywhere be considered to be the sum of the incident field and a "scattered field", $\chi_s(\underline{r})$ with

$$\chi_s(\underline{r}) = \int_V d\underline{r}' G(\underline{r},\underline{r}') \rho(\underline{r}') \chi(\underline{r}'). \tag{7}$$

In practice, it is a set of measurements of $\chi_s$ outside the scattering region, V, that constitutes a physically meaningful data set. The (first) Born approximation is valid when the scattering is not too strong, and states that the scattered field may be expressed as

$$\chi_s(\underline{r}) = \int_V d\underline{r}' G(\underline{r},\underline{r}') \rho(\underline{r}') \chi_0(\underline{r}'). \tag{8}$$

For the case that the incident wave is a plane wave directed along the unit vector $\underline{n}$, i.e. $\chi_0(\underline{r}) = \exp\{ik\underline{n}\cdot\underline{r}\}$, this reduces to

$$\chi_s(\underline{r}) = (1/4\pi) \int d\underline{r}' \exp\{ik|\underline{r}-\underline{r}'|\} \rho(\underline{r}') \exp\{ik\underline{n}\cdot\underline{r}'\} / |\underline{r}-\underline{r}'|. \tag{9}$$

Measurements of the scattered field may be considered to be performed only at very large distances from V (far-field measurement). In the far field, the scattered wave behaves as an outgoing spherical wave, modulated in amplitude and phase by the scattering amplitude, $\Omega$, which has the form (Leeman, 1980).

$$\Omega(k,\underline{n};\underline{m}) = (1/4\pi) \int d\underline{r}' \rho(\underline{r}) \exp\{-ik(\underline{m}-\underline{n})\cdot\underline{r}\} \tag{10}$$

where $\underline{m}$ is a unit vector indicating the direction in which the measurement of the far field scattering is taken.

It is convenient to introduce the notation: $\underline{k}_s \equiv k\underline{m}$; $\underline{k}_I \equiv k\underline{n}$; $\underline{\lambda} \equiv \underline{k}_s - \underline{k}_I$
Equation (10) can then be written as

$$\Omega(\underline{\lambda}) = (1/4\pi) \int d\underline{r}\, \rho(\underline{r}) \exp\{-i\underline{\lambda}\cdot\underline{r}\} \tag{11}$$

This shows that, provided multiple scattering may be ignored, a relationship of the Fourier transform type exists between $\Omega$ and $\rho$. It may be inverted without further ado, to yield:

$$\rho(\underline{r}) = (1/2\pi^2) \int d\underline{\lambda} \Omega(\underline{\lambda}) \exp\{i\underline{\lambda}\cdot\underline{r}\}. \tag{12}$$

Hence, $\rho(\underline{r})$ can be computed if the scattering amplitude can be measured for all values of $\underline{\lambda}$. Indeed, the various ingenious ways which may be devised in order to access all of the $\underline{\lambda}$-space are, in reality, different implementations of the diffraction tomography concept, and not, as often claimed, new types of scatter imaging techniques.

It is interesting to observe that (11) indicates that the measured data ($\Omega$, in this case) can provide information about the Fourier transform of the desired density. It is in this sense that the methods of inverse scatter imaging may be linked to the central slice theorem of conventional reconstruction-from-projections (Leeman, Roberts and Seggie, submitted to Proc. IEE -A).

On introducing the variable, $\Delta \equiv -(\underline{k}_s - \underline{k}_I)^2$, it follows that the scattering amplitude may be regarded as a function of the independent variables $\Delta$ and k, and may thus be written as $f(k;\Delta)$, although it will be convenient to also use the more general notation $f(\underline{k}_s,\underline{k}_I)$. Physically, k may, in principle, take on any real positive value; on the other hand, physical values of $\Delta$ are real and negative, in the range $-4k^2 \leq \Delta \leq 0$ only.

An expansion analogous to (4) may be written for the scattering amplitude. The Lippmann-Schwinger equation states that:

$$f(\underline{k}_s,\underline{k}_I) = -\rho^+(\underline{k}_s-\underline{k}_I)/4\pi - \int d\underline{k}' \cdot \rho^+(\underline{k}_s-\underline{k}') \cdot G^+(k',k) \cdot f(\underline{k}',\underline{k}_I)/2\pi^3 \tag{13}$$

with $^+$ denoting the Fourier transform of the appropriate function. Symbolically, the Lippman-Schwinger equation may be written as

$$f = f_B + f_B G^+ f \tag{14}$$

and may by solved be iteration to give

$$f = f_B + f_B G^+ f_B + \ldots \ldots \tag{15}$$

The Nth term in this expansion is usually referred to as the Nth Born term for the scattering amplitude.

The inversion technique to be described below follows from a consideration of the analyticity properties of the scattering amplitude in the complex Δ domain.


3. THE COMPLEX Δ-DOMAIN

For simplicity, rotational symmetry of the interaction density is assumed

$$\rho(\underline{r}) = \rho(r). \tag{16}$$

More critically, it is also assumed that $\rho$ can be written as the Laplace transform of some general function (or distribution), $\Sigma(\alpha)$,

$$\rho(r) = \int_0^\infty d\alpha \, \Sigma(\alpha) \cdot \exp\{-\alpha r\}. \tag{17}$$

If it is now further assumed that $\Sigma$ is zero for all $\alpha \leqslant \mu$, then it follows, by partial integration, that

$$r\rho(r) = \int_\mu^\infty d\alpha \, \sigma(\alpha) \cdot \exp\{-\alpha r\} \tag{18}$$

with $\sigma$ denoting $d\Sigma/d\alpha$. To ensure convergence in a number of situations, it will prove necessary to demand that $\int d\alpha \cdot |\sigma(\alpha)|$ be finite.

The inversion technique will attempt to construct $\sigma(\alpha)$ from the measured data, i.e. it will aim to uncover the Laplace domain representation of the interaction from the values of the scattering amplitude.

The data acquisition configuration consists of a plane wave incident along the direction $\underline{n}$, with the scattering into all directions being detected, the measured entity actually being the scattering amplitude at all scattering angles, $\theta$, but at fixed incident $\underline{k}$ ($\equiv$ frequency) only. The assumed rotational symmetry of the scattering interaction imposes the same symmetry on the scattering amplitude, which consequently depends on one angular variable only.

It is a relatively simple exercise in Fourier transform evaluation to show that the first Born approximation may be written as

$$f_B = \int d\alpha \ \sigma(\alpha) \cdot [\alpha^2 - \Delta]^{-1}. \qquad (19)$$

The behaviour of $f_B$ in the complex $\Delta$ domain is easy to establish from this integral representation. It is analytic everywhere in the finite complex $\Delta$ plane, except for $\Delta$ real and $\geqslant \mu^2$. The existence of such a branch cut along the real axis of the $\Delta$ plane is associated with the circumstance that the denominator in the integrand of the integral in (19) vanishes for these $\Delta$ values (leading to a singular integral).

The second Born term, $f_{2B}$, has a more complicated structure, but it is still possible to exploit the idea that its analyticity in the complex $\Delta$ domain breaks down when the integrand in its defining integral shows a non-removable singularity. Thus the powerful method devised by Landau (1959) may be invoked to show that $f_{2B}$ is analytic everywhere in the finite $\Delta$ plane, except for a branch cut along the real axis, for $\Delta \geqslant (2\mu)^2$. Indeed, it may be shown quite generally (Leeman, 1970) that the Nth Born term ("NBA") has a similar behaviour, but with its branch cut extending over $\Delta \geqslant (N\mu)^2$. It turns out that the singularities of the scattering amplitude in the finite $\Delta$ plane are the same as those of the Born expansion, a result that may be shown (Leeman, 1970) to be independent of the convergence of the Born expansion.

The discontinuity across the branch cut may be defined as:

$$D(\Delta) \equiv \{f(k;\Delta+i\varepsilon) - f(k;\Delta-i\varepsilon)\}/\pi i \qquad (20)$$

where the usual "$i\varepsilon$" convention is used to imply that the limit as $\varepsilon \to 0$ is always intended.

## 4. THE INVERSION PROCEDURE

An inversion procedure may be devised, which exploits some very interesting properties of the discontinuity across the branch cut. In the range $\mu^2 \leqslant \Delta < (2\mu)^2$, the branch cut arises only from $f_B$. Thus,

$$i\pi D_{12} = \int d\alpha\, \sigma(\alpha) \{ [\alpha^2 - (\Delta+i\varepsilon)]^{-1} - [\alpha^2 - (\Delta-i\varepsilon)]^{-1} \} \qquad (21)$$

where the notation $D_{MN}$ denotes the discontinuity for $(M\mu)^2 \leqslant \Delta < (N\mu)^2$. Invoking the identity

$$1/(x \pm i\varepsilon) = P(1/x) \mp i\pi\delta(x) \qquad (22)$$

where P denotes taking the principal-value integral, it readily follows that

$$\Delta^{\frac{1}{2}} . D(\Delta = \sigma(\Delta^{\frac{1}{2}}). \qquad (23)$$

This fixes the function $\sigma$ in the range $\mu \leqslant \alpha < 2\mu$, thus enabling the following low-resolution version of the object to be computed,

$$r\rho_{12}(r) = \int_{\mu}^{2\mu} d\alpha\, \sigma(\alpha).\exp\{-\alpha r\}. \qquad (24)$$

This finding may be extended further. We note that $D_{23}$ arises from two contributions: (i) the 1BA of the density $\rho_{23}(r)$ - which is, as yet, not known (ii) the 2BA of the density $\rho_{12}(r)$ - which is known from $D_{12}$. The contribution (ii) may be calculated from the known function $\rho_{12}$, and this may be subtracted from $D_{23}$ to give a corrected discontinuity, $D'$. Then, it readily follows that

$$\sigma(\Delta^{\frac{1}{2}}) = \Delta^{\frac{1}{2}} D'(\Delta) \quad \text{for } (2\mu)^2 \leqslant \Delta < (3\mu)^2 \qquad (25)$$

In this way, the function $\sigma$ may be obtained over an even wider range, and an even higher resolution version of the object may be formed, viz. $\rho_{13}(r)$. The extension to higher orders is apparent.

## 5. SUMMARY

The method for recovering the interaction density $\rho(r)$ is summarised, for clarity.

(a) Measure the scattering amplitude at all angles, for a fixed frequency plane wave input.

(b) Calculate, by analytic continuation into the complex $\Delta$-domain, the experimentally determined values, $D_{EXP}(\Delta)$, of the discontinuity across the branch-cut seen to lie on the real $\Delta$-axis, for $\Delta \geq \mu^2$.

(c) Obtain a low resolution version of the object, $\rho_{12}(r)$, from $D_{EXP}$ in the range $\mu^2 \leq \Delta < (2\mu)^2$. Use this to calculate a (subtracted) correction to the values of $D_{EXP}$ in the range $(2\mu)^2 \leq \Delta < (3\mu)^2$.

(d) Obtain a higher resolution mapping, $\rho_{13}(r)$, by utilising the corrected values of $D_{EXP}$ for $(2\mu)^2 \leq \Delta < (3\mu)^2$. Note that the mapping $\rho_{13}$ is obtained by adding a correction term, $\rho_{23}$, to the previous estimate, $\rho_{12}$.

(e) The general structure for obtaining progressively higher resolution mappings is now apparent. Note, however, that the computational effort rises rapidly as higher order corrections are incorporated.

## 6. CONCLUSIONS

It has been demonstrated that inverse scatter imaging methods which are not analogues of conventional CT techniques may, in fact, be formulated. Notwithstanding statements to the contrary, multiple scattering is *not* an obstacle that can be overcome only by ART-like successive approximation techniques (whose convergence has yet to be rigorously established by analytic argument).

The method described here is essentially three-dimensional, and may accommodate more sophisticated physical models (particularly including loss effects), but at the cost of more elaborate data acquisition configurations and computing time.

The method suffers from the disadvantage that it is dependent on noise-sensitive analytic continuation procedures. The (truncated) Laplace domain representation for the interaction is a crucial assumption, and, although reasonably general, may not apply to all objects.

But it has been shown that a well-prescribed, convergent inversion algorithm can be devised, which in principle establishes the feasibility of a unique recovery of the scattering structure (image) of an object by a diffraction tomography technique - at least for a certain class of objects. While the method as developed here may be quite difficult to implement in practice, the calculation does appear to suggest that it may well be feasible to perform diffraction tomography by repeated application of a (first) Born inversion algorithm, not necessarily involving analytic continuation. This may well be the most exciting feature of the results obtained by our approach.


REFERENCES

Landau, L.D. (1959), On analytic properties of vertex parts in quantum field theory, *Nucl. Phys.* 13, pp. 181-192.

Leeman, S. (1970). Non-local potentials and two-body scattering, *Proc. Roy. Soc. Lond.* A 315, pp. 497-516.

Leeman, S. (1980). Impediography revisited, *Acoustical Imaging* 9, pp. 513-520.

## Section 2.2
# Iterative Methods

# POSSIBLE CRITERIA FOR CHOOSING THE NUMBER OF ITERATIONS IN SOME ITERATIVE RECONSTRUCTION METHODS

M. Defrise

*Vrije Universiteit Brussel*
*Belgium*

ABSTRACT

*Two criteria are presented, which provide a choice of the number of iterations for Landweber's recursive algorithm applied to the inversion of an ill-posed linear equation. These stopping rules are shown to regularize the inversion algorithm, and their practical efficiency is discussed. Similar criteria are given to determine the number of singular components to be included in the development of the solution onto the singular system of the problem.*

1. INTRODUCTION

Traditionally, tomographic reconstruction problems are solved through the numerical implementation of explicit analytical inversion formulae for the Radon or X-ray transforms. These "transform" methods allow fast and reliable reconstructions, but are rather rigid and often rely on a very simplified model of the tomographic measurement under study [see e.g. Lewitt, 1983]. Furthermore, the incorporation of prior knowledge is usually restricted to linear -e.g. smoothness- constraints.

A more versatile approach to the reconstruction problem is provided by iterative algorithms. These can be used to solve a very broad class of optimization problems derived from an accurate model of the measurement process, and including virtually any type of constraint concerning the solution [see e.g. Censor, 1983]. Iterative techniques, therefore, are particularly useful when the data sampling is poor or incomplete and (or) when the signal-to-noise ratio is low : that is, whenever the incorporation of prior knowledge is essential to the recovery of a reliable reconstructed image. Recursive algorithms also allow the use of geometrical set-ups for which no explicit inversion formulae are available, as for instance in coded aperture tomography.

The computational requirement of iterative methods is high. It is not, however, their unique drawback: the understanding of the convergence properties of many recursive algorithms is incomplete, and limited to the ideal situation where the modeling of the measurement process is exact, and the data noise-free. Even in this ideal case, the properties of the image estimate obtained after a finite number of iterations are not straightforwardly related to the asymptotic properties predicted by the convergence theorems. Furthermore, in the presence of noisy data, or when an approximate physical model is used, many iterative algorithms are found to diverge from the exact solution when the number of iterations applied exceeds a certain threshold. This behaviour is due to the ill-posedness of tomographic reconstruction problems [Louis and Natterer, 1983], i.e. to the fact that the solution of the problem does not depend continuously on the measured data. Whereas the first iterations ensure the convergence of the lower frequency components of the object, subsequent iterations will eventually amplify the high frequency -and noisy- components of the data, thereby progressively degrading the image. This instability problem can be solved by means of any classical regularization technique [Tikhonov and Arsenin, 1977]. Usually, this means incorporating prior knowledge into the reconstruction problem, thereby defining a modified - regularized - equation. The regularized equation can be solved by means of some adequate iterative algorithm, which will now converge to a stable solution, even in the presence of noisy data.

An alternative regularization technique [Bakushinskii, 1966] consists of stopping an iterative process after an appropriate number of iterations. This number of iterations N plays the role of the regularization parameter appearing in usual regularization techniques, and its evaluation is an important practical issue. From a mathematical point of view, this parameter should be chosen as a function of the signal-to-noise ratio in such a way that the $N^{th}$ iterated solution estimate converges to the exact solution when the noisy data converge to the exact data.
The aim of this paper is to discuss several criteria which can be used as stopping rules for Landweber's recursive algorithm, also known as Simultaneous Iterative Reconstruction Technique (or SIRT).

## 2. MATHEMATICAL FRAMEWORK

We will approach the problem in the more general framework of inversion algorithms for linear ill-posed equations in Hilbert spaces. We consider equations $Kf = g$ where the solution $f$ and the data $g$ belong to real Hilbert spaces $F$ and $G$, and $K$ is a bounded linear operator mapping $F$ onto $G$. We restrict ourselves to the case of invertible compact operators. The singular system of $K$ will be denoted by

$$K u_k = \lambda_k v_k$$
$$K^* v_k = \lambda_k u_k \qquad k=1,2,3,.. \qquad (1)$$

The sets of singular vectors $\{u_k \; k=1,2,3,...\}$ and $\{v_k \; k=1,2,3,...\}$ are orthonormal bases of $F$ and $G$ respectively, and the positive singular values are ordered by decreasing magnitudes : $\lambda_1 \geq \lambda_2 \geq .. \geq \lambda_k \geq \lambda_{k+1} \geq .. > 0$. The solution $f$ can be developed as

$$f = \sum_{k=1}^{\infty} (v_k, g) \frac{u_k}{\lambda_k} . \qquad (2)$$

In practice, one only measures an approximation $g'$ to the exact data $g$; we will assume that an upper bound of the $G$ norm of the measurement error is known :

$$|g-g'| \leq \varepsilon . \qquad (3)$$

As $K$ admits no bounded inverse, the series (2), with $g$ replaced by the noisy data $g'$, becomes meaningless. A stable estimate of the solution $f$ can be recovered by using a "digitally" filtered version of (2) :

$$f'_\alpha = \sum_{k=1}^{\infty} W_k(\alpha) \; (v_k, g') \; \frac{u_k}{\lambda_k} . \qquad (4)$$

For Tikhonov's technique, for instance, the parametric family of filters is $W_k(\alpha) = \lambda_k^2 / (\lambda_k^2 + \alpha)$.

A given prescription for choosing the parameter $\alpha$ as a function of the data g' and of $\varepsilon$ is a "regularizing choice" if

$$\lim_{\varepsilon \to 0} |f'_\alpha - f| = 0. \qquad (5)$$

The existence of a regularizing choice can be guaranteed if the parametric family of filters satisfies the following conditions [Bakushinskii, 1966]:

$$\forall \alpha > 0 : \sup_k (W_k(\alpha)/\lambda_k) = R(\alpha) < \infty, \qquad (6)$$

$$\forall k : \lim_{\varepsilon \to 0} W_k(\alpha) = 1. \qquad (7)$$

This paper is devoted to the definition of practical criteria for the choice of the parameter $\alpha$, in the case of two classical algorithms :
A. *The iterative Landweber algorithm*, which defines successive solution estimates $f'_N$ by

$$f'_{N+1} = f'_N + \tau(K^*g' - K^*Kf'_N) \qquad (8)$$

with $0 < \tau < 2/|K|^2$ and $f'_0 = 0$.
The corresponding filters are

$$W_k(\alpha=1/N) = 1 - (1 - \tau \lambda_k^2)^N, \text{ and} \qquad (9)$$

$$R(\alpha=1/N) \leq \sqrt{2N\tau}. \qquad (10)$$

B. *The truncated development onto the singular system* ("SVD"), obtained by retaining in the series (2) only a finite number of singular components, corresponding to the N largest singular values. Here,

$$W_k(\alpha=1/N) = \begin{matrix} 1 &, k \leq N \\ 0 &, k > N \end{matrix} \qquad (11)$$

$$R(\alpha=1/N) = 1/\lambda_N. \qquad (12)$$

When exact data g are available, both algorithms converge to the solution when N $\to \infty$. In the presence of noisy data, on the other hand, the ill-posedness of the problem reveals itself by an uncontrolled noise amplification when too many singular components are included, or too many iterations applied.

A regularized algorithm can be obtained if an appropriate choice of N is made; it is easy to check, for instance, that the following prescriptions satisfy condition (5) :

A. Landweber :   $N = cst \; \varepsilon^{-2v}$
B. SVD        :   N such that $\lambda_N > cst \; \varepsilon^v$
$\lambda_{N+1} \leq cst \; \varepsilon^v$ for some $v$, $0 < v < 1$.

In section 3, we investigate another criterium, based on Morozov's discrepancy principle. This principle relies on the observation that the estimated solution should not fit the data with an accuracy greater than the accuracy of the measurement. The parameter $\alpha$ will therefore be chosen so as to satisfy :

$$|Kf'_\alpha - g'|^2 = \mu \varepsilon^2 \tag{13}$$

or, in the case of a parameter taking discrete values $\alpha = 1/N$ :

$$|Kf'_N - g'|^2 > \mu \varepsilon^2 \tag{14}$$
$$|Kf'_{N+1} - g'|^2 \leq \mu \varepsilon^2 \tag{15}$$

where $\mu$ is a fixed positive parameter. This criterium is known to yield a regularizing choice of $\alpha$ for Landweber's and SVD algorithms [Vainikko, 1985]. In section 3, we give a simple proof of this theorem. In section 4, we propose an alternative prescription for evaluating N. The idea consists of performing a step N $\to$ N+1 (i.e., one iteration, or the addition of the next singular component) only if we can guarantee that this step will improve the solution estimate.

## 3. THE DISCREPANCY PRINCIPLE

The discrepancy between the solution estimate and the exact solution can be bounded using the triangular inequality

$$|f'_\alpha - f| \leq |f'_\alpha - f_\alpha| + |f_\alpha - f| \qquad (16)$$

where $f_\alpha$ is obtained using the filtered series (4) with the noise-free data g. Using (6), we get:

$$|f'_\alpha - f| \leq R(\alpha)\varepsilon + \left[\sum_{k=1}^{\infty} |W_k(\alpha)-1|^2 |(u_k,f)|^2\right]^{1/2}. \qquad (17)$$

If condition (7) is satisfied - as for the two algorithms considered here - the second term in the r.h.s. will converge to zero if $\varepsilon \to 0$. A criterium for choosing $\alpha$ will therefore be regularizing if

$$\lim_{\varepsilon \to 0} \alpha = 0, \qquad (18)$$

$$\lim_{\varepsilon \to 0} R(\alpha)\varepsilon = 0. \qquad (19)$$

Consider now Morozov's principle. Inequality (14) can be written as

$$\mu\varepsilon^2 < F(\alpha) + 2\varepsilon H(\alpha) + \varepsilon^2 G(\alpha) \qquad (20)$$

with $F(\alpha) = \sum_{k=1}^{\infty} |W_k(\alpha)-1|^2 |(u_k,f)|^2 \lambda_k^2 \qquad (21)$

$$H(\alpha) = \sum_{k=1}^{\infty} |W_k(\alpha)-1|^2 (u_k,f) \lambda_k (v_k, g'-g)/\varepsilon \qquad (22)$$

$$G(\alpha) = \sum_{k=1}^{\infty} |W_k(\alpha)-1|^2 |(v_k, g'-g)/\varepsilon|^2. \qquad (23)$$

We can now solve (20) to get, using Schwarz's inequality $H^2(\alpha) \leq F(\alpha)G(\alpha)$:

$$\varepsilon < \sqrt{F(\alpha)}/(\sqrt{\mu} - \sqrt{G_{max}}) \qquad (24)$$

where $G_{max} = \sup_{k,\alpha} |W_k(\alpha)-1|^2$ and we assume $\mu > G_{max}$.

In order to demonstrate the regularizing character of Morozov's criterium, it is now sufficient to prove (18) and

$$\lim_{\varepsilon \to 0} R(\alpha) \sqrt{F(\alpha)} = 0 . \tag{25}$$

Let us consider the two algorithms under study.

A. *The iterative Landweber algorithm.*
Inequality (15) reads

$$\mu \varepsilon^2 \geq \sum_{k=1}^{\infty} (1-\tau\lambda_k^2)^{2(N+1)} |(v_k, g')|^2 \tag{26}$$

which implies $\lim_{\varepsilon \to 0} N = \infty$. As concerns the condition (25), notice that $|\tau N \lambda_k^2 (1 - \tau \lambda_k^2)^{2N}|$ is smaller than $1/2e$ for $0 < \tau \lambda_k^2 \leq 1$. The series $R^2(\alpha) F(\alpha)$ converges therefore uniformly, and, as $\lim_{N \to \infty} N(1 - \tau \lambda_k^2)^{2N} = 0$ for every k, condition (25) is proved for $\mu > G_{max} = 1$.

B. *The truncated development onto the singular system*
Again, inequality (15) implies $\lim N = \infty$, unless the data g' has only a finite number of singular components, in which case the ill-posedness problem disappears. On the other hand

$$R^2(\alpha) F(\alpha) = \sum_{k=N+1}^{\infty} \lambda_k^2 |(u_k, f)|^2 / \lambda_N^2$$

$$\leq \sum_{k=N+1}^{\infty} |(u_k, f)|^2 . \tag{27}$$

The last quantity converges to zero when N tends to $\infty$. Condition (25) is thus satisfied, and the regularizing property of Morozov's criterium is thereby proved for $\mu > G_{max} = 1$.

Notice finally that the proofs given here are easily extended to the case of the "optimum" regularizing algorithm defined by Bakushinskii as

$$W_k(\alpha) = \begin{array}{ll} 1 & k \leq N \\ \lambda_k/\lambda_N & k > N . \end{array} \tag{28}$$

## 4. AN ALTERNATIVE CRITERIUM

As mentioned in the introduction, the ill-posedness of our problem leads, in the presence of noisy data, to an interruption of the convergence of the iterated Landweber solution towards the exact solution, when the number of iterations exceeds a certain treshold Nopt (the *residual error*, of course, is still continuously decreasing !). Ideally, one should stop the iteration after this Nopt$^{th}$ step, so as to minimize the reconstruction error. Unfortunately, the evaluation of Nopt requires the knowledge of the exact solution !

It is possible, however, given an upper bound $\varepsilon$ on the norm of the noise, to derive sufficient conditions, which can be checked after each iteration step, and which guarantee that one is still in the converging phase of the recursive process. We here derive such conditions for the Landweber and SVD algorithms. They can be shown to guarantee regularization.

### A. *The iterative Landweber algorithm.*

The sufficient condition is derived very simply, by studying the difference between the squared reconstruction error at two iteration steps :

$$|f'_{N+1} - f|^2 - |f'_N - f|^2 = ((f'_{N+1} + f'_N - 2f), (f'_{N+1} - f'_N)) =$$
$$\tau^2 |K^*g' - K^*Kf'_N|^2 + 2\tau((Kf'_N - Kf), (g' - Kf'_N)) \leq$$
$$\tau |Kf'_N - g'| ( 2\varepsilon - (2 - \tau |K|^2) |Kf'_N - g'|). \qquad (29)$$

The l.h.s. is therefore negative, that is, the (N+1)$^{th}$ iterate is better (at least in the sense of the norm of the Hilbert space F) than the N$^{th}$ one, if the following condition is satisfied :

$$|Kf'_N - g'| \geq \mu \varepsilon \qquad (30)$$

with $\mu = 2/(2 - \tau|K|^2)$.

One can then decide to stop the iteration as soon as (30) is no longer satisfied (remember that the residual error $|Kf'_N - g'|$ is a monotonically decreasing function of N). The criterium derived here is by no way a necessary condition : usually, the solution will still be improving during a few iterations, before the noise propagation effect takes over. Notice also that $\mu > 1$. The present criterium is therefore a particular case of

Morozov's criterium and gives a regularizing choice of the number of iterations.

B. *The truncated development onto the singular system*

In the case of the truncated SVD method, a similar condition for choosing the number of singular components is derived by noting that

$$|f'_{N+1} - f|^2 - |f'_N - f|^2 \leq (-|(v_N, g')|^2 + 2\varepsilon|(v_N, g')|)/\lambda^2_N. \quad (31)$$

The introduction of a component N into the SVD series will therefore improve the estimated solution if it satisfies

$$|(v_N, g')| \geq 2\varepsilon. \quad (32)$$

The condition is easily checked in practice since the inner product $(v_N, g')$ must be computed anyway to evaluate the contribution of the $N^{th}$ singular component. We therefore take as estimated solution

$$f' = \sum_{k \, \varepsilon \, S} (v_k, g') \frac{u_k}{\lambda_k} \quad (33)$$

where $S = \{ k : |(v_k, g')| \geq 2\varepsilon \}$.
The proof that $\lim_{\varepsilon \to 0} |f' - f| = 0$ will not be given here.

5. DISCUSSION

The evaluation of an optimal or quasi-optimal value of the regularization parameter is a very important practical issue, which has to be solved when using any regularization technique. In the case of an iterative algorithm - like Landweber's algorithm - the regularization parameter is the number of iterations to be performed. Usually, the decision to stop the recursion is made empirically, e.g. by examination of the successive solution estimates, or by using previous experience with similar objects and signal-to-noise ratios.

The stopping rules proposed in this paper are based on more objective criteria, and are derived using only an upper bound on the norm of the difference between the noisy and exact data. They can be used successfully when little is known about the characteristics of the object and the noise. When more prior knowledge is available, it should be possible to derive more efficient stopping rules; this, however, is difficult from a mathematical point of view, in particular if we wish to introduce non-linear constraints into the problem.

The criteria proposed in this paper have been shown to yield regularizing choices of the number of iteration for Landweber's algorithm, and of the number of singular components for the SVD method. This means that the estimated solution is guaranteed to converge to the exact solution when the data g' converge to the exact data g. In the presence of a fixed, finite noise level, however, the practical efficiency of those criteria can be assessed only by means of numerical simulation. We have performed such simulations for the limited angle tomography problem, known to be severely ill-posed. The Gerchberg-Papoulis algorithm - a special case of the Landweber's iteration - has been used to recover the spectrum of model objects in the so-called "missing cone" [Gerchberg, 1974; Papoulis, 1975]. Plotting the $L^2$ distance between the image estimate and the known model object versus the number of iterations then reveals the usual behaviour : if pseudo random gaussian noise is added to the data, the reconstruction error reaches a minimum after a certain number of iterations - typically 10 to 100 - depending on the signal-to-noise ratio. The number of iterations defined by the criteria of sections 3 and 4 is found to be significantly smaller than the optimal iteration number. This "over-pessimistic" behaviour of the proposed stopping rules is due to their taking into account the worst - and very unlikely - possible noise sample compatible with the upper bound (3).

Numerical experiments have also been performed with one-dimensional problems like the extrapolation of band-limited signals, or the inversion of the Laplace transform; the efficiency of Morozov's criterium appears better for those applications, although the prescribed number of iterations is usually still smaller than the optimal one.

Further study is now in progress to derive objective stopping rules including more restrictive hypotheses about the statistical characteristics of object and noise.

*Acknowledgements*

*It is a pleasure to thank Prof. M. Bertero (Genova) and Dr. C. De Mol (Brussels) for many suggestions concerning the content and presentation of this paper.*

*The author is Research Associate with the National Fund for Scientific Research (Belgium).*


REFERENCES

Bakushinskii, A.B.(1967). A general method of constructing regularizing algorithms for a linear ill-posed equation in Hilbert space, *U.S.S.R Comp. Maths. Math. Phys.* 7, pp. 279-287.
Censor Y. (1983). Finite series-expansion reconstruction method, *Proc. IEEE* 71, pp. 409-419.
Gerchberg (1974). Superresolution through error energy reduction, *Optica Acta* 21, pp. 709-720.
Lewitt R. M. (1983). Reconstruction algorithms : transform methods, *Proc. IEEE* 71, pp. 390-408.
Louis A.K. and Natterer F. (1983). Mathematical problems of computerized tomography, *Proc. IEEE* 71, pp. 379-389.
Papoulis A. (1975). A new algorithm in spectral analysis and bandlimited extrapolation, *IEEE Trans. Circ. Syst.* CAS-22, pp. 735-742.
Tikhonov A.N. and Arsenin V.Y. (1977). *Solutions of ill-posed problems*, Winston & Sons, Washington.
Vainikko G.M. (1983). The critical level of discrepancy in regularization methods, *U.S.S.R Comp. Maths. Math. Phys.* 23, pp.1-9.

INITIAL PERFORMANCE OF BLOCK-ITERATIVE RECONSTRUCTION
ALGORITHMS

Gabor T. Herman and Haim Levkowitz

*University of Pennsylvania*
*U.S.A.*

ABSTRACT

*Commonly used iterative techniques for image reconstruction from projections include ART (Algebraic Reconstruction Technique) and SIRT (Simultaneous Iterative Reconstruction Technique). It has been shown that these are the two extremes of a general family of block-iterative image reconstruction techniques. Here we show that the initial performance of these commonly used extremes can be bested by other members of the family.*

1. BLOCK-ITERATIVE RECONSTRUCTIVE TECHNIQUES

One classification of methods of image reconstruction from projections subdivides them into two categories: transform methods and finite series expansion methods (Herman, 1980). Tutorial articles have been published on both categories; we refer the reader to Lewitt (1983) and Censor (1983). Reconstruction algorithms based on optimization techniques are typically what we call finite series expansion methods. The work reported here also falls into this area.

Much of this work revolves around the so-called Algebraic Reconstruction Techniques (ART). We explain this class of methods using the formulation we used in Herman *et al.* (1984).

Suppose we have a system of equations $Az = Y$, which is written out as

$$\langle a_i, z \rangle = y_i \qquad 1 \leq i \leq LM \qquad (1)$$

where $\langle a_i, z \rangle = \Sigma a_{ij} z_j$. These equations may model the measurement process directly, or they may be derived from an optimization problem. An example of the latter case is when the equations are the normal equations of a regularized least squares minimization criterion, based on Bayesian considerations (Herman, 1980). Note that we think of these LM equations as being subdivided into M blocks each of size L. If we denote $n \pmod{M} + 1$ by $m_n$ (that is, $m_n = 1,2,\ldots,M,1,2,\ldots$ for $n = 0,1,\ldots,M-1,M,M+1,\ldots$), then a generalized version of ART can be described by its iterative step from the n'th estimate $z^{(n)}$ to the (n+1)'st estimate $z^{(n+1)}$ by

$$z^{(n+1)} = z^{(n)} + \lambda \sum_{\ell=1}^{L} \frac{y_i - \langle a_i, z^{(n)} \rangle}{\|a_i\|^2} a_i, \text{ with } i = (m_n - 1)L + \ell. \quad (2)$$

In one iterative step we consider all the equations in one of the blocks. From our previous work (Eggermont et al., 1981) it follows that if the parameter $\lambda$ is within a certain range, then this technique converges to a solution of the system of equations (provided the system is consistent) irrespective of the starting point $z^{(0)}$. Two cases are worthy of special mention.

If the block size L is 1, the method cycles through the equations one-by-one and the technique is (for $\lambda = 1$) the classical method of Kaczmarz (1937), which is the same as the original ART for image reconstruction as proposed by Gordon et al. (1970).

If the number of blocks M is 1, then all equations are considered in each iterative step and the method is (for an appropriately chosen $\lambda$) the classical method of Cimmino (1938), which is very similar to the Simultaneous Iterative Reconstruction Technique (SIRT), as proposed by Gilbert (1972).

Thus the general approach introduced above allows us to discuss both ART-type and SIRT-type techniques within the same framework. In Herman et al. (1984) we investigated whether a multilevel approach, one in which the block size is changing during processing, has an advantage over pure ART or SIRT. The initial results are promising, but more work needs to be done before definitive conclusions can be drawn. Here we report on a different approach, one in which the block size L is kept constant during the iterative process. In an experimental study we considered values of M between 1 and 256 (all powers of 2) and found the initial behavior for M = 32 and 64 to be definitely superior to that for all other values of M.

## 2. THE RELAXED BLOCK-CIMMINO METHOD

To make the concepts introduced in the previous section precise, we appeal to the more general theory of Eggermont *et al.* (1981).

For $1 \leq m \leq M$, we let

$$A_m = \begin{pmatrix} a_{(m-1)L+1}^T \\ \vdots \\ a_{mL}^T \end{pmatrix}, \qquad Y_m = \begin{pmatrix} Y_{(m-1)L+1} \\ \vdots \\ Y_{mL} \end{pmatrix}. \qquad (3)$$

The algorithm (1.10) of Eggermont *et al.* (1981) gives rise to the following iterative step:

$$z^{(n+1)} = z^{(n)} + A_{m_n}^T \Sigma^{(m_n)} (Y_{m_n} - A_{m_n} z^{(n)}) \qquad (4)$$

where $m_n$ is defined as above to be $n \pmod{M} + 1$ and the $\Sigma^{(m)}$ are $L \times L$ matrices, referred to as the *relaxation matrices*.

Theorem 1.3 of Eggermont *et al.* (1981) simplifies to the following statement.

*Theorem.* Let $\Sigma^{(m)}$ satisfy

$$\| A_m^\dagger (I_L - A_m A_m^T \Sigma^{(m)}) A_m \|_2 < 1 \qquad (5)$$

for $1 \leq m \leq M$. If the system (1) is consistent, then (for any $z^{(0)}$) the sequence $\{z^{(n)}\}$ generated by (4) converges to a solution of $Az = Y$. If, in addition, $z^{(0)}$ is in the range of $A^T$, then

$$\lim_{n \to \infty} z^{(n)} = A^\dagger Y. \qquad (6)$$

Even if the system (1) is inconsistent, the sequence $\{z^{(kM)}\}$ converges as $k \to \infty$.

In the statement of the theorem $A^\dagger$ denotes the Moore-Penrose inverse of $A$ (Ben-Israel and Greville, 1974), and thus $A^\dagger Y$ in (6) is the minimum norm solution of the system of equations (1).

We now explicitly define $\Sigma^{(m)}$ and observe that for this choice the iterative step in (4) is the same as that in (2). We let, for $1 \leq m \leq M$,

$$\Sigma^{(m)} = \begin{pmatrix} \dfrac{\lambda}{\|a_{(m-1)L+1}\|^2} & & \\ & \ddots & \\ & & \dfrac{\lambda}{\|a_{mL}\|^2} \end{pmatrix}. \quad (7)$$

With this definition of $\Sigma^{(m)}$ we see that the algorithm whose iterative step is given in (2) is covered by the general theory of Eggermont et al. (1981). To guarantee convergence, $\lambda$ has to be chosen small enough so that (5) is satisfied. Since our interest here is initial (rather than limiting) behavior we do not derive the range of $\lambda$'s which guarantees convergence; for a discussion of such derivations see Eggermont et al. (1981). Our approach for choosing $\lambda$ is based on that of Cimmino (1938).

As mentioned in Section 1, Cimmino's method is described by (2) with $M = 1$ (all equations form a single block). Cimmino's choice of $\lambda$ is $2/L$. This has the following simple geometrical interpretation: the $(n+1)$'st iterate is the mean of the mirror images of the n'th iterate in each of the hyperplanes in (1).

In our block-iterative generalization of Cimmino's method we adopt the same way of choosing $\lambda$, except that in each iterate we deal with only those equations which belong to a particular block (the $m_n$'th block for getting from $z^{(n)}$ to $z^{(n+1)}$. We also introduce the notion of relaxation, which has been found useful in many iterative approaches to image reconstruction (Herman, 1980). Thus we choose

$$\lambda = \frac{2\mu}{L}, \quad (8)$$

where $\mu$ is the so-called *relaxation parameter*.

One approach towards choosing $\mu$, and this is the one we have adopted in this paper, is the following. It is reasonable to start with $z^{(0)}$ as the zero vector. In image reconstruction from projections, the number of measurements, L, is very large. It is over 75,000 in the simple computer experiment we discuss in the next section, but can be well over half a million in real-life computerized tomography. Using Cimmino's method ($\mu = 1$), equation (8) provides a very small $\lambda$ and hence very little change in each of the iterative steps. This accounts for the well-known slowness of the early iterates in SIRT-like methods (Herman, 1980). Choosing a larger $\mu$ can speed things up, but there is the danger of overshooting our target. One reasonable way of choosing $\mu$ is by selecting it so that the components of $z^{(1)}$ have the average value $\bar{z}$, which is the estimated average value of the image based on the data Y. It is well-known (Herman, 1980) that in image reconstruction the data Y determine the average value $\bar{z}$ of the image to be reconstructed accurately and easily. The precise definition of $\bar{z}$ is

$$\bar{z} = \frac{1}{LM} \sum_{i=1}^{LM} \frac{y_i}{\Sigma a_{ij}}. \tag{9}$$

Clearly, the choices that

(i) $z^{(0)}$ is the zero vector,
(ii) the average value of the components of $z^{(1)}$ in the relaxed Cimmino method ($M = 1$) is $\bar{z}$,

uniquely determine $\mu$. In comparing block-Cimmino methods for different block sizes L, the $\mu$ should stay the same, but $\lambda$ should change according to (8). This approach is further discussed after the experimental results.

## 3. EXPERIMENTS

To test the effect of block size in using the block-Cimmino method for image reconstruction from projections we carried out a number of experiments. We used a mathematical phantom which was created based on the actual computerized tomography scan of a cadaver for a previous study (Herman *et al.*, 1982). Fan-beam projection data were generated using the program mini-SNARK (Lewitt, 1982), with the following characteristics: number of projections - 256, number of rays per projection - 295, distance of the source to the detector - 50 cm, distance of the source to the center of rotation - 25 cm, spacing between consecutive detectors - 0.209 cm. Thus the total number of measurements is LM = 256 × 295 = 75,520. For the reconstruction we used a 97 × 97 array of square-shaped picture elements with 0.209 cm sides. Thus the number of unknowns (the number of components of z) is 9,409. Hence the system (1) in this case is highly overdetermined. Since the data (Y) are "real", but the model expressed in (1) is only approximate, the system is also inconsistent (Herman, 1980). We found that the value of the relaxation parameter $\mu$, calculated by the method described in the previous section, is 68.5.

We studied eight different block sizes. In the smallest, one block consisted of all the rays in one projection (L = 295, M = 256), in the others each block was made up from 2, 4, 8, 16, 32, 64, and 256 projections, respectively. (In the last case L = 256 × 295, M = 1, and so we have the relaxed Cimmino method.) Projections were blocked together as follows: with the first projection we blocked projections 1+M,1+2M,...,1+(L-1)M, and similarly for the second, third,..., and M'th projection. Note that since we did not investigate the case L = 1, none of these experiments is ART-type in the pure sense; but the case L = 295 (one projection per block) approaches it and is in fact very similar to the method proposed by Oppenheim (1977).

Figure 1 reports on our results. We plotted the performance of the block-Cimmino method by comparing various measures of merit calculated at the kM'th iterate, for k = 1,2,3,4,5,6,7,8,9. This is fair, since for all values of M, the M'th iterate is obtained by considering all equations exactly once, and so the computational cost of obtaining the (kM)'th iterate (for a fixed cycle number k) is about the same for all M.

We note that there is a significant dependence on the block size when considering the initial performance of the block-Cimmino method. By all our measures of merit, the two extreme block sizes (L = 295 and L = 256 × 295) give the least desirable initial behavior and the intermediate ones (L = 4 × 295 and L = 8 × 295) give the best. In the cases where the number of blocks M is not 1 or 2, the order in which the blocks are processed also influences the initial behavior. We also investigated the case (not reported in Figure 1), when M = 256 and blocks (projections) are processed in the same order as in the case M = 64. Some improvement over what is reported in Figure 1 for M = 256 was observed, but the performance was still much worse than for M = 64.


Legend for Figure 1 (next three pages)

*The performance of the block-Cimmino method for 8 different block sizes compared using three measures of merit calculated at the kM'th iterate, for $k = 1,\ldots,9$. The measures of merit are:*

(a) *The* distance, *which is a normalized 2-norm distance between the phantom and the kM'th iterate. More precisely,*

$$distance = (\Sigma(z_j - z_j^{(kM)})^2 / \Sigma(z_j - \bar{z})^2)^{1/2}.$$

(b) *The* relative error *which is a normalized 1-norm distance between the phantom and the iterate,*

$$relative\ error = \Sigma|z_j - z_j^{(kM)}| / \Sigma|z_j|.$$

(c) *The* residual norm *which is the 2-norm of the difference between the actual measured projection data and the pseudo projection measurement of the current iterate,*

$$residual\ norm = \left(\sum_{i=1}^{LM} (y_i - \langle a_i, z^{(kM)}\rangle)^2\right)^{1/2}.$$

*Note that the actual L used is 295 times the L given in the tables at the top of the figures (the L given there is the number of projections in a block and each projection consists of 295 measurements).*

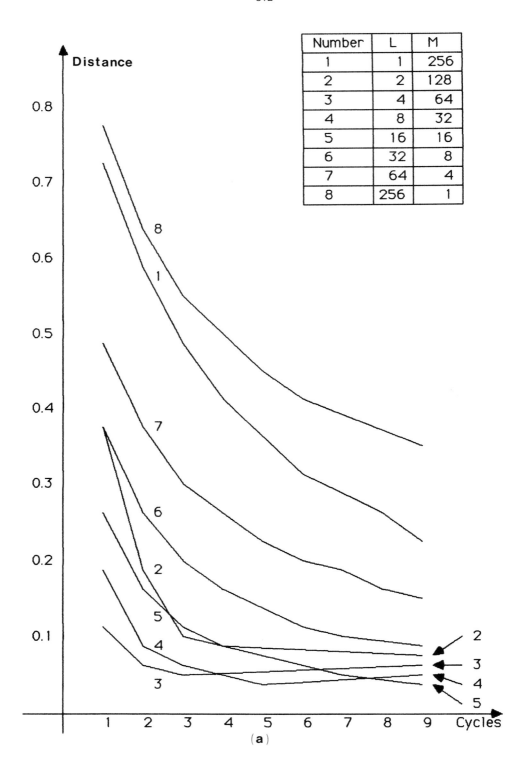

(a)

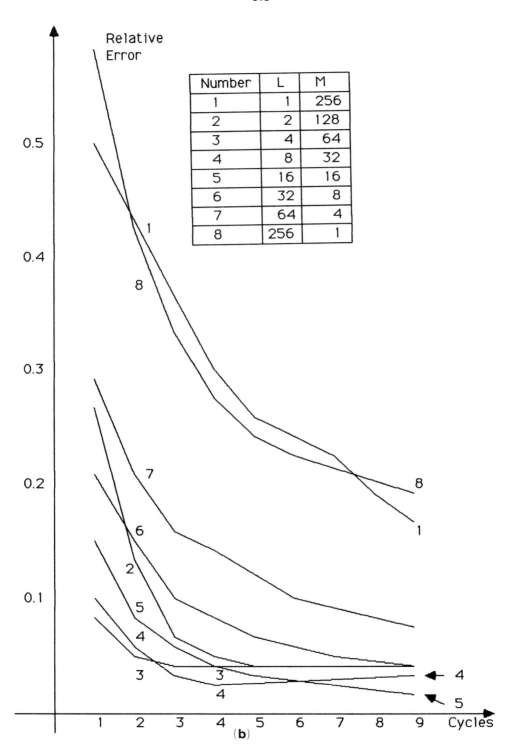

(b)

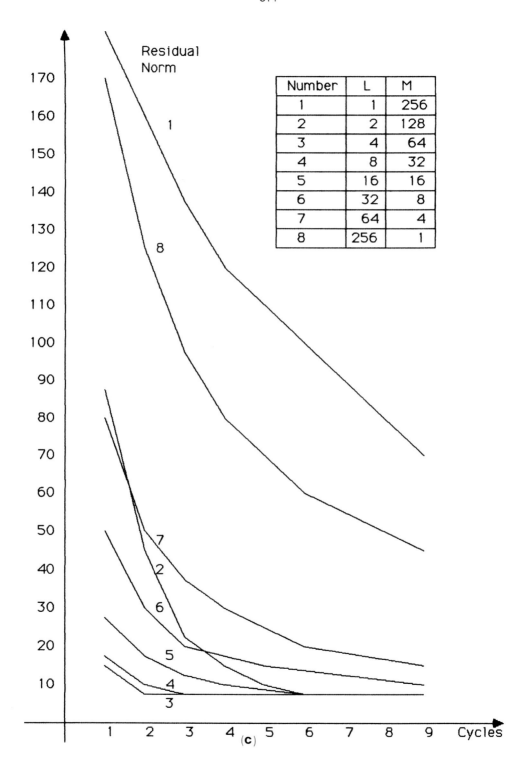

## 4. DISCUSSION

In image reconstruction from projections there have been a number of proposed algorithms that are special cases of the iterative procedure presented above. In this framework we recognize that they only differ from each other in the choice of partitioning the system of equations into blocks. The ART-procedure of Gordon *et al.* (1970) is essentially this process with L = 1 (each equation is a separate block). The procedure of Oppenheim (1977) is essentially this process with all the (parallel) rays in one projection forming a single block. The SIRT-procedure of Gilbert (1972) is essentially this process with M = 1 (all the equations form a single block).

This recognition led us to investigate what happens if we choose different block sizes in the procedure. Our experiment came out with an interesting result: the special choices of the block sizes which give rise to previously suggested iterative reconstruction algorithms resulted in worse initial performance than any of the other choices we tried! Initial behavior is important, since cycling even once through the equations is computationally expensive, and thus the number of cycles should be kept as low as possible.

We complete the paper with two comments.

As reported in the previous section the relaxation parameter $\mu$ in our experiments was 68.5. This implies that our approach of selecting $\mu$ cannot work for small values of L. For L = 1, this choice of $\mu$ would give $\lambda$ = 137, a clearly ridiculous choice, since with ART convergence is guaranteed only if $\lambda < 2$ (Herman, 1980) and $\lambda$ = 137 would yield a rapidly diverging sequence of images. For the smallest value of L we actually used (295), $\lambda < 0.5$ and the experiments indicate improving initial performance. In fact, even though $\mu$ was selected based solely on the largest possible block size (L = 75,520), in all our experiments the average value of the components of $z^{(M)}$ (the image at the end of the first (cycle) was within 3.5% of the desired average $\bar{z}$. This certainly indicates that our approach for selecting $\mu$ is appropriate for all the L's in our experiments.

Finally we wish to point out the applicability of our approach to more sophisticated models of image reconstruction from projections than the one used in our experiments. As explained in Eggermont *et al.* (1981) the variables in (3) and (4) can have many different interpretations, and the approach can be used to solve image reconstruction problems which are couched in the terminology of quadratic optimization (Herman, 1980). Since the numerical aspects of those approaches are not essentially different, we would expect to see a dependence on block size in iterative methods designed for quadratic optimization similar to that demonstrated in this article for the relaxed block-Cimmino method.

*Acknowledgements*

*The research of the authors is supported by NIH grant HL28438. They are grateful to Ms. M.A. Blue for typing.*

REFERENCES

Ben-Israel, A. and Greville, T.N.E. (1974). *Generalized Inverses: Theory and Applications*, John Wiley and Sons, New York.
Censor, Y. (1983). Finite series expansion reconstruction methods, *Proc. IEEE* 71, pp. 409-419.
Cimmino, G. (1938). Calcolo approssimato per le soluzioni dei sistemi di equazioni lineari, *La Ricerca Scientificia (Roma)* XVI, Ser. II, Anno IX 1, pp. 326-333.
Eggermont, P.P.B., Herman, G.T., and Lent, A. (1981). Iterative algorithms for large partitioned linear systems with applications to image reconstruction, *Lin. Alg. Appl.* 40, pp. 37-67.
Gilbert, P.F.C. (1972). Iterative methods for three-dimensional reconstruction of an object from projections, *J. Theor. Biol.* 36, pp. 105-117.
Gordon, R., Bender, R., and Herman, G.T. (1970). Algebraic reconstruction techniques (ART) for three-dimensional electron microscopy and x-ray photography, *J. Theor. Biol.* 29, pp. 471-482.
Herman, G.T. (1980). *Image Reconstruction from Projections: The Fundamentals of Computerized Tomography*, Academic Press, New York.
Herman, G.T., Robb, R.A., Gray, J.E., Lewitt, R.M., Reynolds, R.A., Smith, B., Tuy, H., Hanson, D.P., and Kratz, C.M. (1982). Reconstruction algorithms for dose reduction in x-ray computed tomography, *Proc. MEDCOMP '82*, Philadelphia, pp. 448-455.
Herman, G.T., Levkowitz, H., Tuy, H.K., and McCormick, S. (1984). Multilevel image reconstruction. In: *Multiresolution Image Processing and Analysis*, A. Rosenfeld (ed.), Springer-Verlag, Berlin, pp. 121-135.
Kaczmarz, S. (1937). Angenäherte Auflösung von Systemen linearer Gleichungen. *Bull. Acad. Polon. Sci. Lett. A*, pp. 355-357.
Lewitt, R.M. (1982). Mini-SNARK: Version 2, Technical Report MIPG68, Medical Image Processing Group, Department of Radiology, University of Pennsylvania.

Lewitt, R.M. (1983). Reconstruction algorithms: transform methods, *Proc. IEEE* 71, pp. 390-408.

Oppenheim, B.E. (1977). Reconstruction tomography from incomplete projections. In: *Reconstruction Tomography in Diagnostic Radiology and Nuclear Medicine*, M.M. Ter-Porgossian *et al.* (eds.), University Park Press, Baltimore, pp. 155-183.

# MAXIMUM LIKELIHOOD RECONSTRUCTION IN PET AND TOFPET

Chin-Tu Chen, Charles E. Metz and Xiaoping Hu

*The University of Chicago*
*U.S.A.*

ABSTRACT

*The expectation-maximization (EM) algorithm for maximum likelihood estimation can be applied to image reconstruction in positron emission tomography (PET) and time-of-flight assisted PET (TOFPET). To implement this algorithm, one can employ either the projection data acquired at various angles or the rotationally integrated image. In the case of TOFPET, the latter approach can reduce the required computation time substantially. Three approaches -- the conventional, direct and statistical methods -- that incorporate the effect of photon attenuation in the EM algorithm have been investigated. Preliminary results from computer simulation studies indicate that the direct method can provide a good compromise between image quality and computation time. Three acceleration approaches -- the geometric, additive and multiplicative methods -- that increase the rate of convergence of these EM reconstruction methods have been studied also.*

## 1. INTRODUCTION

Maximum likelihood estimation has been under intensive study in recent years due to its potential versatility in tomographic image reconstruction. Because it can, in principle, take into account in the reconstructed image every physical and statistical process in the sequence of events leading to data collection, the maximum likelihood approach potentially can provide more accurate reconstruction. The EM (expectation-maximization) algorithm for maximum likelihood estimation (Dempster et al., 1977) has received particular attention for its feasibility in executing the calculations for tomographic image reconstruction (Shepp and Vardi, 1982). The application of EM algorithms to image reconstruction in positron emission tomography (PET) and time-of-flight assisted PET (TOFPET) is discussed in this paper.

In the EM algorithm, the 'measured' data set is used in combination with a postulated, unobservable 'complete' data set to facilitate the process of maximizing the likelihood function of the measured data. The actual calculation consists of a series of alternating expectation steps and maximization steps. In theory, this iterative process is equivalent to the original task of maximizing the likelihood function defined on the measured data.

For image reconstruction in PET or TOFPET, the measurements (i.e., 1-D projection profiles in PET or 2-D angular views in TOFPET) at each projection angle are normally used as the 'measured' data. This type of implementation requires angle-by-angle calculations. An alternative is to employ the rotationally integrated image together with associated probability functions that link the photon emission process and these 'measured' data in the EM steps. In the case of TOFPET, this approach reduces the computation time by a factor of 1/M for an image reconstruction using M projection angles.

The virtue of maximum likelihood reconstruction is the possibility of incorporating physical factors such as photon attenuation, scattered radiation, radioactivity decay, detector sensitivity, geometry of the tomograph, detector resolution, TOF resolution, chance coincidence, etc., into a single set of probability functions that is employed subsequently in the reconstruction process. In implementation, the inclusion of these factors is not straightforward, however. More than one approach can be employed in some cases. Therefore, detailed investigations and comparison studies are needed before the merit of the maximum likelihood reconstruction is fully explored. In this paper, only the effect of photon attenuation is discussed as an example to demonstrate the implementation of various approaches.

The major disadvantage of maximum likelihood reconstruction algorithms is their slow convergence rate. To remedy this defect, it is possible to incorporate acceleration schemes to achieve a faster rate of increase of the likelihood function at each iteration. Acceleration methods for the EM reconstruction algorithms can be classified broadly into three general categories: the geometric, additive and multiplicative methods. Some examples of these acceleration techniques are given in this paper.

## 2. THE EM ALGORITHM

The basic idea of the EM algorithm is outlined briefly here. Let Y be the random vector that represents the observed data in a study, with a joint probability density function, or likelihood function, $g_L(Y|\lambda)$, where $\lambda$ is the unobserved parameter vector to be estimated. Maximizing $g_L(Y|\lambda)$ with respect to $\lambda$ yields the parameter vector estimate with which the data are most consistent. The computations associated with this maximization step may be difficult, however. The EM algorithm generally employs a postulated, unobserved random vector X, called the 'complete' data set, which is sampled from a larger or richer space, to facilitate the maximization process. X and its likelihood function, $q_L(X|\lambda)$, can be related to the observed data Y -- now called the 'incomplete' data set -- and its likelihood function, $g_L(Y|\lambda)$, by

$$Y = h(X) \tag{1}$$

and

$$g_L(Y|\lambda) = \int q_L(X|\lambda) \, d\sigma(X), \qquad (2)$$

where h is a many-to-one mapping function that describes the relationship between X and Y, and $\sigma$ is a measure relating $q_L$ to $g_L$. The choice of X is not unique; but in many applications, a natural choice corresponding to the physical process under study is obvious. This will be illustrated in the later sections.

There are two steps at each iteration of the EM algorithm: an expectation (E-) step and a maximization (M-) step. In the E-step, a conditional expectation is formulated by defining

$$Q_E(\lambda|\lambda^n) = E[\ln q_L(X|\lambda) | Y, \lambda^n], \qquad (3)$$

where $E[\cdot]$ denotes the expectation value, $\ln q_L$ is the natural logarithmic likelihood function (LLF) of the 'complete' data set, and $\lambda^n$ are the current ($n^{th}$) estimates. In the M-step that follows, this conditional expectation is maximized with respect to $\lambda$, with $\lambda^n$ held fixed, to yield a formulation for the calculation of the new $[(n+1)^{st}]$ estimates, $\lambda^{n+1}$. Therefore, the original task of maximizing $g_L(Y|\lambda)$ is carried out by many two-step (i.e., E-step and M-step) calculations involving $q_L(X|\lambda)$ in an attempt to approach the maximum of $g_L(Y|\lambda)$ iteratively.

## 3. TWO APPROACHES OF IMPLEMENTATION

Assume that the source distribution function is discrete and that $f_k$ is the $k^{th}$ element. Assume also that the projection data at each angle are discrete, with $p_{ij}$ representing the $j^{th}$ detection element at the $i^{th}$ projection angle. In terms of the EM algorithm described earlier, the parameters to be estimated are the source distribution function elements $\{f_k\}$. One approach of implementation is to employ the projection data measured at each angle, $\{p_{ij}\}$, as the 'incomplete' data set and to devise a 'complete' data set, $\{s_{ijk}\}$, which represents the number of events emitted from $f_k$ and detected in $p_{ij}$. It should be noted that $\{s_{ijk}\}$ cannot be 'measured' or detected physically. Associating $f_k$, $p_{ij}$ and $s_{ijk}$ with the symbols used in the previous section, one has $\lambda=\{f_k\}$, $Y=\{p_{ij}\}$ and $X=\{s_{ijk}\}$. The function relating the complete data set to the incomplete data set [Eq. (1)] can be expressed explicitly for the present case as

$$p_{ij} = \sum_k s_{ijk}. \qquad (4)$$

In PET, $\{p_{ij}\}$ is the data set of 1-D projection profiles. On the other hand, in TOFPET, it represents the data set of 2-D images obtained at individual projection angles. With positron annihilation assumed to be an isotropic process, the expectation value of $s_{ijk}$, given f, can be written as

$$E[s_{ijk}|f] = N_{ijk} = w_{ijk} f_k, \tag{5}$$

where $w_{ijk}$ is the probability that an event emitted from $f_k$ would be detected in $p_{ij}$. In principle, $w_{ijk}$ can incorporate corrections for effects such as photon attenuation, scattered radiation, positron range, chance coincidence, etc. Governed by the Poisson nature of the annihilation process, the probability density function of $s_{ijk}$ is

$$\Pr[s_{ijk}|f] = (N_{ijk})^{s_{ijk}} \cdot \exp(-N_{ijk}) / s_{ijk}!. \tag{6}$$

Because all of $s_{ijk}$ are statistically independent, the likelihood function of the complete data set, $q_L(s|f)$, can be expressed in terms of the joint probability density function as

$$q_L(s|f) = \prod_i \prod_j \prod_k \Pr[s_{ijk}|f]. \tag{7}$$

Substituting Eq. (6) into Eq. (7), one can write the LLF as

$$\ln q_L(s|f) = \sum_i \sum_j \sum_k [-N_{ijk} + s_{ijk} (\ln(N_{ijk})) - \ln(s_{ijk}!)]. \tag{8}$$

Using Eqs. (4) and (8), one can also formulate the LLF of the measured 'incomplete' data set, $\ln g_L(p|f)$, as

$$\ln g_L(p|f) = \sum_i \sum_j [-\sum_k N_{ijk} + p_{ij} (\ln(\sum_k N_{ijk})) - \ln(p_{ij}!)]. \tag{9}$$

In the E-step of the EM algorithm, the conditional expectation of $\ln q_L(s|f)$ with respect to the 'measured' data (i.e., the incomplete data set), p, and the current estimates, $f^{(n)}$, is formulated by

$$Q_E(f|f^{(n)}) = E[\ln q_L(s|f) | p, f^{(n)}]$$

$$= \sum_i \sum_j \sum_k \{ -N_{ijk} + E[s_{ijk}|p_{ij}, f^{(n)}] \ln(N_{ijk})\} + C_1, \tag{10}$$

where $C_1$ is a term independent of f. By noting that all $s_{ijk}$ are independent variables and by using the definition of conditional expectation and Eqs. (4) and (5), one obtains

$$Q_E(f|f^{(n)}) = \underset{i\ j\ k}{\Sigma\Sigma\Sigma}\{-w_{ijk}f_k + [w_{ijk}f_k^{(n)}(p_{ij}/(\underset{m}{\Sigma}w_{ijm}f_m^{(n)}))] \ln(w_{ijk}f_k) + C_1. \quad (11)$$

The M-step that follows is now straightforward. Setting the first derivatives of $Q_E(f|f^{(n)})$ with respect to f equal to zero yields the new estimates

$$f_k^{(n+1)} = [f_k^{(n)} / (\underset{i\ j}{\Sigma\Sigma} w_{ijk})] \cdot [\underset{i\ j}{\Sigma\Sigma} w_{ijk}(p_{ij} / (\underset{m}{\Sigma}w_{ijm}f_m^{(n)}))]. \quad (12)$$

The initial estimate, $f^{(o)}$, is usually set to be a uniform source distribution with non-negative value. Convergence of the estimates obtained in Eq.(12) and convergence to the maximum likelihood estimate have been demonstrated (Shepp and Vardi, 1982; Lange and Carson, 1984). If a non-negative initial estimate is chosen, it is obvious from Eq. (12) that all images produced in subsequent iterations remain non-negative. This property of 'non-negativity' guarantees that negative pixel values, which are physically impossible, will not be produced. Also, it is straightforward to show that the total number of events at each iteration is equal to the sum of the 'measured' data. This important property is called 'self-normalization'.

As indicated in Eq. (12), algorithms which use projection data require angle-by-angle operations. If a rotationally integrated image can be calculated (by backprojection in PET or by rotational averaging in TOFPET) as a preprocessing step before any EM steps, an alternative approach is to employ this rotationally integrated image as the 'measured' or 'incomplete' data (Chen and Metz, 1985). The associated probability functions in this case link the photon emission to this new 'measured' data set. In the case of TOFPET, this type of implementation can save a substantial amount of computation time.

Let $b_j$ represent the $j^{th}$ element of the rotationally integrated image; the incomplete data set is now $Y = \{b_j\}$. A complete data set, $X = \{z_{jk}\}$, can be defined as the number of events emitted from $f_k$ and detected in $b_j$ with an associated probability $u_{jk}$. Again, one should note that $z_{jk}$ is not physically measurable. It should be noted also that the projection-angle variable specified by the index i discussed previously is no longer needed. Paralleling the derivation of algorithms that use projection data, as sketched earlier, one can obtain the LLF of the complete data set as

$$\ln q_L(z|f) = \underset{j\ k}{\Sigma\Sigma}[-N_{jk} + z_{jk}(\ln(N_{jk})) - \ln(z_{jk}!)]; \quad (13)$$

the LLF of the incomplete data set as

$$\ln g_L(b|f) = \underset{j}{\Sigma}[-\underset{k}{\Sigma}N_{jk} + b_j(\ln(\underset{k}{\Sigma}N_{jk})) - \ln(b_j!)]; \quad (14)$$

and the conditional expectation of $\ln q_L (z|f)$ with respect to b and the current estimate $f^{(n)}$ as

$$Q_E(f | f^{(n)}) = E[\ln q_L (z|f) | b, f^{(n)}]$$
$$= \sum_j \sum_k \{ -u_{jk}f_k + [u_{jk}f_k^{(n)} (b_j / (\sum_m u_{jm}f_m^{(n)}))] \ln (u_{jk}f_k) + C_2, \quad (15)$$

where $C_2$ is a term independent of f. Equating the first derivative of $Q_E (f | f^{(n)})$ with respect to f to zero, to obtain a maximum of $Q_E(f | f^{(n)})$, results in

$$f_k^{(n+1)} = (f_k^{(n)} / \sum_j u_{jk}) \cdot [\sum_j u_{jk} (b_j / (\sum_m u_{jm}f_m^{(n)}))] \quad (16)$$

as the criterion for calculating the new parameter estimates with this new approach.

It is easy to demonstrate that the four properties -- i.e., convergence, convergence to the maximum likelihood estimate, non-negativity and self-normalization -- described earlier are preserved. In the case of TOFPET, the computation time required for calculation of Eq. (16) is only approximately 1/M of that required for Eq. (12).


## 4. INCORPORATION OF THE EFFECT OF PHOTON ATTENUATION

Three approaches -- the conventional, direct and statistical methods-- for incorporating the effect of photon attenuation in maximum likelihood image reconstruction are outlined in this section. Effects of other physical factors such as scattered radiation, chance coincidence, etc., are assumed here to be negligible. In principle, similar strategies can be designed to compensate for these factors, but further investigations are needed for their implementation. In the following discussion, emphasis will be placed on algorithms that use projection data. However, the underlying principles are applicable also to algorithms that use a rotationally integrated image.

In the conventional method, an attenuation factor, $A_{ij}$, for the $j^{th}$ detector element at the $i^{th}$ projection angle can be defined as

$$A_{ij} = \exp [ -\int_{ij} \mu(\alpha) d\alpha ], \quad (17)$$

where $\mu(\alpha)$ is the linear attenuation coefficient along the projection ray $\alpha$ and $\int_{ij}$ represents integration along the projection ray passing through $p_{ij}$. $A_{ij}$ can be either estimated from a transmission scan or calculated from an analytical approximation of the physical configuration of the object. Using the conventional correction scheme that is normally employed in PET image reconstruction, one can obtain a modified 'measured' data set, $p'_{ij}$, as

$$p'_{ij} = p_{ij} / A_{ij} \qquad (18)$$

before entering the EM steps. The EM steps that follow are executed in the same manner as indicated in Eqs. (8) - (12), but with $p_{ij}$ replaced by $p'_{ij}$.

In the direct method, the original projection data $\{p_{ij}\}$ are employed, and the attenuation factors are incorporated in the probability functions. If $w_{ijk}$ represents only the geometric relationship between the $k^{th}$ element of the image plane and the $j^{th}$ detector element at $i^{th}$ angle, with the effects of all physical factors, including photon attenuation, ignored, a modified probability, $w'_{ijk}$, that compensates for the effect of photon attenuation can be expressed by

$$w'_{ijk} = w_{ijk} A_{ij}. \qquad (19)$$

The EM steps then proceed as previously described but with the set of modified probability functions given by Eq. (19).

The statistical method employs a binomial model to describe the process of photon attenuation. Let $p''_{ij}$ be the number of events that would be collected in the $j^{th}$ detector element at the $i^{th}$ projection angle, if there were no photon attenuation. The conditional probability of actually detecting $p_{ij}$ events -- considering the effect of photon attenuation -- given $p''_{ij}$ is

$$\Pr(p_{ij} | p''_{ij}) = {}^{p''_{ij}}C_{p_{ij}} (1-A_{ij})^{p''_{ij} - p_{ij}} A_{ij}^{p_{ij}} \qquad (20)$$

where ${}^{x}C_{y} = x!/y!(x-y)!$. Incorporating Eq. (20) into the derivation, Eq. (12) now becomes

$$f_k^{(n+1)} = [f_k^{(n)} / (\sum\sum_{i\ j} w_{ijk})] \cdot (\sum\sum_{i\ j} w_{ijk} \{[p_{ij} + \sum w_{ijl} f_l^{(n)} (1-A_{ij})] / (\sum_m w_{ijm} f_m^{(n)})\}). \qquad (21)$$

The additional term that appears in Eq. (21) but not in Eq. (12) represents those emitted photons that were lost in the process of photon attenuation and never reached the detector array.

Computer simulation studies have been conducted for evaluation of these approaches. The simulated phantom consists of a circular disc 16 cm in diameter, containing three smaller circular discs with 1.25, 2.5 and 5 cm diameter, respectively. A 128x128 matrix with $(0.25)^2$ cm$^2$ pixel size was employed in these studies. The LLF of the 'measured' data, which the EM

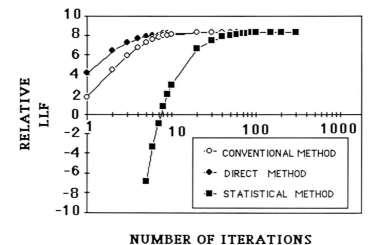

Fig. 1. Comparison of three methods for incorporating the effect of photon attenuation in the EM algorithm.

algorithm is designed to maximize, can be used as a global measure for evaluating the performance of various methods. Figure 1 illustrates the resulting relative LLF values as a function of the number of iterations for all three methods. The direct method obviously performs better than the conventional method for small numbers of iterations. After approximately 200 iterations, the statistical method reaches slightly higher LLF values than those obtained by the other two methods. Although the statistical method is capable of providing slightly better images, the small gain in LLF appears not to be justified by its much slower convergence. These results suggest that the direct method should be the method of choice.

## 5. ACCELERATION TECHNIQUES

'Geometric' acceleration methods use geometric manipulations or geometric information to reduce the computational time required for the production of an acceptable image. For example, a larger pixel size can be used for the first few iterations to increase computational efficiency in the early stage of the iterative process. As convergence is approached, smaller pixel sizes are then employed to refine detailed structures (Politte, 1983). Another technique that can be incorporated into algorithms that use projection data is to use only a subset of the available projection angles at each iteration, thus achieving a reduction in the computation time (Politte,

1983; Tanaka et al., 1986). Including so called 'side-information' in the iterative process so that unnecessary computations are avoided, or employing initial estimates which are better than the uniform distribution in the sense of greater LLF value, can also improve computational efficiency (Snyder, 1984).

Acceleration techniques that employ an additive correction term for up-dating of estimates at each iteration are classified as the additive methods. In the following discussion, the case of algorithms that use rotationally integrated data is shown as an example. Generalization of this approach to other algorithms is straightforward. Equation (16) can be rewritten as

$$f_k^{(n+1)} = f_k^{(n)} + \Delta f_k^{(n)}, \tag{22}$$

where

$$\Delta f_k^{(n)} = f_k^{(n)} \cdot \{[\sum_j u_{jk}(b_j/(\sum_m u_{jm} f_m^{(n)}))/(\sum_j u_{jk})] - 1\}. \tag{23}$$

A modified algorithm with an additive acceleration scheme can be written in the form

$$f_k^{(n+1)} = f_k^{(n)} + \omega^{(n)} \cdot \Delta f_k^{(n)}, \tag{24}$$

where $\omega^{(n)}$ is an overrelaxation parameter for the $n^{th}$ iteration. The value of this overrelaxation parameter is set to be greater than one. Therefore, the additive correction term (or the step size) at each iteration becomes larger, thereby accelerating the iterative process. It is easy to show that the property of self-normalization is preserved by this accelerated algorithm. The non-negativity constraint is not necessarily guaranteed, however. To avoid violation of the non-negativity constraint, special rules can be designed for selecting appropriate values for the overrelaxation parameter (Vardi et al., 1985; Lewitt and Muehllehner, 1986).

The multiplicative methods use multiplicative correction schemes to increase the rate of convergence at each iteration. Again using the case of algorithms that use rotationally integrated data as an example, one can rewrite Eq. (16) in a multiplicative form as

$$f_k^{(n+1)} = f_k^{(n)} \cdot C_k^{(n)}, \tag{25}$$

where

$$C_k^{(n)} = [\sum_j u_{jk}(b_j/(\sum_m u_{jm} f_m^{(n)}))]/(\sum_j u_{jk}). \tag{26}$$

In a multiplicative acceleration algorithm, the correction term can be amplified at each iteration by a power parameter, r, in the form of

$$f_k(n+1) = f_k(n) \cdot (C_k(n))^r. \qquad (27)$$

Since the power factor will preserve the positive polarity of the correction term, C, the non-negativity constraint is automatically fulfilled. On the other hand, the property of self-normalization may be lost. Therefore, a normalization step has to be included at each iteration for rescaling the entire image, so that the total number of events is kept constant. The actual value of the power parameter, r, used in an accelerated algorithm can be selected only on the basis of empirical trials. Using small r values results in ineffective acceleration, whereas using high r values tends to give unstable convergence. It has been suggested that an r value approximately between 2 and 4 will provide satisfactory results (Vermeulen, 1983; Tanaka et al., 1986).

Computer simulation studies have been performed for evaluation of these acceleration techniques. As an example, results from a study in which multiplicative acceleration schemes were used for reconstruction of a circular disc 5 cm in diameter are shown in Fig. 2. It is obvious that rates of convergence are increased when the accelerated approaches are used. For example, in the case of r=2, the LLF value at the 10th iteration for the multiplicative acceleration approach is approximately equal to the value at the 20th iteration for the non-accelerated approach (i.e., r =1). In the case of r=4, the result from 10 iterations using the accelerated algorithm is roughly equivalent to that produced by the non-accelerated algorithm with 40 iterations.

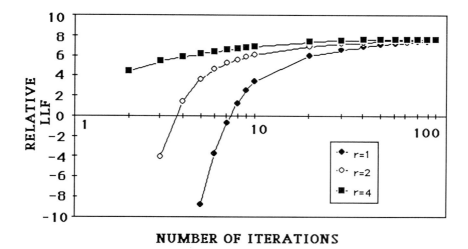

Fig. 2. *Comparison of the non-accelerated EM algorithm (r=1) and two accelerated EM algorithms (r=2,4) that employ the multiplicative method.*

## 6. SUMMARY

Maximum likelihood reconstruction based on the EM algorithm potentially can provide more accurate images in PET and TOFPET. It can be implemented with the use of either the projection data obtained at various angles or of the rotationally integrated image. In the case of TOFPET, the latter approach can reduce the required computation time substantially. Three methods that incorporate the effect of photon attenuation in the EM algorithm have been evaluated. Preliminary results from computer simulation studies indicate that the direct method performs better than the conventional and statistical methods in terms of the balance between reasonable image quality and computation time. Three types of acceleration schemes -- i.e., the geometric, additive and multiplicative methods -- can be employed to increase the rate of convergence of these EM methods. An example was given to demonstrate the effectiveness of the multiplicative acceleration technique.

## REFERENCES

Chen, C.-T. and Metz, C.E. (1985). A simplified EM reconstruction algorithm for TOFPET, *IEEE Trans. Nucl. Sci.* NS-32, pp. 885-888.

Dempster, A.P., Laird, N.M., and Rubin, D.B. (1977). Maximum likelihood from incomplete data via the EM algorithm, *JRSS* 39, pp. 1-38.

Lange, K. and Carson, R. (1984). EM reconstruction algorithms for emission and transmission tomography, *J. Comput. Assist. Tomogr.* 8, pp. 306-312.

Lewitt, R.M. and Muehllehner, G. (1986). Accelerated iterative reconstruction for positron emission tomography based on the EM algorithm for maximum likelihood estimation, *IEEE Trans. Med. Imaging* MI-5, pp. 16-22.

Politte, D.G. (1983). Reconstuction algorithms for time-of-flight assisted positron-emission tomographs, M.S. thesis, Washington University, St. Louis.

Shepp, L.A. and Vardi, Y. (1982). Maximum likelihood reconstruction for emission tomography, *IEEE Trans. Med. Imaging* MI-1, pp. 113-122.

Snyder, D.L. (1984). Utilization of side information in emission tomography, *IEEE Trans. Nucl. Sci.* NS-31, pp. 533-537.

Snyder, D.L. and Politte, D.G. (1983). Image reconstruction from list-mode data in an emission tomography system having time-of-flight measurements, *IEEE Trans. Nucl. Sci.* NS-30, pp. 1843-1849.

Tanaka, E., Nohara, N., Tomitani, T., and Yamamoto, M. (1986). Utilization of non-negativity constraints in reconstruction of emission tomograms. In: *Information Processing in Medical Imaging*, S.L. Bacharach (ed.), Martinus Nijhoff, Dordrecht, pp. 379-393.

Vardi, Y., Shepp, L.A., and Kaufman, L. (1985). A statistical model for positron emission tomography, *J. Amer. Statist. Assoc.* 80, pp. 8-37.

Vermeulen, F. (1983). An improved stochastic reconstruction technique for tomographic imaging. In: *Proc. 2nd Int. Symp. Fundamentals of Tech. Progress in Medicine, Liege*.

# MAXIMUM LIKELIHOOD RECONSTRUCTION FOR SPECT USING MONTE CARLO SIMULATION

Carey E. Floyd, Jr., Stephen H. Manglos,
Ronald J. Jaszczak, R. Edward Coleman

*Duke University Medical Center*
*Durham, USA*

## ABSTRACT

*Reconstructed images for single photon emission computed tomography (SPECT) with quantitative compensation for scatter and attenuation are provided using Inverse Monte Carlo (IMOC): Maximum likelihood estimation with Monte Carlo modeling of the photon interaction and detection probabilities. Quantitative compensation was evaluated by comparing region of interest values for compensated images of line sources scanned in water with line sources scanned in air. Compensation was demonstrated for both $360°$ and $180°$ acquisition. Lesion contrast was investigated for cold spheres in an active background.*

## 1. INTRODUCTION

The source reconstruction problem in single photon emission computed tomography (SPECT) may be viewed as the problem of finding an inverse solution to that photon transport equation which describes the detected photon projection flux. This photon transport equation describes all photon interactions from the point of emission, through all scattering and absorption processes in the body, through all interactions with the detection apparatus, to the time of detection. While a general closed form inverse solution to this complex problem is not obvious (Bell and Glasstone, 1970), a specific solution to the forward photon transport problem with boundary conditions specified by the scattering body geometry and by the parameters and properties of the detection apparatus may be obtained for a specific source distribution using the stochastic estimation techniques of Monte Carlo (Lewis and Miller, 1984; Beck, 1982). Solution of the photon transport problem independently for each element of a normalized complete set of emission volume elements will yield a transport probability matrix. This matrix $T_{ij}$ relates the detected photon flux in a set of discrete detection elements $P_j$ to the source activity in a complete set of discrete basis voxels $S_i$ through the equation

$$T_{ij} S_i = P_j \tag{1}$$

With this mathematical representation of the SPECT problem, reconstruction is achieved through finding a solution to the coupled linear equations.

SPECT imaging is usually count limited, that is, there is a non-negligible uncertainty in the acquired projections due to Poisson fluctuations at low count densities. With the above considerations, the linear system in Eq. (1) will be ill-conditioned and inconsistent, so the solution must be approximated in some sense. We have investigated the behavior of a maximum likelihood estimator using the EM (estimation-maximization) algorithm.

Several researchers have examined the maximum likelihood algorithm for Positron Emission Tomography (PET) (Shepp and Vardi, 1982; Lang and Carson, 1984; Rockmore and Macovski, 1976; Snyder and Politte, 1983; Shepp, Vardi, Ra, Hilial, Cho, 1984; Chen and Metz, 1985; Llacer and Meng, 1985; Llacer, Andrea, Veklerov, Hoffman, 1986; Miller, Snyder, Moore, 1986). The algorithm has been theoretically developed for SPECT (Miller, Snyder, Miller, 1985), has been implemented for experimental SPECT reconstruction both without (Vishampayan, Stamos, Mayans, Kora., Clinthorn, Roger, 1985) and with Monte Carlo techniques (Floyd, Jaszczak, Coleman, 1985; Floyd, Jaszczak, Greer, Coleman, 1986a; Floyd, Jaszczak, Greer, Coleman, 1986b). This algorithm is iterative, uniformly convergent, and provides non-negative images.

Two of the most serious degrading effects in quantitative SPECT reconstructions are scatter and attenuation. Scattered photons have undergone at least one compton scattering before detection. Since the photon is detected along a line which in general does not intersect the original emission site, a simple backprojection will misposition the reconstructed event. Attenuation refers to the removal of photon flux by absorption or by a scattering out of the acceptance angle of the collimators. Both of these processes, scattering and attenuation, introduce inconsistencies in the projections. These inconsistencies may be reduced by including the effects of scatter and attenuation in the system transfer matrix $T_{ij}$ of Eq. (1). The contribution of both of these photon interactions to the T-elements are calculated using Monte Carlo techniques.

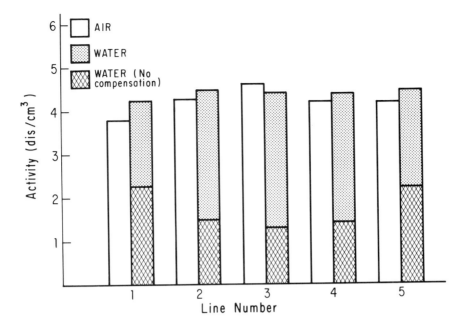

Fig. 1. *Region of interest values for line sources at different radial depths in a cylindrical volume. Open bars to the left were scanned in air. Bars to the right were scanned in water with a) no compensation and b) scatter and attenuation compensation (dotted).*

## 2. METHOD

This preliminary evaluation of IMOC reconstruction for quantitative SPECT imaging will examine two aspects of reconstructed image quality: 1) scatter and attenuation compensation for regions of interest (ROI), and 2) lesion detectability as evaluated by image contrast. All projection data were acquired using the Duke dual headed SPECT system (Greer, Jaszczak, Coleman, 1982) with high resolution collimation and 3.2 mm/pixel sampling. All images were reconstructed onto a grid with equal pixel size (3.2 mm). Photons from the 140 kev emission of Tc-99m were acquired into a 20% energy window (127-154 kev), symmetrically centered at 140 kev. For evaluation, projection data were acquired from a standard SPECT phantom (Data SPECTRUM Corporation, Deluxe SPECT phantom). For the region of interest evaluation, five line sources of activity were placed coaxially in a water filled cylinder of 11 cm radius. The lines were spaced along a

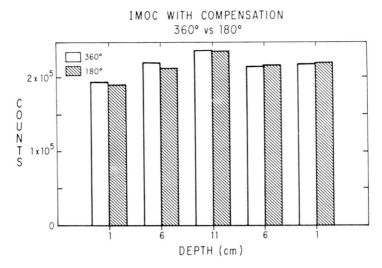

Fig. 2. *Region of interest values for line sources at different depths in a water filled cylinder. Open bars at left for 360° acquisition; hatched bars at right for 180° acquisition.*

diameter of the cylinder 5 cm apart with one located on the axis to provide the depths indicated in the figure. The cylinder was scanned both with and without water to provide projections both with and without the effects due to scatter and attenuation. Compensated reconstructions of the scan in water are compared with reconstructions of the scan in air. For the lesion detectability evaluation, six spheres containing no activity were placed inside a cylinder containing activity. The images reconstructed using IMOC are compared with images reconstructed using Filtered Back-Projection (FBP). The spheres were 3.2 cm, 2.5 cm, 1.9 cm, and 1.6 cm in diameter while the cylinder was 22 cm in diameter. Contrast is calculated as $C = (R - B)/B$ where $R$ is the ROI value in the cold defect and $B$ is the value in the background.

The measurements are expressed as counts per pixel. Thus ideal contrast is represented by -1.0 while no contrast is represented by 0.0. Standard deviations are found for the ROI measurements and are propagated assuming uncorrelated errors.

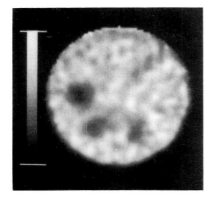 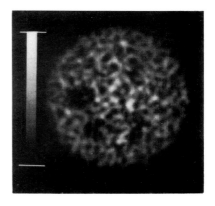

*Fig. 3. Nonactive spheres in active volume, 125,000 counts. IMOC reconstruction with 50 iterations and scatter and attenuation compensation at left. Filtered back-projection with Hanning window cutoff equal to the Nyquist frequency, attenuation and scatter compensation at right.*

IMOC images were reconstructed with 50 iterations. FBP images were reconstructed using a ramp filter with a Hanning frequency window cutoff equal to the Nyquist frequency (Jaszczak, Coleman, Whitehead, 1981). The FBP images were compensated for attenuation using a zero order Chang algorithm (Chang, 1978) with interaction coefficient of 0.15 cm-1 and scatter compensation using the subtraction technique (Jaszczak, Greer, Floyd, Coleman, 1984).

3. RESULTS

The quantitative compensation for scatter and attenuation is demonstrated in Fig. 1 where the region of interest values for the reconstruction in air (no scatter or attenuation) are shown as the open bars to the left. To the right are the reconstructions of the lines in water. The hatched bar shows no compensation while the dotted bar shows compensation for scatter and attenuation. There is good quantitative agreement between the lines in air and the lines in water compensated for scatter and attenuation. Since the scatter and attenuation compensation is an integral component of the reconstruction, compensation is easily achieved for non standard acquisition geometries such as fan beam and cone beam collimation and for less than 360º scans. As shown in Fig. 2, compensation is achieved for 180º

TABLE I
Contrast

Sphere Radius (cm)

|  |  | 3.0 | 2.5 | 1.9 | 1.6 |
|---|---|---|---|---|---|
| 125 K | IMOC | -0.66 (0.12) | -0.51 (0.10) | -0.52 (0.09) | -0.12 (0.10) |
|  | FBP | -0.97 (0.56) | -0.62 (0.44) | -0.56 (0.60) | -0.43 (0.69) |

Table 1. *Contrast measurements with standard deviations in parentheses for nonactive spheres in an active background.*

TABLE 2

Signal to noise (%)

Sphere Radius (cm)

|  |  | 3.0 | 2.5 | 1.9 | 1.6 |
|---|---|---|---|---|---|
| 125 K | IMOC | 10.42 | 7.97 | 8.23 | 1.85 |
|  | FBP | 3.60 | 2.32 | 2.09 | 1.61 |

Table 2. *Signal to noise measurements for nonactive spheres in an active background.*

SPECT acquisition. The bars to the left show results for 360º acquisition while those to the right show equivalent quantitative results for the 180º scan.

Reconstructed images of the cold sphere phantom are shown in Fig. 3 for 125,000 counts. The image to the left in Fig. 3 was reconstructed using 50 iterations of IMOC. The image to the right was reconstructed using FBP with a ramp filter and a Hanning window function with cutoff parameter at the Nyquist frequency. Contrast values with standard deviations are presented for the four largest spheres in table 1. FBP is seen to provide higher contrast but with higher uncertainty. Signal to noise, computed as contrast divided by percent root mean square variation in the background (active) region, is presented in table 2. IMOC is seen to provide better signal to noise than FBP. Visually, the IMOC images have less noise but also have more counts in the cold regions. A true lesion detectability evaluation would seem to require a receiver operator characteristic (ROC) experiment.

## 4. DISCUSSION

Maximum likelihood image estimation has been presented as a reconstruction algorithm for SPECT in an Inverse Monte Carlo technique. Compensation for scatter and attenuation, two of the most significant degrading effects, is provided through Monte Carlo modeling of the system transfer probabilities in the maximum likelihood algorithm. Quantitative compensation was demonstrated for line sources in interacting medium for both 360° as well as 180° acquisition. Lesion detectability was examined through reconstruction of cold spheres in an active medium. Contrast recovery for the IMOC reconstructions at 50 iterations was not as complete as for FBP when scatter compensation was used. Signal to noise values were higher for IMOC however. The question of which technique is more suited for lesion detectability will have to await ROC analysis.

*Acknowledgements*

*The excellent secretarial support of W. Horton is gratefully acknowledged. The aid of K. Greer in acquisition of the data is also acknowledged. This work was supported in part by the National Institutes of Health under Grants CA33541 and CA39251, and by the National Biomedical Simulation Resource, Duke University under USPHS Grant RR01693.*

## REFERENCES

Beck, J.W. (1982). *Analysis of a camera based single photon emission computed tomography (SPECT) system.* PhD Dissertation, 1982 Duke University, (University Microfilms) Ann Arbor, Michigan.

Bell, G. and Glasstone, S. (1970). *Nuclear Reactor Theory*, Van-Nostrand Reinhold.

Chang, L.T. (1978). A method for attenuation correction in radionuclide computed tomography, *IEEE Trans. Nucl. Sci.* NS-25, pp. 638-642.

Chen, C.T. and Metz, C.E. (1985). A simplified EM reconstruction algorithm for TOFPET. *IEEE Trans. Nucl. Sci.* NS-32, pp. 885-888.

Floyd, C.E., Jaszczak, R.J., and Coleman, R.E. (1985). Inverse Monte Carlo: a unified reconstruction algorithm for SPECT, *IEEE Trans. Nucl. Sci.* 32, pp. 779.

Floyd, C.E., Jr., Jaszczak, R.J., Greer, K.L., and Coleman, R.E. (1985). Cone beam collimation for SPECT: simulation and reconstruction, *IEEE Trans. Nucl. Sci.* NS-33, pp. 511-514.

Floyd, C.E., Jaszczak, R.J., Greer, K.L., and Coleman, R.E. (1986). Inverse Monte Carlo as a unified reconstruction algorithm for ECT, *J. Nucl. Med.* 27, pp. 1577-1585.

Greer, K.L., Jaszczak, R.J., and Coleman, R.E. (1982). An overview of a camera-based SPECT system, *Med. Phys.* 9, pp. 455-463.

Jaszczak, R.J., Coleman, R.E., and Whitehead, F.R. (1981). Physical factors affecting quantitative measurements using camera-based single photon emission computed tomography (SPECT), *IEEE Trans. Nucl. Sci.* NS-28, pp. 69-80.

Jaszczak, R.J., Greer, K.L., Floyd, Jr., C.E., Harris, C.C., and Coleman, R.E. (1984). Improved SPECT quantitation using compensation for scattered photons, *J. Nucl. Med.* 25, pp. 893-900.

Lange, K. and Carson, R. (1984). EM reconstruction algorithms for emission and transmission tomography, *J. Comput. Assist. Tomogr.* 8, pp. 306-316.

Lewis, E.E. and Miller, W.F. (1984). *Computational Methods of Neutron Transport*, John Wiley and Sons, New York.

Llacer, J., Andreae, S., Veklerov, E., and Hoffmann, E.J. (1986). Towards a practical implementation of the MLE algorithm for positron emission tomography, *IEEE Trans. Nucl. Sci.* 33, pp. 468.

Llacer, J. and Meng, J.D. (1985). Matrix-based image reconstruction methods for tomography, *IEEE Trans. Nucl. Sci.* 32, pp. 855.

Miller, M.I., Snyder, D.L., and Miller, T.R. (1985). Maximum-likelihood reconstruction for single-photon emission computed tomography, *IEEE Trans. Nucl. Sci.* NS-32, pp. 769-778.

Miller, M.I., Snyder, D.L., and Moore, S.M. (1986). An evaluation of the use of sieves for producing estimates of radioactivity distributions with the EM algorithm for PET, *IEEE Trans. Nucl. Sci.* 33, pp. 492.

Rockmore, A.J. and Macovski, A. (1976). A maximum likelihood approach to emission image reconstruction from projections, *IEEE Trans. Nucl. Sci.* NS-23, pp. 1428-1432.

Shepp, L.A. and Vardi, Y. (1982). Maximum likelihood reconstruction for emission tomography, *IEEE Trans. Med. Im.* MI-1, pp. 113-121.

Shepp, L.A., Vardi, Y., Ra, J.B., Hilal, S.K., and Cho, Z.H. (1984). Maximum likelihood PET with real data, *IEEE Trans. Nucl. Sci.* NS-31, pp. 910-913.

Snyder, D.L. and Politte, D.G. (1983). Image reconstruction from listmode data in an emission tomography system having time-of-flight measurements, *IEEE Trans. Nucl. Sci.* NS-30, pp. 1843-1849.

Vishampayan, S., Stamos, J., Mayans, R., Koral, K., Clinthorne, N., and Rogers, W.L. (1985). Maximum-likelihood image reconstruction for SPECT, *J. Nucl. Med.* 26, pp. 20.

X-RAY CODED SOURCE TOMOSYNTHESIS

I.E. Magnin

*National Institute of Applied Sciences
Lyon, France*

ABSTRACT

*We consider the problem of reconstructing a 3-D object from its 2-D coded radiograph. A new approach to the solution of the problem is presented. The proposed method consists of computing a set of optimal decoding functions using the Kaczmarz algebraic iterative algorithm. To this end, an approximately space-invariant '3-D standard response' is introduced which can be used to characterize any coded source imaging system. Each decoding function corresponds to a specific depth plane inside the object to be reconstructed. The result is a set of 2-D tomograms, each of which is obtained by correlating the coded radiograph with the corresponding decoding function. Two ways of computing the decoding functions are discussed: (i) considering only a single object slice; (ii) treating several immediately adjacent slices (possibly all of them) simultaneously. The proposed reconstruction method can be used for any planar arrangement of discrete sources and is thus capable of comparing the performance of various source point distributions. It is shown that a nine redundant source code provides for better reconstructions than a twelve circular array and a twelve nonredundant array (of same inertia). Finally, the reconstruction of a simulated five planes object using the nine redundant array code is presented.*

## 1. INTRODUCTION

X-ray coded source imaging (e.g. Grant, 1972; Weiss *et al.*, 1979), as well as coded aperture imaging (e.g. Barrett and Swindell, 1981 and for medical applications Ohyama *et al.*, 1984; Van Giessen *et al.*, 1986), permits the representation of the internal structure of an object by an arbitrary number of tomosynthetic slices. A first advantage of the X-ray coded source imaging technique is that a great number of tomograms of the object can be reconstructed from only one 2-D coded radiograph. This property demonstrates the superiority of this new technique over conventional tomography for which each object slice to be reconstructed needs the recording of a new set of projections. A second advantage is that an object can be reconstructed even if its coded radiograph is obtained over a very limited angle of view. The purpose of this paper is to present an original tomosynthesis method able to perform the reconstruction of a 3-D object from its 2-D coded radiograph. The result is a set of 2-D tomograms. In addition, no special autocorrelation properties are required for the source code. Any planar array of discrete X-ray sources such as circular, coherent, redundant, non-redundant or uniformly redundant may be used to provide the coded radiograph of the studied object.

## 2. THE 3-D STANDARD RESPONSE OF A CODED IMAGING SYSTEM

It is necessary to make a set of simplifying hypotheses concerning the physical processes involved in obtaining the coded radiograph of an object, if a tractable mathematical model is desired. We assume straight line propagation of the X-rays, linear interaction with the 3-D object density and we ignore the wavelength-dependent absorption. The statistical nature of the X-ray emission is disregarded and the intensity $I_0(r_s)$ emitted by each source of the code is taken as a constant (of unit value). Any planar array $C(r_s)$ containing m identical punctual sources can therefore be modelled by

$$C(r_s) = \sum_{i=1}^{m} I_0(r_s) \delta(r_s - r_{si}) = \sum_{i=1}^{m} \delta(r_s - r_{si}) \qquad (1)$$

where

$$r_s = (x_s, y_s, z_s) \equiv (x_s, y_s) \qquad \text{source coordinates}$$

$$r_{si} = (x_{si}, y_{si}, z_s) \equiv (x_{si}, y_{si}) \qquad \text{source coordinates of the source i.}$$

Because of the plane parallel geometry (Fig. 1) the z-component of all position vectors is constant in one plane of interest. Making the assumption of linearity and shift invariance of the imaging system, we define the 2-D point spread function (PSF) associated with an object plane located at $z = z_k$ as follows:

$$p_{\delta k}(r_p) = \sum_{i=1}^{m} \delta(r_p - (1 - \gamma_k)r_0 - \gamma_k r_{si}) \qquad (2)$$

where

$$r_p = (x_p, y_p, z_p) \equiv (x_p, y_p) \qquad \text{projection coordinates}$$

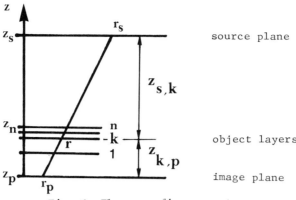

Fig. 1. The recording geometry.

and

$$\gamma_k = -\frac{z_{k,p}}{z_{s,k}}, \qquad \gamma_k \leq 0 \tag{3}$$

is the magnification coefficient related to the depth plane $z = z_k$. The PSF $p_{\delta k}(r_p)$ represents the coded projection of one punctual object of coordinates $r_0 = (x_0, y_0, z_k) \equiv (x_0, y_0)$. When this object belongs to the vertical Oz axis, $r_0 = (0, 0, z_k) \equiv (0, 0)$, the PSF is shifted within the projection plane and becomes

$$\dot{p}_{\delta k}(r_p) = \sum_{i=1}^{m} \delta(r_p - \gamma_k r_{si}). \tag{4}$$

We call $\dot{p}_{\delta k}(r_p)$ the "centered PSF".
Obviously, each depth plane k possesses its own centered PSF $\dot{p}_{\delta k}(r_p)$ which differs from the others by a scaling factor.

### a. Multiplanar object

In order to characterize the general 3-D response of a system in which the source array irradiates a multiplanar object, we introduce a new function $P_s(r_p)$ such that:

$$P_s(r_p) = \sum_{k=1}^{n} \dot{p}_{\delta k}(r_p) = \sum_{k=1}^{n} \sum_{i=1}^{m} \delta(r_p - \gamma_k r_{si}). \tag{5}$$

$P_s(r_p)$ is the sum of the n 2-D centered PSF's given by n punctual objects located on the Oz axis at $z = z_1, z_2, \ldots, z_n$ respectively (Fig. 2a). It can

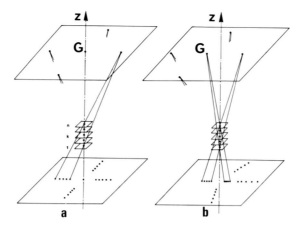

Fig. 2a. The 3-D standard response $P_s(r_p)$ of a coded imaging system irradiating a multiplanar object.
2b. Influence of a shift of the source code on $P_s(r_p)$.

be observed that $P_s(r_p)$ varies when the source code $C(r_s)$ is shifted within its own plane (Fig. 2b). Consequently, to be unique, the 3-D standard response of a given coded source imaging system must be defined for a fixed position of the code within the source plane. We suggest that the center of gravity G of the source array should lie on the vertical Oz axis. In that condition, $P_s(r_p)$ is unique (except for a rotation around the Oz axis) for a given recording geometry.

b. *3-D continuous object*

Let us consider the general case of a 3-D continuous object irradiated by the source code (Fig. 3). We suppose that this object lies inside a sphere of diameter $d = z_n - z_1$, the center of which is on the Oz axis at $z = (z_n - z_1)/2$. In that case, the 3-D standard response of the imaging system can be written:

$$P_s(r_p) = \sum_{i=1}^{m} \int_{\gamma_1}^{\gamma_n} GS_i \, d\gamma \qquad (6)$$

where $S_i$ is the source of coordinate $r_{si}$ and $\gamma_1, \gamma_n$ are the scaling factors related to the object planes $z = z_1$ and $z = z_n$ respectively. As in the previous case where a multiplanar object was considered, $P_s(r_p)$ is unique (except for a rotation around the Oz axis) if the center of gravity G of

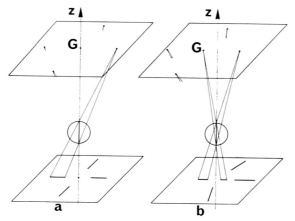

Fig. 3a. The 3-D standard response $P_s(r_p)$ of a coded imaging system irradiating a 3-D continuous object.
3b. Influence of a shift of the source code on $P_s(r_p)$.

the source array is on the Oz axis. The Oz axis is called the "axis of gravity" of the system. It is, by definition, perpendicular to the projection plane.

c. *Experimental situation*

In practical experiments, the axis of gravity of the recording system is generally not exactly perpendicular to the projection plane (Fig. 3b). This implies that the theoretical expression (6) for $P_s(r_p)$ cannot be used to reconstruct an object from its coded radiograph given by the system. In practice, $P_s(r_p)$ is measured experimentally during a calibration step and the recording of the various coded radiographs corresponding to the various objects to be reconstructed are performed afterwards. The 3-D reconstruction of the irradiated objects is then possible.

3. RECONSTRUCTION ALGORITHM

a. *Statement of the problem*

The coded projection of a planar object $\mu_1(r)$ located at $z = z_1$ is ideally given by the convolution equation

$$p_1(r_p) = p_{\delta 1}(r_p) ** \tilde{\mu}_1(r_p) \tag{7}$$

where ** denotes the 2-D convolution operator, $p_{\delta 1}(r_p)$ denotes the PSF of the system associated to the depth plane $z = z_1$ and $\tilde{\mu}_1(r_p)$ is a scaled version (in the projection plane) of the actual object $\mu_1(r)$.

Let us now consider the coded projection of a 3-D object $\mu(r)$. We assume that $\mu(r)$ can be decomposed in a set of n parallel equidistant object layers such that

$$\mu(r) = \bigcup_{k=1}^{n} \mu_k(r). \tag{8}$$

Using linear superposition (of the elementary projections) the actual coded projection of the multiplanar 3-D object can be written as

$$p(r_p) = \sum_{k=1}^{n} p_k(r_p) + N(r_p)$$

$$= \sum_{k=1}^{n} p_{\delta k}(r_p) ** \tilde{\mu}_k(r_p) + N(r_p) \tag{9}$$

where $N(r_p)$ is an additive noise.

The proposed reconstruction method consists of computing a set of optimal decoding functions $p'_k(r_p)$ using the Kaczmarz algebraic iterative algorithm. Each decoding function corresponds to a specific depth plane $z = z_k$ inside the object to be reconstructed. The result is a set of 2-D tomograms, each of which is obtained by correlating the coded radiograph $p(r_p)$ with the corresponding decoding function $p'_k(r_p)$ (see Magnin et al., 1985; Magnin and Goutte, 1985).

These decoding functions differ according to the selected source code. They are independent of the object to be reconstructed.

b. *Reconstruction of the $j^{th}$ object tomogram*

An estimate $\hat{\mu}_j(r_p)$ of the density of the $j^{th}$ object layer can be found by correlating $p(r_p)$ with the corresponding optimal decoding function $p'_j(r_p)$:

$$\hat{\mu}_j(r_p) = p(r_p) \times\times p'_j(r_p) \tag{10}$$

where ×× denotes the 2-D correlation operator. Developing (10) one finds

$$\hat{\mu}_j(r_p) = \tilde{\mu}_j(r_p) ** p_{\delta j}(r_p) \times\times p'_j(r_p)$$
$$+ \sum_{\substack{k=1 \\ k \neq j}}^{n} \tilde{\mu}_k(r_p) ** p_{\delta k}(r_p) \times\times p'_j(r_p) + N(r_p) \times\times p'_j(r_p). \quad (11)$$

A perfect reconstruction of the layer j would exist if the three following conditions could be simultaneously satisfied:

$$p_{\delta j}(r_p) \times\times p'_j(r_p) = \delta(r_p) \quad (12)$$

$$p_{\delta k}(r_p) \times\times p'_j(r_p) = 0 \quad \text{for } k = 1,\ldots,n \quad k \neq j \quad (13)$$

$$N(r_p) \times\times p'_j(r_p) = 0 \quad (14)$$

where $\delta(r_p)$ is the 2-D Dirac impulse.

Theoretically, Eq. (12) represents selecting the "in-focus" j plane for reconstruction and Eq. (13) suppresses all the "out-of-focus" planes. The aim of (14) is to compensate the actual noise. In practice two alternative ways of computing the decoding function $p'_j(r_p)$ are discussed:

*Method 1*

In this approach, the expressions (13) and (14) are implicitly considered as a global noise including the various out-of-focus planes and the optimal decoding function $p'_j(r_p)$ is obtained by solving Eq. (12).
However, $p'_j(r_p)$ must guarantee a good signal-to-noise ratio in the reconstruction $\hat{\mu}_j(r_p)$ of the $j^{th}$ object layer $\mu_j(r)$. That means that $p'_j(r_p)$ must implicitly satisfy Eqs. (13) and (14) to a good approximation. The statistical properties of the noise being fixed, the variance of $p'_j(r_p)$ must therefore be as low as possible.
The solution proposed here is to solve equation (12) using the iterative algebraic algorithm of Kaczmarz, knowing that for an initial value of $p'_j(r_p)$ equal to zero, and without any constraint added, the method converges to a solution with minimum variance (see Durand, 1972).

*Method 2*

In this approach, Eqs. (12) and (13) are solved simultaneously using the iterative Kaczmarz algorithm. It can be noticed, however, that the dimensionality of the linear system to solve grows fast when the number of out-of-focus object planes, which are actually considered, increases. It seems difficult to estimate, a priori, the improvement that is given when $1, 2, \ldots$ or even all the object planes are explicitly considered. In order to get some idea concerning the precision of the reconstruction of a known object and to try and evaluate the performance of method 2, a computer simulation has been conducted.

## 4. RESULTS OF COMPUTER SIMULATIONS

*a. Influence of the out-of-focus planes*

Let us be given a coded imaging system with a nine redundant source code. The object to be reconstructed consists of 5 points located on the Oz axis at $z = z_1, z_2, \ldots, z_5$ respectively. The object plane number 5 ($z = z_5$) is reconstructed according to method 1 and method 2. Fig. 4 shows the mean square error between the actual object (one point) and the reconstructed object using method 1 and method 2 (where all the object planes are explicitly considered), versus the number of iterations necessary to compute the corresponding optimal decoding functions $p'^{0}_{5}(r_p)$ and $p'^{4}_{5}(r_p)$. It can be noticed (Fig. 4) that the quality of the reconstructed tomogram is improved when taking into account all the object slices. The curves corresponding to considering 1, 2, or 3 neighbouring slices in method 2 lie in between the curves 1 and 2, in such a way that the distance between the actual and the reconstructed objects decreases when taking into account more neighbouring slices.

*b. Comparison of various source arrays*

In a second experiment, the same object point is reconstructed for three various source codes (Fig. 5): a redundant array (9 sources), a non-redundant array (12 sources) and a circular array (12 sources). The inertia of these three selected arrays is nearly the same and the recording geometry

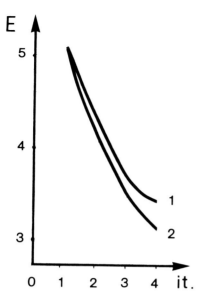
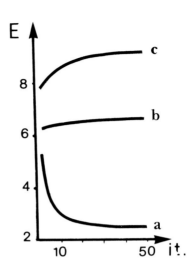

Fig. 4. Mean square error between the actual object (plane 5) and the reconstructed object (arbitrary units), versus the number of iterations necessary to compute the decoding functions. The source code is a 9 redundant array.
curve 1 : method 1
curve 2 : method 2 (all object planes are explicitly considered).

Fig. 5. Mean square error between the actual object (plane 5) and the reconstructed object (arbitrary units), versus the number of iterations necessary to compute the decoding functions, using method 2.
a. 9 redundant array
b. 12 circular array
c. 12 non-redundant array.

of the system remains unchanged. The best results (according to the mean square error criterion) is provided by the nine redundant array (see Fig. 5). The aim of this computer simulation is to show that the comparison of the tomographic capabilities of various source arrays is now possible owing to this new reconstruction method. The study is continuing; no final result concerning the "best source code" can be yet given.

## c. Reconstruction of a heart-like object

The next example illustrates the reconstruction method in the case of a five planes object representing a crude approximation of the heart.

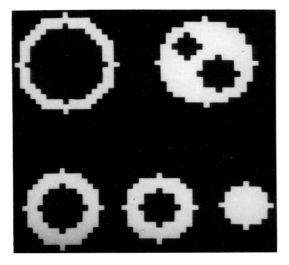

Fig. 6. The five object layers of the multiplanar object.

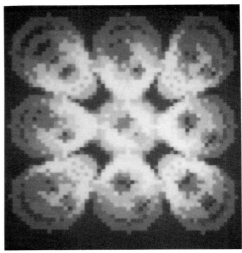

Fig. 7. The simulated coded radiograph obtained with a 9 redundant source code irradiating the multiplanar object.

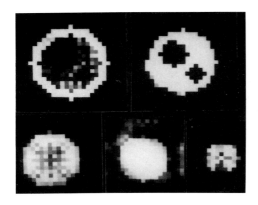

Fig. 8. Reconstruction of the multiplanar object of Fig. 6 (noise free case).

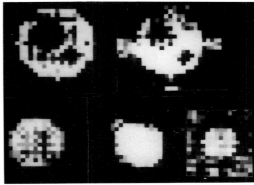

Fig. 9. Reconstruction of the multiplanar object of Fig. 6 (noisy case).

Fig. 6 shows the five layers of the object and Fig. 7 the noise free coded radiograph obtained with a nine redundant square array.

Fig. 8 presents the five reconstructed object layers. The quality of the reconstruction is satisfactory. Fig. 9 presents the five reconstructed object layers when the simulated coded radiograph is degraded by added film grain noise (signal-to-noise ratio: 9,4 dB).

*Acknowledgements*

*The author would like to thank Pr. R. Goutte and Pr. M. Amiel for their helpful discussions and suggestions during the progress of this work. In the same way, the constructive comments and discussions with Dr. M. Viergever were greatly appreciated.*
*This work was in part supported by the Institut National de la Santé et de la Recherche Médicale (INSERM).*


REFERENCES

Barrett, H.H. and Swindell, W. (1981). *Radiological imaging*, Academic Press, New York.
Durand, E. (1972). *Solutions numériques des équations algébriques*, Masson Ed., Paris.
Grant, D.G. (1972). Tomosynthesis: 3-D radiographic imaging technique, *IEEE Trans. Biomed. Eng.* 19, pp. 20-28.
Magnin, I.E., Peyrin, F.C., and Amiel, M. (1985). Reconstruction of object by iterative coded-source image deconvolution, *Signal Proc.* 8, pp. 153-162.
Magnin, I.E. and Goutte, R. (1985). Three dimensional reconstruction in coded-source imaging, *Proc. of the IEEE*, Els. Sc. Pub. Madrid, pp. 221-225.
Ohyama, N., Endo, T., Honda, T., Tsujiuchi, J., Matumoto, T., Linuma, T.A., and Ishimatsu, K. (1984). Coded aperture imaging system for reconstructing tomograms of human myocardium, *Appl. Opt.* 23, pp. 3168-3173.
Van Giessen, J.W., Viergever, M.A., De Graaf, C.N., and Dane, H.J. (1986). Time-coded aperture tomography: experimental results, *IEEE Trans. Med. Im.* 5, pp. 222-228.
Weiss, H., Klotz, E., and Linde, R. (1979). Flashing tomosynthesis: three dimensional x-ray imaging, *Acta El.* 22, pp. 41-50.

SOME MATHEMATICAL ASPECTS OF ELECTRICAL IMPEDANCE TOMOGRAPHY

W.R.Breckon, M.K.Pidcock

*Oxford Polytechnic*
*United Kingdom*

ABSTRACT

*The reconstruction problem for electrical impedance tomography can be formulated as a non-linear inverse problem. We suggest that the problem might be solved numerically using a regularized Newton's method and study the ill-posedness of the linearized problem by numerically calculating the singular value decomposition. For the particular case of a two dimensional disc with pairs of electrodes driven the problem is found to be extremely ill-posed. At realistic signal-to-noise ratios the ill-posedness is reasonably independent of the angular separation between the drive electrodes.*

1. INTRODUCTION

Electrical Impedance Tomography, as a technique for medical imaging, has reached, one might say, its early adolescence of development. Several groups of electrical engineers around the world have developed prototype impedance tomography apparatus and at least one such prototype is shortly to undergo clinical trials. The image reconstruction algorithms used in these machines are fairly crude; based largely on back-projection techniques borrowed from X-Ray CT, they ignore the inherent non-linearity of the electrical impedance problem. Despite their simplicity these algorithms produce useful medical images and it is expected that impedance tomography will prove a valuable clinical tool. Amongst its proposed uses is the diagnosis of cerebral haemorrhage in premature infants.

2. PHYSICAL THEORY

We will assume for simplicity that the biological tissue is an isotropic, Ohmic conductor and that reactance can be neglected. For the validity of these assumptions we refer to Brown (1983). The tissue is probed by applying a small alternating current via an arrangement of electodes in contact with the skin. We will represent the current density vector field by **J** and the conductivity of the tissue by $\sigma$. The electric potential field inside the body will be denoted by $\Phi$. With these assumptions, Ohm's law gives:

$$\mathbf{J} = -\sigma \nabla \Phi. \qquad (1)$$

In the absence of current sources within the body we have

$$\nabla \cdot \mathbf{J} = 0. \qquad (2)$$

The equation relating $\sigma$ and $\Phi$ is therefore the elliptic partial differential equation

$$\nabla \cdot \sigma \nabla \Phi = \frac{\partial \sigma}{\partial x} \frac{\partial \Phi}{\partial x} + \frac{\partial \sigma}{\partial y} \frac{\partial \Phi}{\partial y} + \frac{\partial \sigma}{\partial z} \frac{\partial \Phi}{\partial z} = 0. \qquad (3)$$

If the region of interest is some domain D with boundary $\partial D$ then Eq. (3) holds in the interior of D. In this equation we know neither $\sigma$ nor $\Phi$. The only physical data that we can measure are currents and voltages on the boundary.


3. THE SHEFFIELD APPROACH

In the apparatus developed by Barber, Brown and Seager (1985) (at the Royal Hallamshire Hospital, Sheffield) a typical arrangement is to have a number of electrodes on the skin of the subject. A current from a constant current source is passed betweeen each pair of adjacent electrodes and the voltage difference between other adjacent pairs is measured. This process is controlled by a multiplexer and the measurements are passed to a microcomputer which reconstructs the conductivity distribution in the region between the electrodes and presents a grey-scale picture on a screen.

By analogy with X-Ray CT, each pair of drive electrodes defines a 'projection'. However X-rays propagate in straight lines whereas electric current streamlines are curved, and their shape is dependent on the conductivity. The approach of Barber and Brown is to back-project the voltage differences along iso-potentials and then to take a weighted average over each drive pair. In this algorithm the reconstructed approximation to the conductivity distribution is only a linear function of the boundary measurements. Moreover the linear mapping used is far from the best linear approximation to the true reconstruction mapping.


4.  MATHEMATICAL BACKGROUND

Mathematically the problem of reconstructing the conductivity amounts to identifying the unknown coefficient $\sigma$ in the elliptic partial differential equation (3) from a knowledge of pairs (j,v) of Neumann boundary conditions j (the current density on the boundary) and Dirichlet data v (the potential on the boundary). Recent theoretical work of Kohn & Vogelius (1984a, 1984b) and of Sylvester & Uhlmann (1986a, 1987), has shown the theoretical possibility of distinguishing between any (sufficiently reasonably behaved) conductivities by means of boundary measurement.

While these theoretical results are reassuring, they fall short of answering the more practical question of defining which conducitivity distributions can be distinguished using a given set of possible boundary current distributions. This question still remains largely open although Friedman & Gustafsson (1986) have addressed the problem of what can be determined by just one pair of boundary measurements. They conclude that in a region of uniform conductivity 1 the shape of a (star-shaped) body with conductivity 1+k (where k is a known constant) can be determined by one pair of boundary measurements.


5.  RELATION WITH SCATTERING THEORY

It is interesting to note that Eq. (3) bears a simple relationship to the Schrödinger equation. If $\Psi = \sigma^{1/2}\Phi$ and

$$\eta = (\nabla^2 \log \sigma)/2 + (\nabla \log \sigma \bullet \nabla \log \sigma)/4 \qquad (4)$$

then Eq. (3) implies

$$\nabla^2 \Psi = \eta \Psi. \qquad (5)$$

It is difficult to ascribe any physical significance to the quantities $\Psi$ and $\eta$ but the relationship between Eq. (3) and Eq. (5) is certainly of theoretical interest. Indeed this relationship is used in the proof of the main theorem in Sylvester and Uhlmann (1987). A difficulty, however, in applying any inverse scattering algorithm to the identification of an unknown conductivity would be that once $\eta$ has been calculated one must then solve the non-linear partial differential equation (4) to find $\log \sigma$ and hence $\sigma$ - a process which would be comparatively costly in computer time.

## 6. NUMERICAL ALGORITHMS

The problem of reconstructing $\sigma$ from boundary measurements is a non-linear inverse problem. Like most inverse problems it is not well-posed, that is the solution is extremely sensitive to noise in the data.

To cope with the non-linearity of the problem, an iterative algorithm such as Newton's method can be used. In this method the non-linear operator is first approximated by its linearization, or Fréchet derivative, about some initial guess for the conductivity. The resulting system of linear equations is solved to give an improved estimate for the conductivity and the process is repeated until sufficient accuracy is obtained.

Suppose that a current distribution $j^M$ is applied to the boundary $\partial D$ of the region via a pair of electrodes $M_1$ and $M_2$. Let $\Phi^M$ be the resulting potential which satisfies Eq. (3) in D and the Neumann condition $-\sigma \partial \Phi^M/\partial \nu = j^M$ on $\partial D$, where $\nu$ is the outward pointing normal vector. If another current distribution $j^E$ applied via some different pair of electrodes, $E_1$ and $E_2$, results in a potential which satisfies Eq. (3) in D and the Neumann condition $-\sigma \partial \Phi^E/\partial \nu = j^E$ on $\partial D$, then from the divergence theorem we have

$$\int_D \sigma \nabla \Phi^M \cdot \nabla \Phi^E \, dV = - \int_{\partial D} v j^E \, ds \tag{6}$$

where $v$ is the restriction of $\Phi^M$ to $\partial D$.

We will think of $E_1$ and $E_2$ as the drive electrodes and $M_1$ and $M_2$ as measurement electrodes. In practice no current is passed through $M_1$ and $M_2$ and $\Phi^M$ can be thought of as a 'lead field' in the sense of Geselowitz (1971). Note that if $j^M$ is uniform over $M_1$ and $M_2$ and the total current is unity, then the right hand side of Eq. (6) is simply the voltage difference between the measurement electrodes.

It has been shown by Calderon (1980) that if $\sigma$ is changed to $\sigma + \Delta\sigma$ the resulting change in voltage $\Delta v$ is related to $\Delta\sigma$ by

$$\int_D \Delta v j^M \, ds = -\int_{\partial D} \Delta\sigma \, \nabla\Phi^M \cdot \nabla\Phi^E \, dv + O((\Delta\sigma)^2). \qquad (7)$$

The first term on the right hand side of Eq. (5) is the Fréchet derivative (of the mapping we are trying to invert) applied to $\sigma$. Armed with this expression we can state the algorithm for Newton's method.

Let $\sigma_0$ be an initial guess for the conductivity. Given the $n^{th}$ approximation $\sigma_n$ we solve the Neumann problems

$$\nabla \cdot \sigma_n \nabla \Phi^E_n = 0 \qquad \text{in } D$$
$$-\sigma_n \partial \Phi^E_n / \partial \nu = j^E \qquad \text{on } \partial D \qquad (8)$$

for $\Phi^E_n$ and

$$\nabla \cdot \sigma_n \nabla \Phi^M_n = 0 \qquad \text{in } D$$
$$-\sigma_n \partial \Phi^M_n / \partial \nu = j^M \qquad \text{on } \partial D \qquad (9)$$

for $\Phi^E_n$. We then try to find a correction $\Delta\sigma_n$ which solves the linear equation

$$\int_D \Delta v_n j^M \, ds = -\int_{\partial D} \Delta\sigma_n \nabla\Phi^M_n \cdot \nabla\Phi^E_n \, dv \qquad (10)$$

for each pair of measurement electrodes and each pair of drive electrodes.

Here $\Delta v_n$ is the difference between $\Phi^E_n$ and the measured $v$ on $\partial D$. This correction is used to update the estimate of the conductivity:

$$\sigma_{n+1} = \sigma_n + \Delta\sigma_n. \qquad (11)$$

To implement this algorithm one can calculate $\Phi^E_n$ and $\Phi^M_n$ using the finite element method. The region D is divided into a number of elements and the potentials and conductivity are approximated by polynomials within each of these. A discretized version of Eq. (3) is then solved to find the potentials given the various boundary conditions.

The difficult part is solving the system of equations represented by Eq. (10). To represent Eq. (10) in a more convenient form, let **b** be the vector of differences in boundary voltages (the left hand side in Eq. (10)), let **s** be the vector representing the finite element approximation to $\Delta\sigma$ and let **A** be the discretized version of the integral operator on the right hand side of Eq. (10). The system of equations represented by Eq. (6) is now

$$\mathbf{As} = \mathbf{b}. \qquad (12)$$

Since **A** may not be square we consider the 'least squares normal equation'

$$\mathbf{A}^T\mathbf{As} = \mathbf{A}^T\mathbf{b}. \qquad (13)$$

However, since **A** is the matrix of a first kind integral operator this equation is not well posed. A solution **s** calculated from this equation would be extremely sensitive to noise and some regularization procedure is necessary.

## 7. THE RANK OF THE MATRIX **A**

In this and the remaining sections we consider the particular case of a disc-shaped region D with n equally spaced identical electrodes (see Fig. 1). Although this hypothesis is unnecessary we feel that it will make the exposition clearer.

Let us suppose that we drive a current through electrode $i$ to electrode $i+\alpha$, for $1 \leq i \leq n$. We call $\alpha$ the offset. We can then measure the voltage difference $v$

between electrodes i+j and i+j+1, 1≤j≤n (for convenience we will call electrode n+k, 'electrode k', for any k). These $n^2$ voltage measurements are not linearly independent: two types of dependency exist in all cases except for the special case n = 2m, α = m which we shall consider below.

If $V_{ij}$ is the absolute voltage at electrode i+j, with respect to some fixed reference potential, then $v_{ij} = V_{ij} - V_{i,j+1}$ and we have

$$\sum_{j=1}^{n} v_{ij} = V_{i1} - V_{i2} + V_{i2} - V_{i3} + \ldots + V_{in} - V_{i,n+1} = 0 \qquad (14)$$

for 1≤i≤n. This gives n independent linear equations relating the voltage measurements.

The second type of linear relaton comes from the Reciprocity Theorem (see Gesolowitz (1971)). This principle says: suppose that a current $I_1$ is passed between a pair of electrodes $A_1$ and $B_1$ and the voltage $V_1$ is measured between an identical pair of electrodes $A_2$ and $B_2$. Then if a current $I_2$ is applied to $A_2$ and $B_2$ the voltage $V_2$ between $A_1$ and $B_1$ satisfies

$$V_1 I_2 = V_2 I_1. \qquad (15)$$

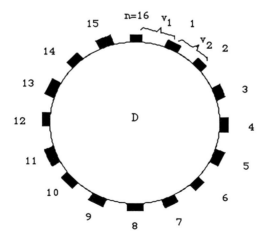

Fig. 1. The case where D is a two dimensional disc with sixteen equally spaced electrodes.

It follows that if we assume that we have identical electrodes, and that a constant current source is used across the drive electrodes we have

$$\sum_{k=0}^{\alpha} v_{i,j+k} - v_{j,i+k} = 0. \tag{16}$$

This is because the voltage between electrode j and j+α is the sum of the voltage differences between the intermediate adjacent electrode pairs. For $1 \leq i < j \leq n$ Eq. (16) are independent of each other giving $n(n-1)/2$ equations.

The $n^2$ possible measurements satisfy, therefore, $n + n(n-1)/2$ relations leaving

$$n^2 - n - n(n-1)/2 = n(n-1)/2 \tag{17}$$

independent readings, in particular 120 for the example of Fig. 1.

An interesting exception to this is the particular case where $n = 2m$ is an even number and one is driving opposite electrodes - that is $\alpha = m$. In this case we have the symmetry relation $v_{ij} = -v_{i+m,j+m}$ for each i and j. Now amongst the $n^2$ variables we have $n^2/2$ relations of symmetry, m sum relations Eq. (14) and only $m(m-1)/2$ reciprocity relations Eq. (16) which are independent of each other. This is because Eq. (16) is equivalent to the same equation with i,j replaced by i+m, j+m. The total number of independent measurements is

$$n^2 - n^2/2 - m - m(m-1)/2 = m(3m-1)/2. \tag{18}$$

For the example of Fig. 1 this works out to be 92. The number of independent voltage difference measurements is the rank of the matrix **A** providing the discretization of the conductivity allows sufficient variation.

## 8. SINGULAR VALUE DECOMPOSITION

To study the ill-conditioning of **A** we introduce the singular value decomposition

$$A = Q\Lambda P^T \tag{19}$$

where $P$ and $Q$ are orthogonal matrices such that $AP = Q\Lambda$ and $A^TQ = P\Lambda$ where

$$\Lambda = \text{diag}(\lambda_1, \lambda_2, \ldots) \tag{20}$$

The numbers $\lambda_i$ satisfy $\lambda_1 \geq \lambda_2 \geq \ldots \geq 0$ and are termed singular values. The columns of $P$ and $Q$ are called the singular vectors of $A$. Equation (13) is equivalent to

$$\Lambda^2 P^T s = P^T A^T b \tag{21}$$

that is, when we express $s$ and $A^Tb$ in terms of the singular vectors (the columns of $P$) we have diagonalized the matrix.

We can use these singular values to study the rank and ill-conditioning of the matrix. As the singular values $\lambda_i$ decrease, the component of $s$ in the direction of the $i^{th}$ singular vector becomes increasingly hard to identify in the presence of noise.

If the matrix $A$ should algebraically have less than full rank, this is displayed by the numerically calculated singular values suddenly decreasing for $i > \text{rank } A$. The fact that they are not zero is an indication of numerical noise.


9. NUMERICAL RESULTS

In a numerical experiment a 129-node, 120-element finite element partition of the two dimensional disc was used. The model had 32 nodes on the boundary and each of the 16 electrodes spanned the element side between two adjacent nodes. The matrix $A$ was calculated for a number of different values of the drive offset $\alpha$. The singular values of $A$ were then calculated and $\log_{10} \lambda_i/\lambda_1$ was plotted against i.

Both the phenomena mentioned above were displayed here (see Fig. 2). The singular values decay rapidly in all cases. All fall to zero after 120 as

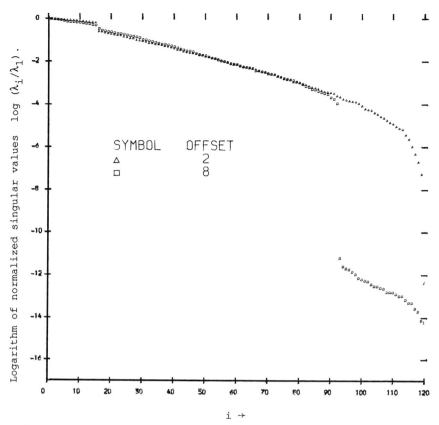

Fig. 2. The normalized singular values of the linearized problem in decreasing order plotted on a logarithmic scale. The curves are for offsets 2 and 8. On this scale, the cases of offset 3 to 7 are virtually indistinguishable from offset 2. Note the decline at the 92nd singular value for the case of offset 8.

predicted (and so could not be plotted on the logarithm scale). For an offset of 8, corresponding to the special case of $n = 2m$, $\alpha = m$, a rapid drop (9 orders of magnitude) is observed as predicted above. The singular values, which have been normalized so the largest is unity, should be compared with the ratio of noise to signal. The singular value decomposition should be truncated after the first k terms where $\lambda_k$ is the first singular value below the noise ratio. It can be seen from Fig. 2 that for realistic noise levels (i.e. more than $10^{-4}$) the singular values are very similar whatever the drive offset and that the symmetry effect of drive offset 8 is irrelevant. We can conclude that the amount of information contained in the data is fairly independent of the offset but that priority is to reduce the ratio of noise to signal.

## 10. CONCLUSIONS

We have shown how the reconstruction problem for Electrical Impedance Tomography can be formulated as a non-linear inverse problem. We have suggested how the inverse problem might be solved numerically using a regularized Newton's method and studied the ill-posedness of the linearized problem by numerically calculating the singular value decomposition. For the particular case of a two dimenstional disk with pairs of electrodes driven, the problem was found to be extremely ill-posed. The angle between the electrodes does not significantly affect the information contained in the data but it is important to reduce the ratio of noise to signal.

*Acknowledgements*

*The finite element program used incorporated ideas and subroutines from the Numerical Algorithms Group Finite Element Library. Our thanks to its authors - in particular Dr C. Greenough of the SERC Rutherford Appleton Laboratory for his help.*

REFERENCES

Barber, D., Brown, B., and Seagar, A. (1985). Applied Potential Tomography: possible clinical applications, *Clin. Phys. P.* 6, pp.109-121.

Brown, B.H. (1983). Tissue impedance methods in: *Imaging with Non-Ionizing Radiation*, D.F. Jackson (ed.), Surrey Univ.Press, pp. 85-110.

Calderon, A.P. (1980). On an inverse boundary value problem. In: *Seminar on Numerical Analysis and its Application to Continuum Mechanics*, Soc.Brasileira de Matematica, Rio de Janeiro, pp. 65-73.

Friedman A. and Gustafsson B. (1987). Identification of the conductivity coefficient in an elliptic equation, submitted for publication.

Gesolowitz, D.B. (1971). An applicaton of electrocardiographic lead theory to impedance plethsmography, IEEE *Trans. Bio-Med. Eng.*, BME-18, pp. 38-41.

Kohn, R.V. and Vogelius, M. (1984a). Determining the conductivity by boundary measurement, *Comm. Pure App. Math.* 37, pp. 289-298.

Kohn, R.V. and Vogelius, M. (1984b). Determining the conductivity by boundary measurement II, interior results, *Comm. Pure App. Math.* 39, pp. 644-667.

Sylvester, J. and Uhlmann, G. (1986). A uniqueness theorem for an inverse boundary value problem in electrical prospection. *Comm. Pure Appl. Math.* 339, pp. 91-112.

Sylvester, J. and Uhlmann G. (1987). A global uniqueness theorem for an inverse boundary value problem, to appear in *Annals of Maths*.

## Section 2.3
# Display and Evaluation

HIERARCHICAL FIGURE-BASED SHAPE DESCRIPTION
FOR MEDICAL IMAGING

Stephen M. Pizer, William R. Oliver,
John M. Gauch, Sandra H. Bloomberg

*University of North Carolina*
*Chapel Hill, USA*

ABSTRACT

*Medical imaging has long needed a good method of shape description, both to quantitate shape and as a step toward object recognition. Despite this need none of the shape description methods to date have been sufficiently general, natural, and noise-insensitive to be useful. We have developed a method that is automatic and appears to have great hope in describing the shape of biological objects in both 2D and 3D.*
*The method produces a shape description in the form of a hierarchy by scale of simple symmetric axis segments. An axis segment that is a child of another has smaller scale and is seen as a branch of its parent. The scale value and parent-child relationship are induced by following the symmetric axis under successive reduction of resolution. The result is a figure- rather than boundary-oriented shape description that has natural segments and is insensitive to noise in the object description.*
*We extend this method to the description of grey-scale images. Thus, model-directed pattern recognition will not require pre-segmentation followed by shape matching but rather will allow shape properties to be included in the segmentation itself.*
*The approach on which this method is based is generally applicable to producing hierarchies by scale. It involves following a relevant feature to annihilation as resolution is reduced, defining the component that is annihilating as a basic subobject, and letting the component into which annihilation takes place become its parent in the hierarchy.*

## 1. OBJECT DEFINITION VIA HIERARCHICAL SHAPE DESCRIPTION

A common task in medical image processing and display is the definition of the pixels or voxels making up a particular anatomic object. With such a definition the object can then be displayed or analyzed, full scene analysis can begin, or parameters of image processing on the object or its image region can be chosen. The weaknesses of common methods of object definition are well-known. Image noise has major effects on the object definition. Separate objects are inadvertently connected when they have similar properties. And the global variation of such features as image intensity across an object undermines definitions based on these features. These weaknesses follow from the locality of decisions on boundary or region specification, and the inability of the methods to take into account global expectations about the objects being defined. Thus, detail due to noise or normal variations interferes with the determination of an object's global properties, and global information cannot be brought to bear until after a tentative segmentation (definition of an object or subobject) has occurred.

To avoid these weaknesses, we are developing methods that
(i) model expected objects hierarchically by scale so that detail is seen as a property of a subobject that does not destroy the description of the object at a larger scale,
(ii) use descriptors that capture global object properties that commonly go under the name *shape*, but at the same time can capture typical intensity properties such as level and profile, and
(iii) operate in a way that does not require finding the object before it can be described.

In our approach, shape is described using the symmetric axis transform (SAT), a descriptor that is global by depending not on the object boundary but on the figure (included pixels or voxels) and that together with a multiresolution approach induces a hierarchical subdivision of an object into meaningful objects and subobjects. Furthermore, we suggest a form of this approach that allows 'intensity shape' to be captured as well. Thus, an object is defined by computing a description of an image or image region and matching that description against a predefined description of the object, possibly together with its environment.

We begin by facing the problem of creating an adequate shape description of an object whose boundary has already been defined and then go on to see how the method can be extended to describe images or image regions defined only by the intensity values of their pixels or voxels. In all cases the method is discussed in two dimensions but applies directly to three dimensions, not slice by slice, but by replacing two dimensional elements (pixels) by three dimensional elements (voxels) and two dimensional distance by three dimensional distance.

After reviewing other work on shape description in Section 2, we review the symmetric axis transform and its properties in Section 3. Then in Section 4 a general approach for generating hierarchies by scale using multiple resolutions is presented, and it is applied to the symmetric axis to produce a hierarchical description of shape. Section 5 covers details of the multi-resolution symmetric axis transform approach such as the means of reducing resolution and of following the axis as resolution is reduced. In Section 6 the relation between the proposed shape description and other methods is discussed, and extensions are suggested. Finally, in Section 7 a method applicable to grey-scale images is presented.


2. SHAPE DESCRIPTIONS


Many methods of shape description have been previously proposed, among them many focusing on the description of boundary curvature (Koenderink, 1985; Zahn and Roskies, 1972; Richards and Hoffman, 1985), some based on a list of somewhat ad hoc features, some focusing on description of deformation from a primordial shape such as an oval (Bookstein *et al.*, 1985; Leyton, 1984; Leyton, 1986b), some focusing on description of the object figure, i.e., the area (or volume in 3D) of the object (Blum and Nagel, 1978; Nackman and Pizer, 1985), and some focusing on both boundary and figure (Brady and Asada, 1984). All except those based on the feature list have difficulty with the effects of noise in the figure or boundary definition and with separating detail from more essential shape characteristics, and some, e.g. the symmetric axis transform, have especial difficulties in this regard. Many produce descriptions of questionable naturalness.

To handle detail and noise naturally, one is led to a representation of the object as a hierarchy of segments at successively smaller levels of scale. For example, a human face might be described in terms of regions and subregions as in Figure 1. An advantage of this approach is that detail that may be noise is relegated to the lower parts of the hierarchy, and if it is to be seen as noise rather than important detail, it can be ignored without disturbing the description at higher levels of scale. Another advantage is that it allows top-down (large scale subobjects first) matching of models and descriptions of data objects. This approach of producing and using a hierarchical description has been taken with attractive results in regard to grey-scale image description by multiresolution methods which focus on

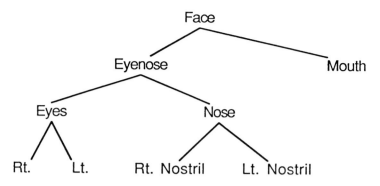

Fig. 1. *A hierarchical description of the human face.*

intensity extrema (Crowley and Sanderson, 1984; Pizer et al., 1986).

The symmetric (or medial) axis transform (SAT) (Blum and Nagel, 1978) has elegant properties in inducing segmentation of shapes into natural components, but its major flaw has been its sensitivity to noise in the boundary or figure specification. A related problem has been that no measure of the closeness of shapes fell out from the SAT, because there was no way to discern how to group parts of the axis into major components or how to measure the importance of a component. In this paper we show how the multi-resolution and symmetric axis transform approaches can be married, producing a noise-insensitive hierarchical shape description with the natural segments that had been the original promise of the symmetric axis transform.

3. THE SYMMETRIC AXIS TRANSFORM

The symmetric or medial axis (SA) of a 2D object is intuitively the set of points within the object figure that are medial between the boundaries. More precisely, the SA is the locus of the centers of all maximal disks in the object, where a maximal disk is a disk entirely contained within the object figure but which is not contained by any other such disk. Figure 2 shows an example.

The SA forms a graph (a tree if the object has no holes). Segmenting the

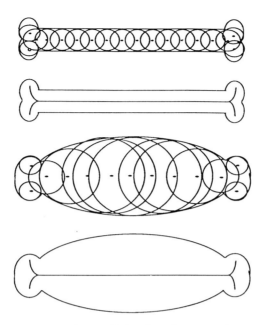

*Fig. 2. Maximal disks in simple objects and the corresponding symmetric axis.*

SA at branch points produces so-called simplified segments. Associated with each simplified segment is a part of the object figure made up of the union of the maximal disks of the points on the simplified segment. The object segmentation thus induced is frequently very natural. For example, in Fig. 2 the bone shape is segmented into a rod and four knobs.

Associated with each point on a simplified segment is the radius of the maximal disk at that point. The SA together with this radius function of position on the axis is called the *symmetric axis transform (SAT)*. The radius function can be analyzed in terms of curvature properties (Blum and Nagel, 1978) which characterize the behavior of the width of the object at that point, e.g., as flaring or cupping. One of the attractive properties of the SAT is that it separates these width properties from the curvature properties of the axis. These two sets of curvature properties can be used to further segment the SA and thus the object.

Nackman and Pizer (1985) have shown how the ideas of the SAT and the associated width and axis curvatures generalize to three dimensions. The 'axis'

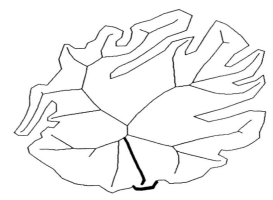

*Fig. 3. Sensitivity of the SA to figure noise in an image of a glomerulus. Note, for example, the large ratio of boundary to axis arc length in the region marked in bold.*

becomes a locus of the centers of maximal balls, which in general is a branching surface. This surface can be subdivided into simplified segments at the branch curves, and again a natural subdivision is frequently produced.

The major weakness of the SAT is its sensitivity to properties of the detail of the object boundary. That is, changes in the figure or its boundary that are small in terms of distance can produce major changes in the SA (see Fig. 3). A boundary feature that has a short arc length may result in a long symmetric axis branch, which moreover distorts the branch to which it is connected. The result is not only that branches that describe only detail are difficult to discern as such but also that major branches are split in such a way that a portion of axis that should naturally be viewed as a unit (a *limb* of the SA tree) is broken into unassociable portions. This weakness of the SAT has been so great as to destroy interest in it despite its otherwise elegant properties. Attempts to ameliorate it by pre-smoothing the boundary or by analysis, after SAT calculation, of properties such as axis arc length to boundary arc length ratios have foundered on the arbitrary thresholds that had to be imposed ("One man's noise is another man's detail."). However, we suggest that the imposition of a scale-based hierarchy on the symmetric axis segments solves these problems and thus allows one to take full advantage of the attractive properties of the SAT of inducing segmentation strongly related to our sense of shape and of separating width curvature from axis curvature.


4. THE SYMMETRIC AXIS BRANCH HIERARCHY

A useful paradigm for creating scale-based hierarchies for describing a complex distribution of components, such as an image or image object, is to find an important component feature that smoothly changes as the underlying distribution is blurred and that annihilates after an appropriate amount of blurring, then becoming part of another component. This approach has been fruitful with grey-level images, where, in a generalization of the pyramid approach, Koenderink (1984) and Pizer *et al.* (1986) have suggested following intensity extrema under blurring until they annihilate. There, the amount of blurring necessary for a particular extremum to annihilate is taken as the scale of the extremum; in the process a region surrounding the extremum is associated with the extremum, producing its *extremal region*; in addition, when upon annihilation an extremum melts into another, the former is associated with the latter as its child in the hierarchy. The result is that the image is described by a tree of extremal regions, each labeled with a scale, where larger scale regions have tree descendants that are smaller scale regions contained by it.

We can apply this paradigm to the problem of describing object shape by focusing on the branches of the symmetric axis. We have found empirically that reasonable methods of smoothing of the object boundary cause the branches of the symmetric axis to change smoothly, such that at certain levels of smoothing a branch will disappear (see Fig. 4). According to the general approach laid out above, we associate with the branch, as a measure of its scale, the amount of resolution reduction necessary to achieve annihilation, and we say that the annihilating branch is a subobject of the branch into which it disappears. This process is continued until only a branch-free SA remains.

Every annihilating branch can be traced back to the part of the SA of the original, unblurred object from which it was smoothly generated. In the case of all but the branches at the frontier of the tree, these SA parts consist of a limb of the original SA, i.e., a sequence of simplified seg-

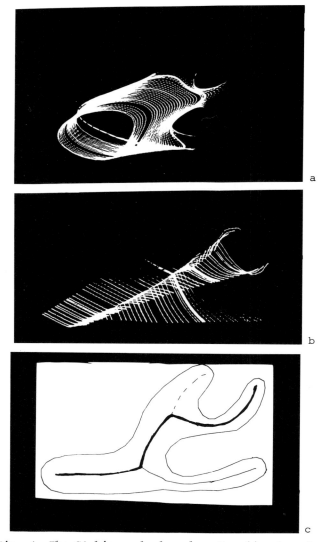

*Fig. 4. The SA hierarchy based on B-spline boundary smoothing a) stack of boundaries, b) stack of SA's, c) induced description.*

ments from which twigs have been removed at the one or more branch points where smaller scale branches that annihilated earlier were attached. The result is that the method has the property of defining naturally associated axis components from the associated segments in the sequence.

The complete multiresolution process defines a tree (hierarchy) of limbs and twigs (axis portions) in the original SA. The root is the portion of the axis to which the final branch-free SA traces back, and descendants of the axis portion at any node are the portions of axis which annihilated into that axis portion. The axis portion at each node is either a single SA simplified segment or a limb made up of a sequence (without branching) of simplified segments.

With every axis portion in the SA tree there is the original radius function on that axis portion. The union of the maximal disks centered at each point on the axis portion and with radii given by the respective radius function value is a subobject associated with the axis portion. The description tree can then be thought of as a tree of subobjects, of decreasing scale (but not necessarily area or volume) as you move down the tree. Each node in the tree (subobject) can be labeled by its scale together with properties describing the width (radius function) curvature and the axis curvature.

These ideas generalize straightforwardly to three dimensions. The axis components at nodes in the tree are simple surfaces, and the subobjects associated with a node are corresponding unions of maximal balls. Axis curvature and width curvature properties, as described by Nackman and Pizer (1985), as well as scale, label each node.

Examples given in Fig. 5 suggest that this method produces natural descriptions. Furthermore, our experience is that objects that we see as similar, such as outlined human skulls viewed laterally, produce similar descriptions. Problems with this description arise from four facts. First, with some types of resolution reduction the topology of the figure and of the axis is not maintained: for example, a simplified segment of SA can split into two, or two can join into one. Second, shape features related to boundary concavities are not directly represented by this approach. In Section 5, where types of resolution reduction are discussed, we show how including other axes of symmetry in the representation seems to handle both of these problems.

The third problem is that the sensitivity of the symmetric axis to small changes in the boundary can cause implementation difficulties in axis segment following. In particular, axis segments that do not exist at one resolution level can be artifactually created, or those that decrease smoothly

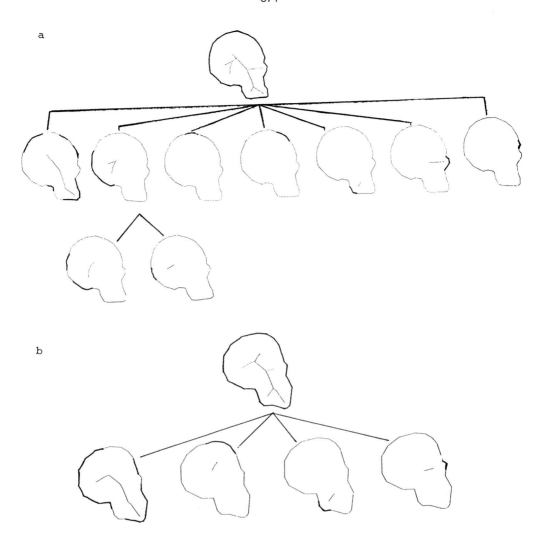

Fig. 5. *Shape descriptions produced by the SA hierarchy. The descriptions above are lateral views of two different human skulls. Each node of a description tree shows a component of the SA and shows in bold the part of the boundary corresponding to that component. The leftmost child of a node is its principal component axis. The other children of that node represent axis complexes branching from the principal component, arranged in order of decreasing scale.*

in size for an underlying smooth boundary are artifactually removed at some resolution levels, only to reappear at later levels. Therefore, following the axis segments across steps is made difficult. This problem is discussed in the next section as well.

Finally, there is one case in which similar objects have dissimilar descriptions. When the two branches emanating from a branch point are similar in scale in that when one annihilates the other is also almost gone, a small change in the scale (length or width) of one of the branches can change which of these two branches annihilates first and thus change which is considered part of the limb and which the attached twig. The tree changes that result from such small object changes are predictable and are discussed further in Section 6.

## 5. RESOLUTION REDUCTION AND SYMMETRIC AXIS FOLLOWING

What method should be used to continuously reduce the object resolution to produce the multiresolution stack of symmetric axes that induce the SAT hierarchy? The natural first thought is to focus on the boundary of the object to be described by applying some smoothing operator to its curvature. The result of such an approach, in which boundary points at one level of resolution were used as control points for a B-spline which forms the boundary at the next level, is shown in Fig. 4. Koenderink and van Doorn (1986) discuss why it is preferable to focus on first blurring the figure and then computing a consequently smoother boundary from the result, rather than directly to smooth the boundary. In essence the argument is that figure properties better capture the global relationships which we call shape than do boundary properties, which are too local. This very argument is the basis of the appeal of the symmetric axis method of shape description over methods based on describing boundary curvature.

Koenderink and van Doorn's (1986) suggestion of the means for figure-oriented resolution reduction starts by treating the figure as a characteristic function, i.e., an image which is 0 outside the figure and 1 inside. They then would convolve the result with an appropriate Gaussian and compute a new figure boundary as a level curve in the result. The authors suggest that the level curve be taken at some fixed intensity, but this requires the choice of some arbitrary intensity, and the choice affects the shape description. Moreover, the approach of choosing a fixed intensity level causes the figure to shrink as resolution is reduced, so that after some amount of resolution reduction the figure disappears entirely. While there are many means of level choice that avoid this disappearance, we take our cue from the accepted definition of shape that it is what is left after normalization for size and orientation. Thus, we normalize to constant size at each amount of resolution reduction, choosing the figure-defining level such that the area (in 2D, volume in 3D) of the figure remains constant.

Resolution reduction based on figure blurring does indeed behave more intuitively correctly than direct boundary smoothing. Furthermore, it does have the additional advantage that it is in principle directly applicable to grey-scale object representations and not just characteristic function representations. However, there are two difficulties with this type of resolution reduction. First, topology is not maintained: connected components can split under blurring, disconnected components can join, indentations in the figure can turn into holes and vice-versa, and holes can disappear (see Fig. 6). Second, with an implementation using a piece-wise linear boundary the smoothness of the SA branch disappearance is more affected

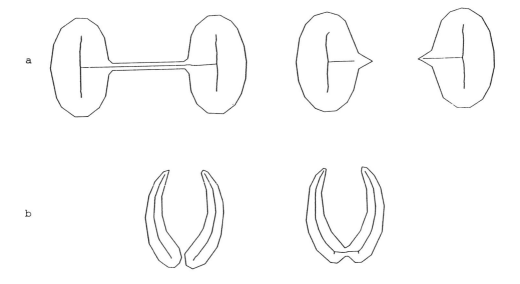

Fig. 6. a) Splitting, and b) Joining of the SA under resolution reduction.

than with direct boundary smoothing. Let us discuss each of these difficulties in turn.

From the point of view that the figure is essential and the boundary is derivative, the non-maintenance of topology is no problem at all. It is easy to argue that the object figures shown in Fig. 6a are indeed close and that it is natural that a decrease in resolution should cause the isthmus between the two disks to be broken (eventually to be rejoined at yet lower resolution). Similarly, the object figures in Fig. 6b should naturally combine as the resolution is lowered; other natural transitions are the closing of two nearby points of land around a bay to form a lake, the melting of a narrow strip between a lake and the sea to form a bay, and the drying up of a lake. We see it as unnatural to insist that topology be maintained under resolution reduction; instead, we must arrange our shape description not to be too sensitive to topology.

On the other hand, we are more disturbed by nonsmooth appearances or disappearances of large pieces of the SA. These appear to happen under some of the changes in topology listed above. For example, when two ellipses, each with a horizontal major axis and one just above the other, are blurred, a vertical segment of SA will appear nonsmoothly as the two ellipses join (see Fig. 7a). We can avert many of these difficulties by including the external symmetric axis or the global SA as part of the SA.

The external SA of a figure is the SA of its complement. If we take the overall SA as the union of the internal and external SA (see Fig. 7b), we find first that with resolution reduction as a piece of internal SA breaks, a corresponding pair of external axis pieces come together, and second that the overall SA now reflects concavities in the figure boundary directly. We therefore suggest that an improved shape description can be obtained by following this overall SA under resolution reduction. Note that as the resolution is reduced the object eventually becomes ovoidal and the external part of the SA becomes null.

The global SA (Blum, 1979) is formed by the locus of the centers of all disks tangent to two or more disconnected regions on the figure boundary. The ordinary (first order) SA and the external SA are subsets of the global SA. We have observed that when a new segment of ordinary SA appears non-

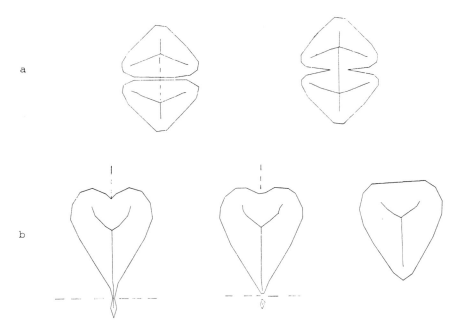

Fig. 7. The change under resolution reduction of a) The ordinary (———) and one other (- - -) component of the global symmetric axis of two nearby joining figures, and b) The internal (———) and external (- - -) SA of two splitting figures.

smoothly as two pieces of figure join or a hole is eliminated, it actually forms smoothly from a piece of global SA that is transformed into ordinary SA. On the other hand, when a hole fills in, the ordinary SA segment loop around the hole does disappear nonsmoothly. Further research is needed to catalogue these transitions and to determine the usefulness of computing the global SA.

The other disadvantage of resolution reduction based on figure blurring, as compared to direct boundary smoothing, is that following SA branches to annihilation is more difficult to implement using a piecewise linear boundary approximation. In our implementation the histogram of the image at each level of resolution is used to find the intensity such that the number of pixels with greater or equal intensity is equal to the original figure area (or volume). Points on the isointensity contour at that intensity are connected by linear segments, with the points selected by the recursive splitting method described in Ballard and Brown (1982, Algorithm 8.1). Our SAT algorithm essentially computes the Voronoi diagram of this collection of line segments and takes the SA to be the curves separating the regions in the Voronoi diagram, less those separating curve segments that touch the object boundary. The latter are removed because they are an artifact caused by using a piecewise linear approximation to a smooth object boundary. The result is an SA that is piecewise made up of linear and parabolic components.

The computed axis depends on the boundary approximation. In particular, small pieces of SA are either artifactually added or omitted depending on the approximation, and the added pieces are not necessarily small compared to SA pieces reflecting the 'true' boundary at the given level of resolution, especially since the 'true' branches get small as resolution is reduced. The artifactual addition or omission of pieces that are unrelated across stages of resolution reduction complicates following the SA branches to annihilation.

This difficulty does not occur as greatly with direct boundary smoothing, as the boundary pieces at one step are related to those at the next. Nevertheless, the fundamental attractiveness of resolution reduction by figure blurring leads us to cope with its difficulty rather than resort to direct boundary smoothing. We believe that we need either to develop a method of figure-blurring-based resolution reduction in which the piecewise linear approximation at one step has related pieces to those in the previous step, or to replace the piecewise linear boundary approximation by a smooth piecewise approximation and develop a method of calculating the SAT for such approximations.

In our results to date, including those shown in Fig. 5, the matching of SA segments from level to level has been done by hand, though the large majority of the matchings can be correctly made by a straightforward algorithm.

## 6. RELATIONS AND EXTENSIONS OF THE MULTIRESOLUTION SAT

Brady and Asada (1984) have defined another form of an axis of symmetry that they call *smoothed local symmetries (SLS)*. They combine boundary curvature and figural properties in their definition. We find the boundary curvature aspects unattractive, as they involve an arbitrary degree of boundary smoothing, but the SLS, defined as the set of smooth loci of centers of chords that have angular symmetry with the boundaries they touch, is a set of axes that have some advantages (and some disadvantages) over the SA. We note here only that the multiresolution approach to inducing a hierarchy on the SA could equally well be applied to the SLS (in our modified definition).

The hierarchical SA or SLS descriptions have a relation to shape descriptions based on boundary curvature, such as the codons of Richards and Hoffman (1985) or the boundary deformations of Leyton (1986b). Codons describe simple convexities or concavities of the boundary. Leyton (1986a) has shown that these simple boundary segments have simple SA segments, i.e., correspond to the outermost branches of the overall SA. That work begins to relate the boundary and figure points of view, and more mathematics in this direction would be valuable.

Leyton's and Richards' methods, as well as others, describe the boundary as a sequence of curvature features or deformations to obtain them, but they frequently have difficulty determining an order of features or deformations to be applied. The multiresolution SAT method described above could be used to induce the order by scale for these methods, if the descriptions produced by latter are deemed superior to those produced by the multiresolution SAT.

Whether the multiresolution SAT is used directly to produce a shape description or as an order-inducing auxiliary to another method of producing a shape description, its weakness of having small changes in the image produce discontinuous change in the description must be dealt with. Recall that this behavior results from a close decision in deciding which of two branches or limbs emanating from a branch point forms the branch and which forms part of the branching limb. This behavior can happen at any branch point at which the two SA branches have almost equal scale. For example, in Fig. 5a the axis piece corresponding to the base of the skull is more prominent, while in Fig. 5b the axis piece corresponding to the back of the skull is more prominent. It seems to be an application-dependent question whether trees differing in one or more of these close decisions should be considered instances of the same shape or of different shapes. In fact, it might be decided that the 'close call' case is a single shape, independent of which way the call goes, while each of the cases, in either direction, of a 'distinct difference' in scale between the respective branches might be called two yet different shapes. Of course, this decision would require an arbitrary threshold to be established between each of the shapes, and it also would require some means to be developed to define the scale of the non-annihilating branch that becomes part of the limb.

An alternative seems to be to let an annihilation cause both the annihilating branch and the 'other' branch at the fork both to be declared as branches. The problem here is which of the (normally) three branches incident to a branch point to declare as the limb and which the two branches. Without a means to make such a declaration, the strength of the method in associating axis segments into a limb is lost. With such a means, one could detect when two branches are nearly equally strong and then produce alternative description trees corresponding to each of these branches annihilating first. Further research is needed on this question.


## 7. THE MULTIRESOLUTION SAT FOR GREY-SCALE IMAGES

Finally, we move on to the problem of creating a shape-based description of intensity-varying images or objects. With a method for creating such a description we can avoid the preliminary step of defining the pixels in the object before the shape is described, so the shape description can contribute to the object definition. Furthermore, the description can reflect the levels and 'shape' of changes in the intensity dimension as well as spatial shape.

A common way to visualize a grey-scale image is as the surface in three dimensions defined by the image intensity function. This has two benefits. First, the 'shape' of the grey-scale image can be described by a three-dimensional shape description method. And second, the familiar tools of differential geometry can be used to study properties of grey-scale images and the shape description.

Our first idea was to describe the image surface using the 3D SAT. This approach has the problem that the intensity dimension is being treated as commensurate with the spatial dimensions; some choice must be made as to what intensity change is equivalent to what spatial distance. Unfortunately, two different choices may result in very different shape descriptions.

Thus, we must design a shape description which treats the intensity and spatial dimensions separately.

To meet this requirement, we look at the image intensity function as a collection of isointensity contours. Since the level curves for the intensity function provide a complete representation of the image, the collection of SAT's of all of these level curves will also provide a complete representation. We describe each isointensity contour with the 2D SA of the region with intensity greater than or equal to that of the contour. Thus the whole image is described by a collection of 2D SA's obtained by varying the intensity defining the contour (see Fig. 8). We then continuously blur the image and follow the annihilation of major components of this pile of SA's to induce a hierarchical description of the image based on a natural subdivision. Let us examine some of the properties of our description.

First, let us examine the connection of SA's from one intensity to the next, at a fixed degree of blurring. Consider functional surfaces which have a finite number of discréte critical points. Any selected intensity level $i$ determines a collection of isointensity contours at that intensity. Except at levels at which critical points occur, the set of contours varies smoothly with $i$. Since an SA varies smoothly with the region it represents, the SA's of the regions corresponding to the two sets of contours are also very similar. For this reason, the set of SA's for all intensities in the (2D) image will form a branching surface in three dimensions. We call this structure the *SA-pile* for a grey-scale image, and we call its branches *SA-sheets*. For a 3D image the SA-pile appears in four dimensions and its sheets form a three-dimensional hyper-surface in this space.

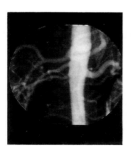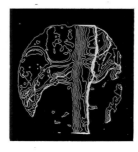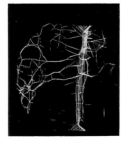

*Fig. 8. A digital subtraction angiogram and corresponding contour pile and SA-pile.*

To visualize the behaviors of SA-piles, view the image as terrain and the regions whose SA's we are computing as horizontal cross-sections through this solid mass. Thus near a maximum a cross-section is a closed region, while near a minimum a cross-section is a region with a hole in it.

For an image with only one smooth oval bright spot (maximum), the SA-pile consists of only one SA-sheet. For images with bright spots of a more complex shape, there may be a branching structure of SA-sheets (very much like the branching structure in the 2D SA for non-oval 2D objects). What about more general images? We know that the critical points of the surface cause catastrophic changes in the isointensity contours. Contours appear and disappear at local maximum points and local minimum points, and contours connect at saddle points. The SA-sheets also change drastically at these points.

Local extrema are the first two types of critical points. Consider the terrain near a local maximum. As illustrated in Fig. 9a, the SA-sheet at this position ends at the intensity of the hilltop. Now consider the terrain near a local minimum. As illustrated in Fig. 9b, the SA forms a loop around the hole formed by the minimum, which instantaneously transforms into an SA without a loop as we move below the intensity of the minimum. That is, the SA-pile consists of a cylindrical sheet which instantaneously turns into a simpler sheet at the local minimum. Following our discussion in Section 5, this abrupt appearance of the simpler sheet can be avoided if the global SA rather than just the first-order SA is used.

The third type of critical point, the saddle point, comes in three generic varieties: passes between two hilltops and between two pit-bottoms (see Fig. 9c) a pass between a hilltop and a hole in the hillside (see Fig. 9d), and a pass between a pit-bottom and a peak on the pitside (see Fig. 9e). These figures illustrate that as we move from below near these critical points, the components of the SA-piles behave as follows. In the first and third cases an SA-sheet tears at the saddle point. In the second case a tear is formed in an SA-cylinder, turning the cylinder into an ordinary sheet. In summary, the SA-pile consists of a branching structure of possibly torn and SA-sheets and cylinders.

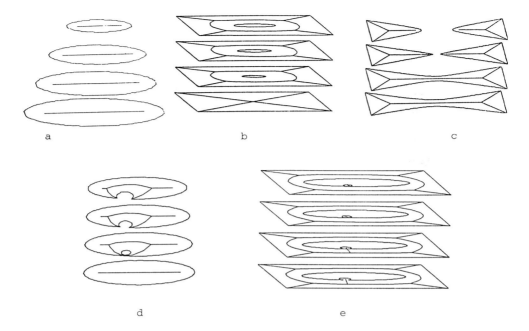

*Fig. 9. Components of SA-piles near intensity critical points: a) Maximum; b) Minimum; c) Saddle point between two hills and between two pits; d) Saddle point between hill and hole in its side; and e) Saddle point between pit and peak on its side.*

Now we must address the problem of imposing a natural hierarchy on the branching sheets in our description. We know that Gaussian blurring of an image annihilates extrema and saddle points in pairs (Koenderink, 1984). Because the contours associated with the image will simplify with blurring, the corresponding SA-pile will also simplify. Our experience to date confirms the inference that the process of blurring will cause successive disappearance of SA-sheets and SA-cylinders. We then impose a hierarchy based on the order of disappearance of SA-sheets and SA-cylinders, in a way analogous to the multiresolution SA of defined objects described in Section 4.

We have implemented this process using subroutines that convolve an image with a Gaussian, threshold the image at a selected intensity, and compute an SA by connecting centers of maximal disks in the thresholded region as one follows the boundary of the region. The results of successively larger threshold intensities are displayed as a pile of SA's on a vector graphics screen. Similarly, the successive isointensity contours can be displayed as

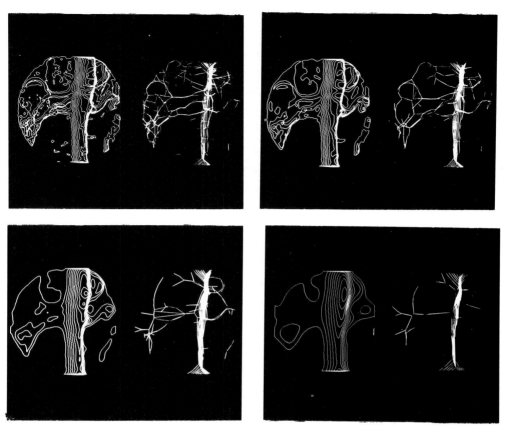

*Fig. 10. A sequence of blurred digital subtraction angiograms, their contour piles, and their SA-piles.*

a pile. Figure 10 shows the contour piles and associated SA-piles of a sequence of blurred versions of a digital subtraction angiogram. The predicted behavior can be seen.

Further mathematical investigation and algorithmic development of the multi-resolution grey-scale SA-pile is required. The usefulness of the external and global SA deserves attention. In addition, the relationship of this method to natural vision might be enhanced if the original image was first transformed into 'perceived intensity' by reflecting the effect of edges and intensity diffusion (Grossberg and Mingolla, 1985).

## 8. SUMMARY AND CONCLUSIONS

In closing, we suggest that the multiresolution symmetric axis transform is a main challenger in the possibilities for shape description. However, more work is needed in a number of areas. Mathematical study is needed of how symmetric axes behave under figure- or boundary-based resolution reduction and how SA-piles behave under Gaussian blurring. Algorithms must be developed to connect SA's into an SA-pile and to follow branches of SA's and SA-pile through resolution reduction to annihilation. This will require methods of computing the SA that limit the creation of artifactual branches and the premature omission of real branches.

In addition, we hope to have focused attention onto the multiresolution approach of following features to annihilation and thus inducing scale-based hierarchies. This appears to be a powerful method of deriving descriptions that are suitable for model-directed pattern recognition.

*Acknowledgements*

*We are indebted to Fred L. Bookstein, Jan J. Koenderink, and Lee R. Nackman for useful discussions. We are grateful to Sharon Laney for manuscript preparation and to Bo Strain and Karen Curran for photography. The research reported in this paper was carried out under the partial support of NIH grant number R01-CA39060 and NATO Collaborative Research Grant number RG. 85/0121.*

## REFERENCES

Ballard, D.H. and Brown, C.M. (1982). *Computer Vision*, Prentice-Hall, Englewood Cliffs, p. 234.

Blum, H. (1979). 3-D Symmetric axis co-ordinates: An overview and prospectus, draft of talk at Conf. on Representation of 3D Objects, Univ. of Pennsylvania.

Blum, H. and Nagel, R.N. (1978). Shape description using weighted symmetric axis features, *Pattern Recognition* 10, Pergamon, Oxford, pp. 167-180.

Bookstein, F., Chernoff, B., Elder, R., Humphries, J., Smial, G., and Strauss, R. (1985). *Morphometrics in Evolutionary Biology*, Special Publication 15, The Academy of Natural Sciences of Philadephia.

Brady, M. and Asada, H. (1984). Smoothed local symmetries and their implementation, Tech. Rept., Artificial Intelligence Laboratory, Massachusetts Institute of Technology.

Crowley, J.L. and Sanderson, A.C. (1984). Multiple resolution representation and probabilistic matching of 2-D gray-scale shape, Tech. Rept. No. CMU-RI-TR-85-2, Carnegie-Mellon Univ.

Grossberg, S. and Mingolla, E. (1985). Neutral dynamics of perceptual grouping: textures, boundaries, and emergent segmentations, *Perception and Psychophysics* 38, pp. 141-171.

Koenderink, J.J. (1984). The structure of images, *Biol. Cyb.* 50, pp. 363-370.

Koenderink, J.J. (1985). The internal representation of solid shape based on the topological properties of the apparent contour. In: *Image Understanding*, W. Richards (ed.), MIT Press, Boston.

Koenderink, J.J. and van Doorn, A.J. (1986). Dynamic shape, Tech. Rept., Department of Physics, Univ. of Utrecht.

Leyton, M. (1984). Perceptual organization as nested control, *Biol. Cyb.* 51, pp. 141-153.

Leyton, M. (1986a). A theorem relating symmetry structure and curvature extrema, Tech. Rept., Department of Psychology and Social Relations, Harvard Univ.

Leyton, M. (1986b). Smooth processes on shape, Tech. Rept., Department of Psychology and Social Relations, Harvard Univ.

Nackman, L.R. and Pizer, S.M. (1985). Three-dimensional shape description using the Symmetric Transform, I: Theory, *IEEE Trans. PAMI* 7, pp. 187-202.

Pizer, S.M., Koenderink, J.J., Lifshitz, L.M., Helmink, L., and Kaasjager, A.D.J. (1986). An image description for object definition, based on extremal regions in the stack. In: *Inf. Proc. in Med. Imaging*, S.L. Bacharach (ed.), Martinus Nijhoff, Dordrecht, pp. 24-37.

Richards, W. and Hoffman, D.D. (1985). Codon constraints on closed 2D shapes, *Computer Vision, Graph., Im. Proc.* 31, pp. 265-281.

Zahn, C.T. and Roskies, R.Z. (1972). Fourier descriptors for plane closed curves, *IEEE Trans. Comp.* 21, pp. 269-281.

# GIHS:
# A GENERALIZED COLOR MODEL AND ITS USE FOR THE REPRESENTATION OF MULTIPARAMETER MEDICAL IMAGES

Haim Levkowitz and Gabor T. Herman

*University of Pennsylvania, Philadelphia, USA*

ABSTRACT

*It is often desirable to display medical images of different parameters in correlation to each other. Multiparameter color images can be used to represent such information. We discuss the most common color models in computer graphics, presented in a new unifying framework. A new generalized color model, GIHS, is presented of which the existing color models are special cases realized by particular choices of parameter values. The use of the GIHS model for the representation of multiparameter images in general and some medical applications are discussed.*

1. INTRODUCTION

Many techniques and modalities are currently being used in medical imaging, both clinically and for research purposes, for the imaging of various parameters in the human body relating to *anatomy* (such as *Computerized Tomography (CT), Magnetic Resonance Imaging (MRI)*), *physiology* or *function* (such as *Positron Emission Tomography (PET), MRI, Cerebral Blood Flow (CBF)*), and *biochemistry* (such as *MRI*).

Functional and biochemical modalities do not contain sufficient information to precisely identify the various anatomical sites. It would be most desirable to have images that would include useful information from all (or several) of the modalities simultaneously and would thus convey the information in a much more comprehensive fashion than each of the modalities viewed separately.

The outcome of the different modalities are several three (or higher) - dimensional scenes, each representing a different parameter, such that volume elements (*voxels*) in corresponding locations of different scenes correspond to the same location in the patient's body. Together they form a multidimensional *multiparameter* scene.

In displaying information in such a scene, the usual single parameter black-and-white monitor is not enough to incorporate all the parameters. The mapping of the relevant information in the multidimensional multiparameter scene onto commonly used display devices so as to enhance contrast between parts of the scene that need to be distinguished (such as a lesion and normal tissue) is a matter of great interest (Jaffe, 1985).

Color is a *tristimulus* entity, that is, the specification of a single color requires the specification of three different components (Young, 1802; Helmholtz, 1866). Thus, it enables us to represent several parameters simultaneously.

In order to use color for the representation of multiparameter information it is necessary to use a color model whose coordinates are perceptually orthogonal. In such models it is easy to determine the values of each of the coordinates of a viewed color independently of the others. Out of the color models that are based on orthogonal coordinates, some are straightforward for interaction with color reproduction hardware but not easily interpreted by viewers (like the *RGB* model for color monitors and the *CMY* model for printing), others are better suited to describe color to people but cannot be easily implemented using a computer (such as the *Munsell Book of Color* (Munsell, 1976)), yet others are suitable for the measurement of color differences but use coordinates that are practically meaningless to people (like the *CIELUV Uniform Color Space* (CIE, 1978)). Moreover, the coordinates of some of these models have not been proven to be perceptually orthogonal. The *perceptual primaries* — *intensity, hue*, and *saturation* — are well suited for human interaction and for computer implementation. This article describes a generalized model based on the perceptual primaries and its potential uses for the display of multiparameter image information.

Section 2 describes the most commonly used color models in computer graphics in a new unifying framework. Section 3 presents a new *generalized intensity, hue, and saturation* model, *GIHS*. The models presented in Section 2 are special cases of this generalized model, realized by particular choices of parameter values. Section 4 discusses the use of the GIHS color model (and its existing special cases) for the display of multiparameter images and describes some medical applications. Section 5 presents some conclusions.

## 2. COLOR MODELS FOR COMPUTER GRAPHICS

In computer graphics there is a need to specify colors in a way that is compatible with the hardware and comprehendable to the user for specification and recognition of the colors and their components. These two requirements are referred to as *hardware* and *user* oriented requirements, respectively. Unfortunately they cannot be fulfilled within the same model. Thus, it is preferred to let humans interact with user oriented models and then transform them into the model of the hardware used.

*a. Hardware oriented models: The color monitor, the Colorcube, and the RGB model*

The most commonly used color display monitors work on the principle of additive mixtures to create different colors. They have three electron guns, for the three additive primaries R(ed), G(reen), and B(lue). The intensity of each gun is controlled independently between zero and a maximum voltage. The gamut of an RGB color monitor can be represented by the cube of all colors $0 \leq (r,g,b) \leq 1$, referred to as the *RGB Color Cube*, or just the *Colorcube*, see Fig. 1a.

*b. User oriented color models: The intensity, hue, and saturation (IHS) family*

The colorcube specifies colors in a straightforward way but lacks an intuitive appeal. For example, it is difficult for the user to specify or estimate the exact amounts of red, green, and blue necessary to represent a dark brown.

An easier way to make such estimates corresponds to the way artists mix, and specify, colors using *tints*, *shades*, and *tones* of pure colors (achieved by mixing in white, black, or both). The most common user oriented models in computer graphics are based on this notion of color mixing. We refer to them as the *IHS (intensity, hue, saturation) family of models*. To relate their three coordinates with the mixing notions of tints, shades, and tones one has to note that decreasing saturation is 'tinting' (corresponding to increasing whiteness), and decreasing intensity is 'shading' (corresponding to increasing blackness). 'Toning', is the combination of both.

The three most common IHS models have been developed by different people at different times and thus use somewhat different terminology (for intensity and in the order of the components in the name). We show here, however, that they can all be described in the same framework as different transformations of the RGB colorcube to the perceptual coordinates intensity, hue, and saturation. We describe the general framework, emphasizing the details that realize each one. The details of the actual models can be found in the cited literature. In all the three models:

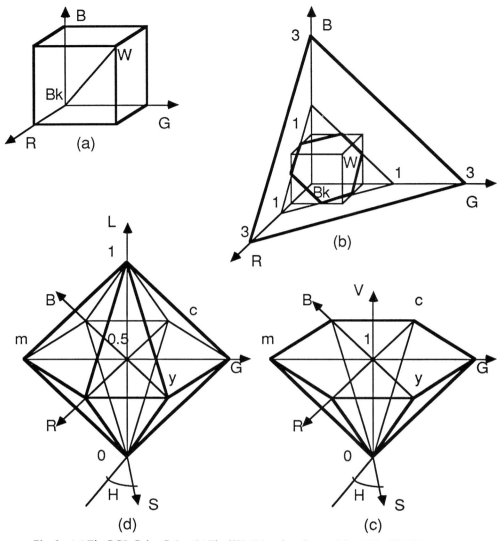

Fig. 1. (a) The RGB Color Cube, (b) The HSL Triangle color model, (c) The HSV Hexcone model, (d) The HLS double-hexcone model. (R = Red, G = Green, B = Blue, H = Hue, S = Saturation, L = Lightness, V = Value, c = cyan, m = magenta, y = yellow, Bk = Black, W = White.)

- Cylindrical coordinates are used.

- The *intensity* axis is represented by the main diagonal of the cube. An *intensity function* is defined and *constant-intensity* surfaces are formed, each of which intersecting the intensity axis at a single point, representing its intensity.

- The constant-intensity surfaces are projected onto planes perpendicular to the intensity axis and intersecting it at their intensity points. Their projections define their shape (such as a triangle or a hexagonal disk). Stacking these projected surfaces in the order of their intensities (0 at the bottom, 1 at the top) yields a three-dimensional body which is the *color solid* of the particular model.

- The *hue* of a color is represented by (some approximation to) the angle between the projections of the color vector and a predefined origin axis (traditionally the red axis) onto the planar projection of the constant-intensity surface in the color solid.

- The *saturation* of a color is represented by the ratio of the length of the projection of the color vector onto the planar projection of the constant-intensity surface to the maximal possible length along the same direction.

- Colors represented by points on the main diagonal of the colorcube (with equal amounts of red, green, and blue) are grays (sometimes referred to as *achromatic colors*). They are represented by points on the intensity axis of the color solid, their saturation is 0, and their hue is undefined.

The models differ from each other in the way their intensity functions are defined and thus in the way constant-intensity surfaces are formed. These two factors determine the shape of the color solid that represents each model.

## - The HSL Triangle

The triangle model (Smith, 1978) defines the intensity L (for lightness) of a color (r,g,b) to be

$$L = (r + g + b)/3, \quad (1)$$

where the division by 3 serves only to normalize the intensity into the range [0,1]. A constant-intensity surface L is the plane

$$\{(r,g,b) : (r + g + b)/3 = L\}. \quad (2)$$

For $0 \leq L \leq 1$, these planes are perpendicular to the main diagonal of the colorcube and parallel to each other and when limited to only non-negative values of r, g, and b they form triangles of increasing size starting at the origin (L = 0) and ending at the triangle (3,0,0), (0,3,0), (0,0,3), see Fig. 1b. The gamut of interest is the gamut of *realizable colors*, which is the intersection of the generated color solid with the RGB colorcube.

## - The HSV Hexcone

The HSV Hexcone model (Smith, 1978) is derived from the colorcube by defining the intensity of a given color (r,g,b) (called *value* and denoted V) to be

$$V = \max(r, g, b). \tag{3}$$

The locus of all the colors with at least one component that equals V are the points

$$\{(r,g,b) : \max(r, g, b) = V\}. \tag{4}$$

Limiting the coordinates such that $0 \leq r, g, b \leq 1$ yields the surface that consists of the three 'visible' faces of the subcube originating at the origin with side length V. We call this subcube a *V-subcube*.

The planar projections of the visible faces of the V-subcubes are corresponding hexagonal disks. Stacking the disks in their V value order yields the HSV hexcone, see Fig. 1c.

### - The HLS double-hexcone

The HLS Double-Hexcone developed by Tektronix, see description in (Foley and van Dam, 1982), is derived from the colorcube by defining the intensity function L of a color $(r,g,b)$ to be

$$L = [\max(r, g, b) + \min(r, g, b)]/2. \tag{5}$$

This intensity function gives rise to a set of *L-subcubes* such that the main diagonal of each of them is a segment of the main diagonal of the colorcube with the following starting and ending points:

$$\begin{cases} (0,0,0) \to (2L, 2L, 2L) & \text{if } 0 \leq L \leq 0.5, \\ (2L-1, 2L-1, 2L-1) \to (1,1,1) & \text{if } 0.5 < L \leq 1. \end{cases} \tag{6}$$

For each L-subcube the constant-intensity surface is the collection of points

$$\{(r,g,b) : [\max(r, g, b) + \min(r, g, b)] = 2L\}. \tag{7}$$

Limiting the color coordinates to $0 \leq r, g, b \leq 1$ yields a surface that consists of the six triangles resulting from connecting the point $(L, L, L)$ — located in the middle of the main diagonal of the L-subcube — to the six corners of the subcube that are not connected by the main diagonal. The planar projections of these triangles are again hexagonal disks, similar to the ones in the HSV hexcone. The only differences are that the largest disk corresponds now to $L = 0.5$ and the disks for both $L = 0$ and $L = 1$ (the black and white points) are single points. Stacking the disks vertically in the order of their L values yields the HLS double-hexcone, see Fig. 1d.

## 3. GIHS: A GENERALIZED INTENSITY, HUE, AND SATURATION MODEL

The various IHS models presented in the previous section have related properties and are derived from the colorcube using similar concepts. This gives rise to a generalization that describes a whole class of models. The existing models are special cases of the general model that are realized by special parameter values. We discuss now two levels of generalization.

*a. A first-order generalization*

The first-order generalization uses piece-wise planar constant-intensity surfaces. For each color (r,g,b) we define

$$\min = \min(r,g,b), \quad \mathrm{mid} = \mathrm{mid}(r,g,b), \quad \max = \max(r,g,b), \tag{8}$$

and three weights $w_{min}$, $w_{mid}$, $w_{max}$, such that $w_{min} + w_{mid} + w_{max} = 1$. Then, the intensity function is defined as

$$I = w_{min} \cdot \min + w_{mid} \cdot \mathrm{mid} + w_{max} \cdot \max, \tag{9}$$

and a constant-intensity surface for a given intensity I is given by the locus of points

$$\{(r,g,b): w_{min} \cdot \min + w_{mid} \cdot \mathrm{mid} + w_{max} \cdot \max = I\}. \tag{10}$$

These are the six planes that result from the six combinations of the order of magnitude of r, g, and b.

Different values of $w_{min}$, $w_{mid}$, and $w_{max}$ realize the different existing models introduced above, see Table 1.

|                    | $w_{min}$ | $w_{mid}$ | $w_{max}$ |
|--------------------|-----------|-----------|-----------|
| HSL-Triangle       | 1/3       | 1/3       | 1/3       |
| HSV-Hexcone        | 0         | 0         | 1         |
| HLS-Double-Hexcone | 1/2       | 0         | 1/2       |

*Table 1. The values of the three weights that realize the computer-graphics color models*

By changing the values of the three weights a continuum of models can be achieved. It is possible to implement any one of them using a single algorithm by the mere specification of a particular set of weights.

*b. Higher-order generalizations*

It is possible to define a generalization based on topological notions, where different color models are homeomorphic deformations of each other. The deformations are parametrized and by changing the values of the parameters different deformations (and thus, different models) are achieved. Each particular deformation defines a particular shape of the constant-intensity surfaces. The surfaces must cover the entire volume of the colorcube in an order-preserving fashion. That is, they intersect the achromatic axis in the order of their intensities and they do not intersect each other.

The first-order model described in the previous section is a special case with piece-wise planar surfaces. An example of another special case is the intensity function

$$I = [(r^2 + g^2 + b^2)/3]^{1/2}, \tag{11}$$

which results in surfaces of octants of I-subspheres (similar to the subcubes introduced before). Here, the gamut of interest is the intersection of the solid generated by the collection of these surfaces with the RGB colorcube.


4. GIHS IN MULTIPARAMETER MEDICAL IMAGING

The GIHS color model can be utilized to display multiparameter images, generally as well as in several medical applications (Levkowitz and Herman, 1986). We first discuss the general approach without specifying a particular model or application. It is applicable with any particular model that is realizable within the GIHS framework.

*a. Multiparameter image display using the GIHS model*

Multiparameter images can be generated from (up to three) separate input images by coding them as the intensity, hue, and saturation components of the color. If there are only two parameter images, $p_1$ and $p_2$, and we wish to correlate $p_2$ to $p_1$, then $p_1$ should be assigned to the intensity component and $p_2$ should be assigned to hue. The reasons for this particular assignment are the following: areas where $p_1$ is zero will not appear in the output image (intensity zero), while areas where $p_1$ is non-zero but $p_2$ is zero will still appear in a color conveying this information. The choice of the hue component is made due to the higher perceptual resolution of hue (humans can distinguish about 128 different hues, but only about 20 different levels of saturation (Foley and van Dam, 1982)). The saturation should be kept constant, at levels high enough to ease the distinction between colors but low enough to avoid eye-fatigue due to refocusing (Murch, 1984). Typical levels are above 0.6 and below 1.

An alternative possibility is to assign $p_2$ to both hue and saturation. In such a case, areas were $p_2$ is zero will be displayed in gray-scale while areas where $p_2$ is non-zero will be enhanced in color. This approach is limited, though, to two parameters only.

If there is a third image, it can be mapped onto the saturation component.

### b. Potential applications of GIHS in multiparameter medical imaging

We present now some specific applications of the GIHS color model in multiparameter medical imaging.

#### - Anatomy and function images

In medicine it is often desired to correlate function to anatomy. In this case, anatomy ($p_1$) is assigned to intensity and function ($p_2$) is assigned to hue (and, possibly, to saturation). Thus, areas without structure will not appear in the output image (intensity zero), while areas of structure but no activity will still appear in a color that corresponds to lack of activity.

Reported experiments with this scheme, in the medical field, show meaningful and informative pictures. For an example, see Faintich (1985).

#### - Magnetic Resonance Images (MRI)

MRI is a multiparameter modality by itself. Several different parameters can be measured and imaged using this modality. It is a relatively new imaging modality and many aspects of it are still in their research stage. Thus, the ways various parameter images can be utilized optimally need to be studied (Jaffe, 1985).

*- Measuring blood flow using Doppler Ultrasound and MRI*

The ability to see the flow of blood in an anatomical context can provide a lot of knowledge about blood dynamics and the vascular system in general.

It is possible to combine blood flow and anatomy using a Doppler Ultrasound dual image and color to create an image of flow. The color assignment depends on the direction of the signal (with respect to the transducer) and its velocity.

Echo-signals from stationary objects are assigned to gray-scale intensities relative to their amplitudes, creating an ultrasound image of anatomy. Signals from moving objects are assigned to colors as follows. The directionality is assigned to hue — relative to the transducer. For example, red can be used for movement away from the transducer and cyan for movement towards the transducer. Speed is assigned to saturation. Thus, signals from stationary objects (speed 0) will show the anatomy in gray-scale (saturation 0), while signals from moving objects will show flow in colors where the hue represents directonality and the saturation represents speed.

While direction information in Doppler Ultrasound is limited exclusively to directions to and away from the transducer, MRI provides the potential of measuring flow with 360° angle information (Axel, 1986). With such information, the GIHS approach can be used to its full power.

## 5. CONCLUSIONS

We have discussed the desirability of displaying multiparameter medical images. Color has been suggested as a way to achieve this goal. We have shown that the most commonly used color models in computer graphics can be described within a unifying framework. A new generalized color model, GIHS, has been suggested, that realizes the existing intensity, hue, and saturation color models, as well as potential additional ones. We have discussed the use of the GIHS model for the display of multiparameter images as well as some of its potential application in multiparameter medical imaging.

*Acknowledgments*

*The research of the authors is supported by NIH grants HL28438 and RR02546. The support of the organizers of the conference to help the attendance of the first author is gratefully acknoweledged.*


REFERENCES

Axel, L. (1986). Personal communication.
Commission Internationale de l'Eclairage (CIE) (1978). CIE recommendations on uniform color spaces-color difference equations, psychometric color terms. *CIE Pub.* (15, (E-1.3.1) 1971/(TC-1.3) 1978, Supplement No. 2), pp. 9-12.
Faintich, M. (1985). Science or art? *Comp. Graph. World*, April 1985, pp. 81-86.
Foley, J.D. and van Dam, A. (1982). *Fundamentals of Interactive Computer Graphics*, Addison-Wesley, Reading.
von Helmholtz, H.L.F. (1866). *Handbuch der Physiologischen Optik*, Voss, Hamburg.
Jaffe, C.C. (1985). Color displays: widening imaging's dynamic range, *Diagn. Im. 7*, pp. 52-58.
Levkowitz, H. and Herman, G.T. (1986). Color in multiparameter multidimensional medical imaging, *Color res. and appl. 11*, pp. S15-S20.
Munsell Color Company (1976). *The Munsell Book of Color*, Munsell Color Company, Baltimore.
Murch, G.M. (1984). Physiological principles for the effective use of color, *IEEE Comp. Graph. Appl. 4*, pp. 49-54.
Smith, A.R. (1978). Color gamut transform pairs, *Comp. Graph. 12*, pp. 12-19.
Young, T. (1802). On the theory of light and colours, *Philosoph. Trans. Royal Soc. London 92*, pp. 12-48.

# THE EVALUATION OF IMAGE PROCESSING ALGORITHMS FOR USE IN MEDICAL IMAGING

M. De Belder

*Agfa-Gevaert N.V.*
*Mortsel, Belgium*

ABSTRACT

*The evaluation of image processing algorithms implies the determination of the receiver operated characteristics (ROC's) for typical details. This is in principle an experimental procedure. An alternative approach consists in calculating the ROC's by using a model for the human visual system. A simple form of such a model is proposed. It is shown that realistic results can be obtained. The main conclusion to be made from this approach is perhaps that model calculations show more clearly the limitations of image processing.*

1. INTRODUCTION

The newcomer in the field of image processing is overwhelmed by the enormous variety of algorithms found in the literature. The choice between the algorithms is not always easy, it will be dictated as well by the ease of implementation as by the benefit to be derived from its use.

This paper will exclusively be concerned with the evaluation of image processing algorithms (IPA) from the point of view of their usefulness, and this with particular reference to medical imaging.

The ultimate test for any improvement in an imaging system is the determination of the ROC characteristic for a particular diagnosis as discussed e.g. by Lusted (1968). For reasons of convenience it is often indicated to determine the ROC for the detection of certain artificial details simulating typical diagnostic details; see for example Rossmann (1969), De Belder et al. (1975) and Loo et al. (1985).

It is well known that the current approach to produce artificial details experimentally is time consuming. Therefore, we have sought for alternative methods to evaluate IPA's. Clearly image quality criteria - if they are any good - should be able to do just that. However, none of the published image quality criteria yields ROC's with the typical shape found for visual stimuli. This is in particular true for the image quality criterium proposed by De Belder et al. (1971). This image quality criterium was based on a simplified model of the human visual system. It was assumed that perceptibility depends only on the signal to noise ratio at some location in the brain, obtained after introducing internal visual noise and linear filtering by the visual system.
Realistic ROC curves can be obtained by introducing some additional features to this model.

2. MODEL DESCRIPTION

The calculation of a ROC involves, just as in the experimental approach, a number of trials in which a noisy image, whether containing a signal or not, is presented to the system (Fig. 1)

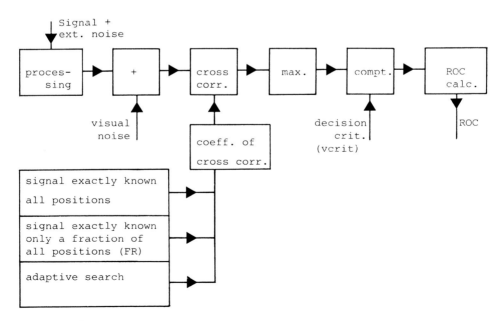

Fig. 1. Proposed model for the human visual system.

The noisy background is generated independently for each image. Firstly the image is processed using the algorithm to be evaluated. Next the human visual system is simulated. The main assumptions made in the model refer to this step. Just as in the model which served as the basis for the image quality criterium, internal visual noise is superimposed on the incoming data (see also Burgess et al., 1981 and Loo, 1985).

It is further assumed that detection and recognition involve cross-correlating the incoming data with some memorised function, the imprint. This idea was first explicitly introduced by Kabrisky (1966); see also Burgess (1985) and Pelli (1985).

An essential hypothesis finally consists in introducing uncertainty in the human decision process. The suggestion that uncertainty is an important element in detection and recognition tasks was first made by Tanner (1961). Green and Swets (1974) showed that it is only by introducing uncertainty in ROC theory that the characteristic asymmetry of ROC curves for visual tasks can be explained; see also Pelli, 1985.

In our model, uncertainty can be introduced in a number of different ways.

(i) The observer is uncertain about the location of the detail and cross correlates the imprint with the data at a limited number of random positions. This assumption is very easily modeled.

(ii) Perhaps a more realistic assumption is that the observer has no clear recollection of the shape of the detail, and makes a number of trial cross correlations with imprints from a given set.

(iii) Maybe still more realistic is the mechanism whereby the observer, again being uncertain about the shape of the detail, as well as about the position, proceeds adaptively by applying small changes to the imprint (both in position and shape) so as to maximize the cross correlation.

Though assumptions 2 and 3 seem more realistic they have not yet been implemented in our model.

The model in its actual form yields as the output of the cross-correlator the value $z_i$.

$$z_i = \sum_F [S(x) + N(x)] \cdot I(x - x_i) \cdot dx \tag{1}$$

where S represents the signal, N the noise, I the imprint, and F the field of observation. The value $z_i$ can be considered as the result of one of a series of successive fixations. The position $x_i$ is chosen at random from all possible positions in the field of observation.

The actual quantity on which the decision is based is the maximum value of $z_i$ for a given number of fixations, Z.

$$Z = \underset{i}{\text{MAX}}(z_i). \tag{2}$$

It is interesting to note that in the case where the imprint - apart from a constant factor - is identical to the signal, and for a large number of fixations, Z can be considered as the result of a matched filtering operation.

Once the probability distributions P(Z/n) for noise alone and P(Z/s) for signal plus noise are known, the ROC can be calculated in the standard manner.

However, in order to make the bridge to the experimental approach (see e.g. Todd-Pokropek (1980)), let us define the vector VCRIT whose components correspond to various levels of confidence. To each component $VCRIT_j$ of this vector there corresponds an operating point $Q_j$ on the ROC curve with coordinates

$$P_{FP}^j = \int_{VCRIT_j}^{\infty} P(Z/n) \cdot dZ \quad , \quad P_{TP}^j = \int_{VCRIT_j}^{\infty} P(Z/s) \cdot dZ \quad . \tag{3}$$

## 3. THE EFFECT OF IMAGE PROCESSING ON THE ROC CURVE

It is evident that image processing will change the ROC curve for a particular type of detail. In order to describe the effect of image processing, it is tempting to characterise the ROC curve by a single parameter, and the surface under the curve has often been used for this purpose.

It appears however that the ROC curves generated in this study cannot be considered as members of a one-parametric set - as would e.g. result in a Gaussian noise - equal variance situation. In particular our calculations in a number of cases yield ROC curves for the processed images crossing the ROC curves for the reference images. It should be mentioned that crossing of ROC curves has been observed before (Lams et al., 1986.

Apparently, another ranking method is needed. Ideally, the observer should choose an operating point maximising the expected value of the pay-off. As was shown by Lusted (1968), this is equivalent to maximising the expression

$$P_{TP} - uP_{FP} \qquad (4)$$

where $u = (P_N \cdot C_{FP})/(P_p \cdot C_{FN})$

$P_N$ = the probability of occurence of noise alone

$P_p$ = the probability of occurence of signal plus noise

$C_{FP}$ = the cost of making a false positive decision in stead of a true negative

$C_{FN}$ = the cost of making a false negative decision in stead of a true positive

We suggest the maximum expected value of the pay-off as a measure of system performance, and in most diagnostic situations this should lead us to focus our attention on the strongly curved portion of the ROC near the $P_{FP} = 0$ axis.

In practice the observer can only hope to approach this optimal performance after considerable training. We may assume that he will initially choose an operating point corresponding to some confidence level, as defined by one of the components of VCRIT.

Fig. 2 illustrates a case where the effect of image processing is clearly beneficial, since the working points are shifted in such a way that lower false positive rates are combined with higher true positive rates.

Fig. 3 illustrates a somewhat more critical situation. Though the surface under the ROC is increased by image processing one can imagine diagnostic situations where the new working points actually yield a smaller pay-off than the corresponding points of the original image.

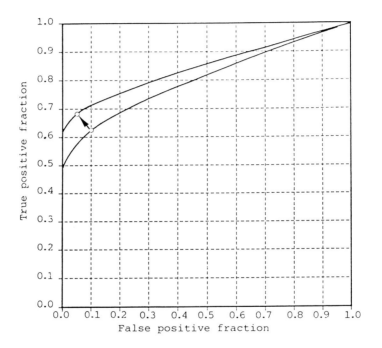

*Fig. 2. Situation where image processing shifts the working points toward lower false positive rates and higher true positive rates.*

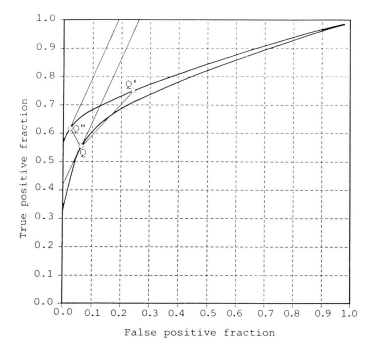

*Fig. 3. Though the surface under the ROC curve is increased by image processing the working point Q is shifted in such a way, that the diagnostic quality of the image may be lowered. In a diagnostic situation where the slope of the line with constant pay-off is 2, the new working point Q' gives a lower pay-off. It is only by retraining that a better working point, nearer to the optimum Q" can be found.*

The relevant quantity here is the slope of the line with constant pay-off, u, which was defined earlier. After a suitable training period the radiologist can be expected to shift his working point from Q' to Q'', which will give a better pay-off. This implies basing his decisions on a lower threshold. The fact that the working point corresponding to a given threshold is shifted toward higher false positives as well as higher true positives implies that the observer "sees more". We suggest that it is very important to make distinction between these two theoretical situations described in Figs. 2 and 3, where the ROC curves of both the original and the processed images look alike, but where the diagnostic implications are totally different.

## 4. CALCULATIONS AND RESULTS

*a. Programme limitations*

The results we shall present here were obtained with a one-dimensional prototype of our model, written in APL. The relative slowness of this programme imposed a limitation on the number of pixels per image. We think, however, that neither this limitation nor the restriction to one-dimensional images affects the validity of our results. More important is the statistical aspect of these simulations. Neglecting this aspect has led us to erroneous conclusions in the initial phase of this project. To detect certain small effects as many as 2000 indepentent noisy images had to be generated.

*b. Details*

Since the model in its actual form can handle only one-dimensional images we chose as details the one-dimensional analogues of the two-dimensional details we use currently in our experimental work. These are shown in figures 4 to 6.
The one-dimensional details will be called resp. the absorbing circle, the modulated Gaussian, and the localized edge. These details were located in the centre of the field of observation, but in view of the way the image is cross-correlated with the imprint we are effectively simulating a situation where the position of the detail is unknown.

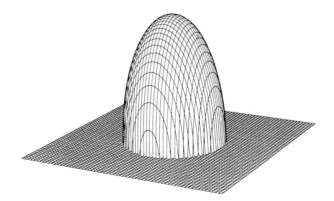

*Fig. 4. Absorption distribution in a spherical detail.*

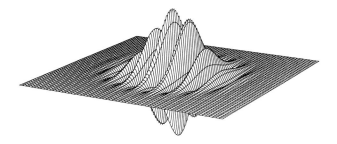

*Fig. 5. Absorption distribution in a modulated Gaussian detail.*

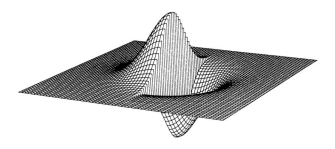

*Fig. 6. Absorption distribution in a localised edge.*

c. *Image processing algorithms*

The IPA's we used were
- (i) smoothing with a kernel of size 3,
- (ii) unsharp masking,
- (iii) matched filtering.

In all cases the kernels were normalised so as to introduce no grey scale transformation.

The smooting operator used the kernel 0.2  0.6  0.2.

The kernel for unsharp masking is represented in Fig. 7. Due to the normalisation it is completely defined by the enhancement factor K and the mask size S. For matched filtering the kernel was identical to the detail.

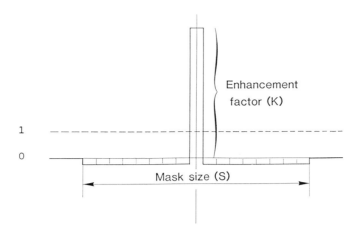

Fig. 7. Significance of the two parameters defining the unsharp masking algorithm. By requiring that the grey scale be unchanged the kernel is completely specified by mask size and enhancement factor.

d. Results and discussion

In this section we present a selection of the results of our calculations.
In Fig. 8 the effects on a coarse detail - the absorbing circle - of the three linear operators described above are presented. Fig. 9 shows an enlarged portion of the diagram. Smoothing hardly affects the ROC which is somewhat surprising. On the other hand the effects of matched filtering and of unsharp masking are very pronounced. They are however of a totally different nature, as will be visible from the shift of the operating points corresponding to a given confidence level. We suggest that it is impossible to make a choice between these two algorithms without specifying the diagnostic situation so as to define the slope of the line with constant pay-off.
Furthermore, in the case of unsharp masking retraining may be required. Figs. 10 to 12 show the effect of unsharp masking on the three types of detail we mentioned in section 4.c.
The influence of unsharp masking on these details is totally different. However, taking as a figure of merit the pay-off for a given slope u it is possible to formulate a general conclusion.

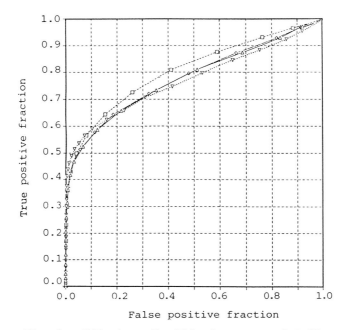

*Fig. 8. Effect on the ROC of a coarse detail (an absorbing circle) of three linear operations. (i) smoothing ($\Diamond$) (ii) matched filtering ($\Box$) and (iii) unsharp masking ($\nabla$). The ROC for the untreated image is also shown ($\triangle$).*

Table 1 shows the increase in pay-off for a slope $u = 2$. It appears that for all types of detail a large mask size combined with a moderate enhancement factor is to be preferred.

| K | S | Coarse detail | H.F. detail | Edge |
|---|---|---|---|---|
| 2 | 21 | 60 | 16 | 30 |
| 2 | 41 | 65 | 17 | 33 |
| 3 | 21 | 58 | 13 | 30 |
| 3 | 41 | 65 | 17 | 29 |
| Reference | | 59 | 10 | 27 |

*Table 1 : Effect of unsharp masking on pay-off for $n = 2$ (arbitrary units)*

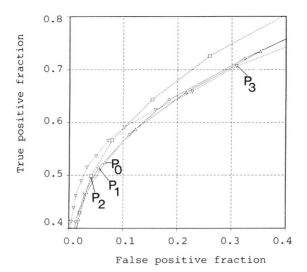

Fig. 9. Enlarged portion of fig. 8, showing how the working point $P_0$ of the reference curve is shited by the three IPA's. The resulting working points $P_1$, $P_2$ and $P_3$ correspond resp. to smoothing, matched filtering and unsharp masking.

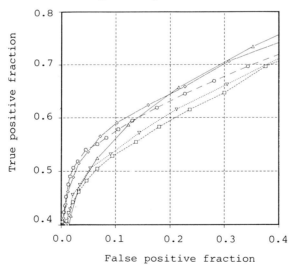

Fig. 10. Enlarged portion of ROC diagram, showing the influence of unsharp masking on the visibility of coarse details (absorbing circles). The ROC curves were calculated with the proposed model for the human visual system.
△ : reference, ◇ : $K=2$, $S=41$,
○ : $K=3$, $S=41$, ▽ : $K=3$, $S=21$,
□ : $K=3$, $S=21$.

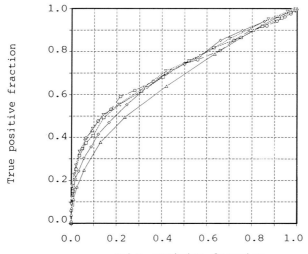

*Fig. 11. ROC curves for modulated Gaussians, computed with the proposed model for the human visual system. The images were processed with the unsharp masking algorithm.*
△ : reference, ▽ : K=2, S=41, ◇ : K=3, S=41, □ : K=2,S=21, ⌒: K=3,S=21

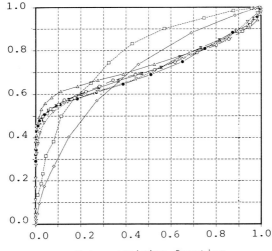

*Fig. 12. ROC curves for localised edges computed with the proposed model for the human visual system. The images were processed with the unsharp masking algorithm. (i) low contrast details.*
◇: reference □: K=2,S=21
The symmetry is due to the fact that the crosscorrelation was calculated at a larger number of locations. (ii) high contrast details.
▽: reference △: K=2,S=41, ○: K=3,S=41, ●: K=2, S=21, ⊠: K=3,S=21.

## 5. CONCLUSION

A simple model was constructed allowing to make predictions about the detectability of details in a noisy background. The model is based on a number of assumptions about the human visual system and the processing of visual data in the cortex. The model allows the computation of the ROC curves and the characteristic asymmetric shape of these is obtained in a natural way. The model was applied to a number of combinations (detail-linear filter).

Within the limitations of the model it is possible to formulate some general conclusions about the possible effects of image processing, as well as to optimize the parameters of a given algorithm.

## REFERENCES

Burgess, A.E., Wagner, R.F., Jennings, R.J., and Barlow, H.B. (1981). Efficiency of human visual signal discrimination, *Science* 214, pp. 93-94.

Burgess, A.E. (1985). Visual signal detection. III On bayesian use of prior knowledge and cross correlation, *J.Opt.Soc.Am.* 2, pp. 1498-1507.

De Belder, M., Bollen, R., and Duville, R. (1971). A new approach to the evaluation of radiographic systems, *J.Phot.Sc.* 19, pp. 126-131.

De Belder, M., Bollen, R. and Van Esch, R. (1975). Medical X-ray Photo-Optical Systems Evaluation *SPIE* 56, pp. 54-63.

Green, D.M., and Swets, J.A. (1974). *Signal detection theory and psychophysics*, Krieger, Huntington.

Kabrisky, M. (1966). *A proposed model for visual information processing in the human brain*, Univ. of Illinois Press, Urbana.

Lams, P.M., and Cocklin, M.L. (1986). Spatial resolution requirements for digital chest radiographs, *Radiology* 158, pp. 11-19.

Loo, L.N., Doi, K. and Metz, C.E. (1985). Investigation of basic imaging properties in digital radiography, *Medical Physics*, 12, pp. 209-214.

Lusted, L.B. (1968). *Introduction to medical decision making*, Thomas, Springfield.

Pelli, D.G. (1985). Uncertainty explains many aspects of visual contrast detection and discrimination, *J.Opt.Soc.Am.A/2*, No.9., pp. 1508-1530.

Rossmann, K. (1969). Image quality, *Radiologic Clinics of North America* VII, No.3, pp. 419-422.

Tanner, W.P. (1961). Physiological implications of psychophysical data, *Ann. N.Y. Acad. Sci.* 89, pp. 752-765.

Todd-Pokropek, A. (1983). *Fundamental algorithms for computer graphics*, Springer-Verlag Berlin XVI-1042.

# FOCAL LESIONS IN MEDICAL IMAGES: A DETECTION PROBLEM

J.M.THIJSSEN

*University of Nijmegen*
*The Netherlands*

## ABSTRACT

*The problem of detecting a signal in a noisy background is highlighted from the point of view of communication systems, sensory perception studies and assessment of the intrinsic quality of medical imaging equipment. The concepts of "psychometric curves" (PMC) and of "receiver operating characteristics" (ROC) are introduced.*
*The likelihood ratio approach is discussed to illustrate the techniques for constructing optimum detectors for communication systems, but at the same time to find the best figure of merit to perform the quality assessment. The concepts of "detectability index" and of the "area under ROC" are shown to be the most suitable figures of merit which can also profitably be employed in medical imaging experiments with human observers. Several medical imaging modalities are discussed and the methods to specify the detectability index from measurable characteristics of the imaging performance and of the intrinsic noise of the equipment are presented.*

## 1. THE DETECTION PROBLEM

### a. Introduction

The observer is assumed to have the task to either detect the presence of a signal, or to differentiate between the presences of either of two signals. In both cases the decision by the observer is influenced by noise. The noise can be due to the image generating system, to some uncertainty in the observer's decision making, or to both. Assumptions regarding the properties of the observer noise (internal noise) can only be deduced indirectly from the observer's behaviour in a detection experiment (cf. Thijssen and Vendrik, 1968). Generally it is assumed that this noise has a Gaussian distribution. A further assumption has to be made about the dependence of the internal noise on signal strength, i.e. whether or not the internal noise is additive, or multiplicative. In simple detection tasks the noise is additive (Thijssen and Vendrik, 1968), but recent evidence indicates that for more complex stimuli the multiplicative noise model would be more adequate (Burgess, 1986).

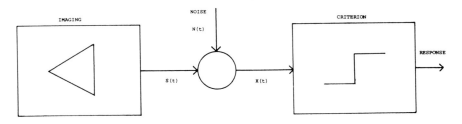

Fig. 1. Block diagram of detection model in medical imaging.

The block diagram of the detection problem is drawn in Fig. 1. A signal s(t) is generated by an imaging device and noise n(t) is added (or multiplied). So the detector receives the observation variable x(t)

$$x(t) = s(t) + n(t) \tag{1}$$

which in the frequency domain becomes

$$X(f) = S(f) + N(f). \tag{2}$$

The detector has to decide upon observation x(t) whether or not a signal s(t), specified exactly, and occurring or not, during a known observation interval, is present.
This description equally applies to the spatial domain, where t has to be replaced by space coordinates and f by spatial frequency.

b. *Psychometric curve (PMC)*

In a detection experiment stimuli are presented many times and the

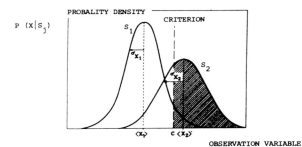

Fig. 2 Probability density functions of observations variable x, positive response probability $P(+|s_j)$ for exceeding criterium C.

fractions of correct and incorrect decisions are estimated. The decision problem the observer is faced with is illustrated by Fig. 2. The observer has to employ a certain decision level, the criterion C, to be able to decide whether an observation is due to signal $s_1$, or $s_2$. Because of the overlap of the probability distribution functions corresponding to the two possible signal levels a certain fraction of the responses inevitably will be incorrect. Because the response probabilities add up to one two by two, two probabilities are sufficient to characterize the decision process. Here the false positive and the true positive probabilities are chosen, which correspond to the area under the curves in Fig. 2 to the right of the criterion C (hatched regions).

$$\text{FPF} = P(\text{FP}) = \int_{C}^{+\infty} (2\pi\sigma_1^2)^{-1/2} \exp(-(x-\langle x_1 \rangle)^2/2\sigma_1^2)dx \tag{3}$$

$$\text{TPF} = P(\text{TP}) = \int_{C}^{+\infty} (2\pi\sigma_2^2)^{-1/2} \exp(-(x-\langle x_2 \rangle)^2/2\sigma_2^2)dx \tag{4}$$

where a Gaussian probability density function has been assumed, for both conditions $s_1$ and $s_2$. Formulas (3) and (4) can be simplified to a standard Gaussian p.d.f., $G(0,1)$:

$$P(\text{FP}) = \int_{Z_1}^{+\infty} (2\pi)^{-1/2} \exp(-x^2/2)dx \tag{5}$$

$$P(\text{TP}) = \int_{Z_2}^{+\infty} (2\pi)^{-1/2} \exp(-x^2/2)dx \tag{6}$$

where $z = (c-\langle x_i \rangle)/\sigma_i$ = the normal deviate.

In psychophysics these probabilities are often plotted vs. signal strength $s_i$, assuming a linear transfer from $s_i$ to the decision variable $x_i$ employed in the brain of the observer (which is also the present assumption, but most probably not true). The resulting graph is called the "Psychometric Curve", PMC.

As is shown in Fig. 3a the horizontal position of the PMC's is dependent on the criterion level employed by the observer. This property has completely undermined the classical concept of sensory thresholds (cf. Green and Swets, 1966). In Fig. 3b the ordinate is scaled in normal deviates z, which corresponds to the use of normal probability paper.

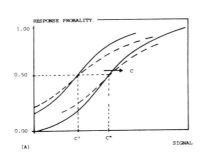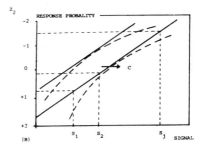

*Fig. 3 a) Psychometric curve's (PMC's) with linear probability ordinate.
Drawn: $\sigma_1 = \sigma_2$, dashed: $\sigma_1 \neq \sigma_2$
b) As in a), ordinate is normal deviate.*

equal variance case: $\sigma_1 = \sigma_2$
In this case the PMC's are parallel, both in Fig. 3a, and in Fig. 3b.
unequal variance case: $\sigma_1 \neq \sigma_2$
Now the shape and the slope of the curves in Fig. 3a change, depending on the criterion level. In Fig. 3b the lines do not remain straight, but the slope gradually increases with increasing criterion level (cf. Thijssen and Vendrik, 1968).

*c. Receiver Operating Characteristic (ROC).*

It will be evident from the discussion so far that PMC's are not suited to express the detectability threshold of a signal, because the positive response probabilities not only depend on signal strength but strongly on the criterion level as well. Therefore, a more adequate figure of merit is needed. Before deriving such a measure it is necessary to introduce the ROC, which is done by Fig. 4a.

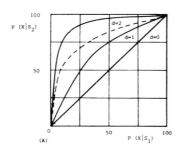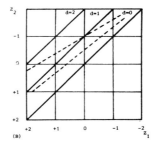

*Fig. 4 a) Receiver Operating Characteristics (ROC's) with linear probability coordinates. Drawn: $\sigma_1 = \sigma_2$, dashed: $\sigma_1 \neq \sigma_2$
b) ROC's with normal deviate coordinates.*

Here the P(TP) is plotted vs. P(FP), both on linear scales. If the criterion level changes both probabilities change in such a manner, that the operating point on the ROC is always in the range to the left of the diagonal. As can be deduced from Fig. 2, the higher the criterion level C, the lower the response probabilities are, as indicated by the arrow in Fig. 4a. It can easily be shown that the ROC is symmetric around the negative diagonal in the equal variance case (drawn ROC in Fig. 4a) and non-symmetric in the unequal variance case (dashed). If, like before, the response probabilities are replaced by normal deviates the ROC's become straight lines (Fig. 4b), with a unity slope for the equal variance case (drawn) and a non-unity slope otherwise (dashed). The ROC on z-coordinates can be generally described as

$$z_2 = az_1 + b. \tag{7}$$

The constants a and b can be determined with Formulas (5,6):

$$z_2 = (\sigma_1/\sigma_2) z_1 - (\mu_2 - \mu_1)/\sigma_2. \tag{8}$$

For equal variance, and $n(t) = G(0,\sigma)$: $a = 1$, $b = (s_2 - s_1)/\sigma$.
The figure of merit describing a ROC, and the detectability is called the *detectability index*, d'. Several definitions can be found in literature:

1. d' (after Green and Swets, 1966)

$$d' = (s_2(t) - s_1(t))/\sigma \tag{9}$$

In literature the detectability index is often referred to as the "signal to noise ratio": SNR, which term will be used in this paper as well. It is evident from Formula (8) that in the unequal variance case two parameters remain. Therefore, the search for a single figure of merit can not be successful. Many authors have tried to solve this problem and three of the proposals are discussed now (cf. Simpson and Fitter, 1973)

2. $d'_e$ (after Egan and Clarke, 1966)
This index summarizes the ROC by its cross section with the negative diagonal: It can be shown that (for $z_2 = -z_1$):

$$d'_e = (s_2(t) - s_1(t))/(1/2\ (\sigma_2 + \sigma_1)) \tag{10}$$

hence the average standard deviation is involved.

3. $d'_a$ (after Schulman and Mitchell, 1966)
This index summarizes the ROC by its orthonormal distance to the origin:

$$d'_a = (s_2(t) - s_1(t))/(1/2\ (\sigma^2_2 + \sigma^2_1))^{1/2}. \tag{11}$$

Now the square root of the average variance is involved.

4. P(A) = area under the ROC

$$P(A) = \int_0^1 (1-P(FP))\ dP(TP). \tag{12}$$

It can be shown (see Section 3) that the last figure is identical to the probability of correct response in a two alternative forced choice experiment. The question which of these figures of merit is to be preferred when a single criterion independent measure is to be used to describe the performance of an observer, or alternatively the quality improvement of an image processing technique, has been answered by historical development. It appears, that in medical imaging often P(A) is employed when psychophysical experiments are reported. However, when the intrinsic quality of imaging devices is being specified one generally employs the SNR concept, which is equivalent to d' of Formula (9), for the equal variance case (cf. Wagner and Brown, 1985).

## 2. LIKELIHOOD RATIO APPROACH

a. *Time domain*

Like before, signals $s\nu(t)$ are involved, but now the bandwidth B and the duration T are considered. The signals are sampled at times k, with $k \in (o,T)$:

$$x(t) = \sum_{k=1}^{2BT} x_k = \sum_{k=1}^{2BT} (s_k + n_k). \tag{13}$$

The sampling interval for 2BT samples (according to the sampling theorem) is 1/2B. The noise is considered to be additive, Gaussian and white, with spectral density $N_o$, hence

$$\sigma^2 = N_o B. \tag{14}$$

The likelihood $\ell$ is defined as:

$$\ell(x_i(t)) = \prod_{k=1}^{2BT} p(x_{i,k}) = \prod_{k=1}^{2BT} (2\pi\sigma^2)^{-1/2} \exp(-(x_k - s_{i,k})^2/2\sigma^2) \tag{15}$$

where the stochastic independence of white noise samples is employed and $x_i(t)$ is the observation when signal $s_i$ is present.
The likelihood ratio of two possible conditions $x_1$ and $x_2$ is:

$$Y_{1,2} = \frac{\ell(x_2(t))}{\ell(x_1(t))} = \prod_{k=1}^{2BT} \exp(-((x_k - s_{2,k})^2 - (x_k - s_{1,k})^2)/2\sigma^2). \tag{16}$$

The decision strategy then is to decide that signal $s_2$ is present, if $Y'_{1,2} \geq \beta$.
Since the logarithm is a monotonic function, thresholding the logarithm of Formula (16) is equivalent to thresholding $Y'_{1,2}$, so:

$$Y_{1,2} = \ln Y'_{1,2} = \sum_{k=1}^{2BT} ((x_k - s_{2,k})^2 - (x_k - s_{1,k})^2)/2\sigma^2 \tag{17}$$

and the decision that $s_2$ is present has to be taken if $Y_{1,2} \geq \ln \beta$.
If $s_1 = 0$, the decision problem reduces to a detection of a signal s in a noisy background and Formula (17) becomes:

$$Y_{1,2} = \sum_{k=1}^{2BT} x_k s_k / \sigma^2 - \sum_{k=1}^{2BT} s_k^2 / 2\sigma^2. \tag{18}$$

The second term on the righthand side of Eq. (18) being a constant, the detection strategy becomes a cross correlation of the expected signal ("signal known exactly" case) with the observed variable x(t).

The response probabilities for the general case of Formula (17) can be specified now, when it is realized that, because n(t) is Gaussian white noise, $Y_{1,2}$ has also these properties. Hence,

$$P(TP) = \int_{\ln\beta}^{+\infty} P(Y_{1,2}|s_2) = \int_{\ln\beta}^{+\infty} (2\pi\sigma_Y^2)^{-1/2} \exp(-(Y_{1,2} - \mu_{Y2})^2/2\sigma_Y^2) \qquad (19)$$

$$P(FP) = \int_{\ln\beta}^{+\infty} (2\pi\sigma_Y^2)^{-1/2} \exp(-(Y_{1,2} - \mu_{Y1})^2/2\sigma_Y^2) \qquad (20)$$

where

$$\mu_{Yi} = \sum_{k=1}^{2BT} (s_{2,k} - s_{1,k})^2 (-1)^i / 2\sigma^2 \qquad (21)$$

$$\sigma_Y^2 = \sum_{k=1}^{2BT} (s_{2,k} - s_{1,k})^2 / \sigma^2. \qquad (22)$$

Inserting (21) and (22) into the definition of d' (Formula (9)) yields

$$d' = \sum_{k=1}^{2BT} (s_{2,k} - s_{1,k})^2 / (\sum_{k=1}^{2BT} (s_{2,k} - s_{1,k}) \sigma^2)^{1/2} \qquad (23)$$

which can be written as

$$d' = (2\Delta E_S/N_0)^{1/2} \qquad (24)$$

where

$$\Delta E_S = (\sum_{k=1}^{2BT} (s_{2,k} - s_{1,k})^2)/2B \qquad (24a)$$

and

$$\sigma^2 = N_0 2B.$$

## b. Frequency domain

For some applications it is more convenient to consider the detection problem in the (spatial) frequency domain (cf. Section 5).
So, for $X(f) = S(f) + N(f)$:

$$X(f) = (2\pi)^{-1/2} \int_{-\infty}^{\infty} x(t) \exp(-i2\pi ft) \, dt. \tag{25}$$

The continuous case will not be discussed in this chapter, although in imaging generally only non-periodic "signals" are considered. The problem that in digital computers only sampled versions of the original images can be processed is solved by the quasi-periodic treatment of sampled signals in discrete Fourier transform algorithms. Hence,

$$X(f_n) = \sum_{k=1}^{2BT} x_{k,n} = \sum_{k=1}^{BT} (a_k \cos(2\pi kn/BT) + b_k \sin(2\pi kn/BT)) \tag{26}$$

where $a_k = (2/BT) \sum_{n=1}^{BT} x_{k,n} \cos(2\pi kn/BT)$  (26a)

and $b_k$ similarly with a sine in the sum.
The Fourier coefficients are assumed to have a Gaussian, white noise distribution. Therefore the coefficients $a_p$ and $b_p$ are independent with a $G(0,\sigma)$ distribution, where $\sigma^2 = N_0/T$.
The likelihood for the $a_k$ series then becomes:

$$\ell_a = \sum_{k=1}^{BT} (2\pi N_0 T)^{-1/2} \exp(-(X_k - S_{k,a})^2/2(N_0/T)) \tag{27}$$

and

$$Y_{1,2}(a) = \sum_{k=1}^{BT} ((-(X_k - S_{k,a_2})^2 + (X_k - S_{k,a_1})^2)/2(N_0/T)) \tag{28}$$

$$Y_{1,2}(a|S_2) = \sum_{k=1}^{BT} ((S^2{}_{k,a_1} - S^2{}_{k,a_2}) - 2N_k(S_{k,a_2} - S_{k,a_1}))/(2N_0/T). \tag{29}$$

From Formula (29) it becomes clear that the cross correlator mentioned in

Chapter 2.1 can be replaced by a matched filter, yielding the cross-term of Np and Sp.

Like has been shown in Section 2.1 for the temporal domain, the d' (=SNR) in the frequency domain becomes for $Y_i(a) = Y_{1,2}(a|S_i)$:

$$d' = \sum_{k=1}^{BT} (S_{k,a_2} - S_{k,a_1})^2 / (\sum_{k=1}^{BT} (S_{k,a_2} - S_{k,a_1})^2/(N_0/T))^{1/2} +$$

$$+ \sum_{k=1}^{BT} (S_{k,b_2} - S_{k,b_1})^2 / (\sum_{k=1}^{BT} (S_{k,b_2} - S_{k,b_1})^2/(N_0/T))^{1/2}. \quad (30)$$

## 3. PSYCHOPHYSICAL PROCEDURES

### a. Yes-No procedure

The observer is confronted with the situation that in an observation he has to decide which of two possible alternatives is presented. So in a simple detection experiment whether a condition is present or not. This detection task has been treated extensively for communication systems in the fifties (e.g. Peterson et al. 1954, Swets, 1964). The various stimulus-response conditions are listed in Table 1. The true response probabilities can only be approximately estimated from the response fractions, so P(TP) TPF, etc. where TPF = true positive fraction. It will be clear:

$$TPF = 1-FNF \text{ and } FPF = 1-TNF. \quad (31)$$

|  |  | Radar |  | Psychophysics |  | Medical Imaging |  |
|---|---|---|---|---|---|---|---|
| response |  | 1 | 2 | - | + | - | + |
| sign. $s_1$ |  | correct reject | false alarm | true negative | false positive | specificity | 1 - spec. |
| $s_2$ |  | miss | hit | false negative | true positive | 1 - sens. | sentivitity |

Table 1. Response categories of detection tasks in various disciplines.

Often a single measure is used to quantify the observer's potential to detect the signal (i.e. a pathological condition), cf. Metz (1978): the diagnostic accuracy, (DA) of responses:

$$DA = (\#correct\ responses)/(\#cases) = (TPF) \times (PF) + (TNF) \times (NF) \qquad (32)$$

where PF and NF are the a priori fractions of both conditions in the total set of images. Equation (32) implies, that the diagnostic accuracy may remain constant even if the response fractions change and the reverse it may change due to different a priori fractions if the response fractions are identical. It is questionable, therefore, whether it is an adequate measure.

The response fractions TPF and FPF can be employed to estimate one of the figures of merit obtained from a ROC. For the equal variance case either $d'$, or $P(A)$ could be chosen. However, from a single set of fractions only a single point at a ROC is found. Since it is not known a priori whether the equal variance case applies a procedure has to be found to increase this number of points obtained from an experiment. Before proceeding in this matter, it might be illustrative to compare the DA to $d'$ as a figure of merit. If one finds for the a priori condition NF = 0.83, PF = 0.17:
    TPF = 0.40, FNF = 0.04, then DA = 0.87 and $d'$ = 2.0
and TPF = 0.70, FNF = 0.10, then DA = 0.87 and $d'$ = 1.8.
So the diagnostic accuracy remains constant, although the $d'$ indicates a worse SNR in the second case. If one finds the response fractions as in the first case, for a different a priori condition: NF = 0.40, PF = 0.60, then: DA = 0.78 and $d'$ = 2.0
From these examples it can be concluded that the diagnostic accuracy is sensitive to the criterion level to be used by the observer (because that is the change induced by different a priori probabilities), and not only on the SNR (cf. Swets, 1986). The latter is the preferable figure of merit for the detectability of the pathology, but sometimes the area under the ROC is to be preferred (Eq. 12).

### b. Rating procedure

As stated above the ROC analysis induced the need to let the observer employ several criteria in a single experiment. The way to do this is to

| response | sure neg. <br> - - | neg. <br> - | undecided <br> 0 | positive <br> + | sure pos <br> ++ |
|---|---|---|---|---|---|
| signal $s_1$ <br> signal $s_2$ | $P(--\|s_1)$ <br> $P(--\|s_2)$ | $P(-\|s_1)$ <br> $P(-\|s_2)$ | $P(0\|s_1)$ <br> $P(0\|s_2)$ | $P(+\|s_1)$ <br> $P(+\|s_2)$ | $P(++\|s_1)$ <br> $P(++\|s_2)$ |
| criterion | | $C_1$ | $C_2$ | $C_3$ | $C_4$ |

Table 2. Response probabilities in rating experiment.

ask the observer to rate his answer at every trial according to some certainty scale. The result will be that for each stimulus condition several response categories are present, and a number of points of a ROC become available. An example of such a rating scale is given in Table 2. The table indicates that the observer has to handle (n-1) criterion levels to be able to make a rating in n categories. This also implies that (n-1) points of the ROC are determined. The procedure to follow is: first plot $P(++|s_2)$ vs. $P(++|s_1)$, then accumulate these probabilities with the second row from the right, i.e. $(P(++|s_2) + P(+|s_2))$ and $P(++|s_1) + P(+|s_1))$, to find the second pair, etc. This procedure corresponds to finding always the total area under the probability density curve to the right of any criterion (cf. Fig. 1). Having estimated the points to construct a ROC the problem arises how to find the best fit in a statistical sense. Although it has been shown that a straight line is to be expected if one employs z-coordinates (normal deviates), linear curve fitting based on least-squared error methods is not allowed because of unequal variances in z-space. A method to make a fit in probability space, and a computer programme to perform it, has been made available by Dorfman and Alf (1969). This method is to be preferred and can be extended to yield the area under the ROC: P(A).

c. *Two-alternative forced choice (2AFC) procedure.*

This procedure is best described for a simple detection task: signal, or pathology, present or not. The observer in every trial is sequentially,

or simultaneously, confronted with two stimuli (images) one of which
contains the signal. His task is to decide (forced) in which of the two
stimuli. The response probability for this condition, i.e. fraction of
correct answers, P(A) is given by:

or:
$$P(A) = \int_{-\infty}^{\infty} p(x|s) \left[\int_{-\infty}^{x} p(\xi|n) d\xi\right] dx \qquad (33)$$

$$P(A) = \int_{0}^{1} [1-P(FP)] \, dP(TP) \qquad (34)$$

So, this probability is identical to the area under the ROC for the
yes-no and rating procedures.

## 4. DECISION GOALS AND COSTS

a. *Optimize expected cost (maximize benefit)*

In the simple yes-no procedure the overall costs of the experiment can be
specified. Instead of response fractions the probabilities will be
employed in the following.
The expected value of the costs of a decision strategy E(C) is given by:

$$E(C) = (C_{tp} P(TP) + C_{fn} P(FN)) P(P) +$$

$$+ (C_{tn} P(TN) + C_{fp} P(FP)) P(N) \qquad (35)$$

where $C_{tp}$, etc. are the costs of a true positive, false negative,
etc.; decision P(P) and P(N) are the total a priori probabilities of a
positive case or a negative case to be present.
By using Formula (31) this can be rewritten as

$$E(C) = (C_{tp} - C_{tn}) P(P) P(TP) - (C_{tn} - C_{fp}) P(F) P(FP) +$$

$$+ C_{fp} P(P) + C_{tn} P(F). \qquad (36)$$

The last two terms in Formula (36) are constants with respect to the observer's behaviour. Optimization of expected costs means setting E(C) to zero, hence

$$\frac{dP(TP)}{dP(FP)} = \frac{(C_{tn} - C_{fp}) P(N)}{(C_{fn} - C_{tp}) P(P)} = \beta. \qquad (37)$$

It should be realized that the left-hand term in Formula (35) is the first derivative of the ROC and equals the likelihood ratio in the operating point, given by the criterion level. This formula also quite clearly shows how the decision criterion should be adapted to the expected, or known, a priori probabilities P(N) and P(P). As discussed in Chapter 3.1 this, however, does not indicate a change in the detectability of the stimulus.

b. *Maximize correct response fraction* (*diagnostic accuracy*)

The diagnostic accuracy correct responded fraction (CRF) has been defined by Formula (32) and is now rewritten in response probabilities:

$$P(CR) = P(TP) P(P) + P(TN) P(N) \qquad (38)$$

or:

$$P(CR) = P(TP) P(P) + (1-P(FP)) P(N). \qquad (39)$$

For equal costs of all responses, the P(CR) is optimized if P(CR) = 0, so:

$$\frac{dP(TP)}{dP(FN)} = \frac{P(N)}{P(P)} = \beta. \qquad (40)$$

So like before an operating point is found on a ROC, which result could have been derived from Formula (37) by setting costs equal to one.

c. *Neyman-Pearson objective*.

This objective is described as: maximize P(TP) for a prefixed value of P(FP) = k. It can be shown that the likelihood ratio approach satisfies the Neyman-Pearson objective (cf. Green and Swets, 1966). The response strategy is to answer positively if the likelihood ratio equals, or exceeds a threshold value :

$$\frac{dP(TP)}{dP(TN)} \geq \beta \qquad (41)$$

so:

$$P(TP) \geq P(TN) + \text{constant} \qquad (41a)$$

hence:

$$P(TP) \geq k + \text{constant} \qquad (41b)$$

which confirms the above statement.

## 5. DETECTION THEORY IN MEDICAL IMAGING

*a. Introduction*

Although it is not generally realized two important developments in this context should be mentioned. The first is the application of ROC concepts to medical decision making in general (e.g. Patton, 1978) and to evaluation of diagnostic methods and image processing techniques in particular (cf. Swets et al., 1979, Swets and Pickett, 1982). The second is the employment of the SNR and the ideal observer concepts to objectively quantify the inherent performance characteristics of imaging equipment (cf. Wagner and Brown, 1985). This chapter is mainly devoted to the latter aspect, but the question: how "ideal" is a real human observer will be addressed as well. But first of all the concept of an ideal observer has to be elucidated: the ideal observer makes optimal use of the information, without decreasing, therefore, the signal to noise ratio (SNR). In statistical terms the ideal observer achieves the Neyman-Pearson objective (Green and Swets, 1960) and meets the Cramer-Rao bound for minimum variance of the estimation (cf. Wagner and Brown, 1985).

*b. Detection of focal lesion*

In the following it is first assumed that the detection problem can be specified as follows: to detect a lesion of known size and location in a (noisy) image. The noise is generally inherent in the imaging modality, e.g. photon noise in X-ray and scintigraphic (γ-ray) images, thermal noise in MRI images and speckle in echographic images. So both the background intensity and the lesion intensity are specified statistically. The probability density of the intensity is given by:

$$p(I) = (1/\langle I \rangle) \exp(- I/\langle I \rangle) \tag{42}$$

where $\langle I \rangle$ is the average over a large region.
If a lesion of diameter D is considered, it can be assumed that m independent spatial samples are present within this region (e.g. photons, speckles). The p.d.f. of the intensity of the lesion then is:

$$p_m(I) = (m/\langle I \rangle)^m I^{m-1} \Gamma(m)^{-1} \exp(- mI/\langle I \rangle) \tag{43}$$

(gamma p.d.f., with $\Gamma(m) = (m-1)!$).
So the logarithm of the likelihood ratio $\gamma_{1,2}(I)$ becomes:

$$\gamma_{1,2}(I) = m \ln(\langle I_1 \rangle/\langle I_2 \rangle) + mI((\langle I_2 \rangle - \langle I_1 \rangle)/(\langle I_2 \rangle \langle I_1 \rangle)) \tag{44}$$

where $\langle I_1 \rangle$ is the average background intensity level.
Therefore, thresholding $\gamma$ is equivalent to thresholding the last term in this formula, except for a constant between brackets, i.e. $\gamma' = mI$, which is exponentially distributed (cf. Formula (42)), so:

$$\mu_{\gamma'} = m \langle I \rangle$$
$$\sigma^2_{\gamma'} = m \langle I \rangle^2 \tag{45}$$

and:
$$d'_a = SNR = m^{1/2}(\langle I_2 \rangle - \langle I_1 \rangle)/((\langle I_2 \rangle^2 + \langle I_1 \rangle^2)/2)^{-1/2} \tag{46}$$

or:
$$d'_a = C_I M^{1/2} \tag{47}$$

where $C_I$ is the intensity contrast as defined by comparison of (46) with (47).
A more general approach of the detection index can be found in the literature (e.g. Wagner and Brown, 1985). For the one-dimensional case

the following definition is given:

$$(d')^2 = \int_t dt (\Delta s(t))^2 / \left( \int_t dt \int_{t'} dt' \, \Delta s(t) \, R(t-t') \, \Delta s(t) \right) \tag{48}$$

where t is a 1-D time, or space, coordinate and the autocorrelation R is given by

$$R(t-t') = \langle n(t) \, n(t') \rangle \xrightarrow{FFT} N(f) \tag{49}$$

In the frequency domain Formula (48) becomes:

$$(d'_{qi})^2 = \left( \int_f df |\Delta S(f)|^2 / \left( \int_f df \, N(f) |\Delta S(f)|^2 \right) \right) \tag{48a}$$

where $N(f)$ is the Fourier transform of $R(t)$.
This formula gives the d' (i.e. SNR) for the so-called "quasi-ideal" observer, i.e. the observer who is not able to "whiten" the coloured noise. For the ideal observer this formula becomes (Thomas, 1969):

$$(d'_i)^2 = \int_f df |S(f)|^2 / N(f) \geq (d'_{qi})^2. \tag{50}$$

Experimental evidence points to an even lower efficiency by human observers, i.e. the real $d'_r$ as compared to the $d'_{qi}$ is of the order of 0.7 (Burgess et al., 1981). A suggestion to explain this was given by Wagner and Brown (1985): a similar decrease of d' is observed if the spatial derivative is presented to an ideal observer. It is well known that some differentiation occurs in the retina, so this suggestion may be taken seriously.

c. *Clinical examples.*

(i) *Echographic imaging* (Smith et al., 1983)
Imaged quantity: echolevel $v(ct) = (v(z))$ (c = sound velocity)
Processed quantity: intensity $I = v^2$

$$SNR_i = (D^2/(S_{cx}S_{cz}))^{1/2} \, C_I \tag{51}$$

or:

$$SNR(I) = m^{1/2} \, C_I \tag{52}$$

where  D = diameter of lesion
       $S_{cx}$ = 0.87 $\lambda_c$F/d = speckle correlation size in lateral
       $S_{cz}$ = 0.26/$\sigma_f$ = speckle correlation size in axial direction
       $\lambda_c$ = wavelength at centre frequency of transmitted pulse
       F = focal length of transducer
       d = diameter of transducer
       $\sigma_f$ = bandwidth of employed transducer
       C = intensity contrast (cf. Formula (46))

Smith et al. (1983) found that the C of the human observer was not inversely proportional to lesion diameter D, but to $D^{-0.8}$ (Fig. 5). Their conclusion was that the human observer had a lower efficiency than

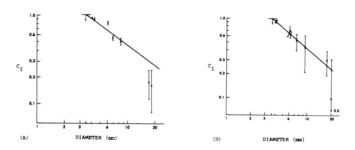

*Fig. 5. Threshold contrast of lesion in echographic images vs. lesion diameter, human observers (Smith et al, 1983).*

*(ii) NMR imaging*

Measured quantity: proton spin density (volts per unit of area, $\Delta v$)

$$SNR_v = \Delta V\ D\ (\pi/4)^{1/2}(XY)^{1/2}(4kT\ R_e\Delta f_t F)^{1/2} \qquad (53)$$

where XY = slice area           $R_e$ = effective electrical resistance
      k = Boltzmann's constant  $\Delta f_t$ = bandwidth of receiver
      T = temperature (Kelvin)  F = noise figure of front-end amplifier.

this thermal noise (0.1 eV) far outweights the photon noise ($10^{-8}$ eV). In addition noise due to the algorithm has to be considered if a projection reconstruction technique is employed (see also following section iii).

*(iii) X-ray, optics, γ-ray)*

Measured quantity: # photons, n

$$SNR_n = C_n(\pi D^2/4)^{1/2} \tag{54}$$

where $C_n = \Delta n/n$
This formula is too simple, because the aperture limitation by the imaging device has not been included. So, more generally (Wagner and Brown, 1985):

$$SNR = (\int_f df \, |\Delta S(f)|^2 \, G^2 \, MTF^2(f)/N(f))^{1/2} \tag{55}$$

where G = macroscopic transfer function of imaging system
MTF = modulation transfer function of imaging system.
For X-ray images the formula is written as:

$$SNR = (\int_f df \, |\Delta S(f)|^2 \, NEQ(f))^{1/2} \tag{56}$$

where NEQ(f) = noise equivalent quanta = $G^2 \, MTF^2(f)/N(f)$ (cf. Dainty and Shaw, 1974).
If this concept is applied to CT imaging, when using the line integrals of the measured quantity, the NEQ has to be multiplied by a filter function πf. In addition to this in CT imaging the employed reconstruction algorithms cause a blur, which can be introduced in Eq.(55) by an additional $MTF_{alg}(f)$ (Wagner and Brown, 1985).


## 6. DISCUSSION

This paper has been confined to the "signal known exactly" detection situation, i.e. both size (shape) and position of the lesion to be detected are considered to be a priori known to the observer. Several reasons can be conceived to expand this situation to more general approaches in case of medical images.
-the human observer may not be able to make adequate use of the a priori knowledge

-the diagnostician generally does not know whether a lesion is present, and if present at which position in the image
-a detectable lesion can be in a range of possible sizes, or even might have to be specified statistically

The first expansion can be described as introducing a spatial "jitter" in the expected position of the lesion. The consequence of employing a cross-correlator is that the mask position in image space is stochastically defined. The problem has been dealt with by Van Trees (1971), cf. Burgess (1986), and this author showed, that in addition to the "normal" cross-correlation (matched filter) a second correlation has to be performed, which is quadratic and covers an extended region of expectance of the lesion.

A second expansion can be split up in two conditions: either position not known, or size not known, to the observer. The position not known condition has been dealt with in classical signal detection theory: "signal known exactly, except for phase", or signal at one of k possible locations. The latter approach will be followed here, while assuming that the diagnostician subdivides the region of interest in k-subregions of equal areas. So, one has to deal with a k-alternative forced choice experiment, with simultaneous display (e.g. Green and Swets, 1966). If the possibility of: "no lesion present at all" is included the mathematics to derive an expression for the correct response probability of "detection and localization" is somewhat complicated but can be found in literature as well (Hershman and Lichtenstein, 1967: Starr et al., 1975: Judy and Swensson, 1981). For a Gaussian probability density function of equal variance the observed correct response fraction can be used to estimate d' from the published tables. If the size of the lesion is not known either, one should consider the possibility that the observer employs a set of matched filters, or cross correlators, in parallel covering the whole range of possible sizes. This description might be well suited to cover the actual situation, when an observer performs a diagnostic task on medical images. Moreover, it fits with the concept of parallel narrow-band filtering by the human visual system (cf. Burgess et al., 1981).

The third approach of a signal specified statistically (cf. also Green and Swets 1966, ch. 6.5) is the most general one and it was treated by Barrett et al. (1985). These authors introduced the Hotelling trace concept in the field of assessment of imaging systems. The parameter employed is a generalized signal to noise ratio, which for the signal known exactly case is equivalent to the detectability index d'.

In this paper the model for the detection problem (cf. Fig. 1) may be just a first approximation to the actual situation. It has been established that an internal noise, i.e. somewhere in between the physical stimulus source and the final detection stage within the brain, is involved additively (cf. Thijssen and Vendrik, 1968, 1978, Burgess et al., 1981). This noise may be considered as a kind of disturbing "background" activity which is always present in the brain. In addition to that also fluctuations in the criterion level(s) a subject has to employ in yes-no and rating procedures might be involved. This latter source of noise when criterion independent, might be modelled in the already mentioned internal noise. Because the observer compares an actual observation with a criterion, the total variance of the noise is just the sum of the variances. However, carefully designed experiments showed that the detectability index d' obtained with yes-no, or rating, experiments corresponded fairly well with the d' obtained from 2-AFC experiments, where criterion noise cannot occur (cf. Green and Swets, 1966). So, for trained observers criterion noise will not be important in the signal known exactly condition. For other conditions it might be true, but experimental evidence is lacking.

If the image contains noise itself, it could be shown by Burgess et al. (1981) that above a certain level (approximately when the noise becomes visible) a multiplicative noise effect occurs. So, the noise level apparent from the d' value exceeds the effect to be expected from the external noise by a fixed fraction. This observation is of great importance to medical images in which photon noise or speckle is involved. It might be an alternative, or an addition, to the explanation proposed by Wagner and Brown (1985) for the reduced efficiency of human observers as compared to the quasi-ideal observer (cf. Chapter 5.1), i.e. the spatial differentiation by the visual system.

In this contribution much emphasis has been placed on the employment of

ROC's to assess the observer's behaviour, or the quality of medical images. If it can be shown that the noise is additive it suffices to perform a yes-no experiment (i.e. a single point at the ROC) to estimate the detectability index, d', i.e. from sensitivity and specificity only. Otherwise two to four points should be estimated in a rating procedure, which of course is more laborious. Then either one of the other d' indices (cf. chapter 1.3) should be employed, or alternatively the area under the ROC, P(A). The assessment of equipment performance can proceed just by estimating a d'-index, involving the average variance preferably, i.e. $d'_a$, or in more general sense the signal to noise ratio (Wagner and Brown, 1985) and the Hotelling trace (Barrett et al., 1985).

*Acknowledgements*

*This work is supported by a grant from the Foundation for Technological Research-STW, and was carried out within the framework of the Concerted Action Programme on Ultrasonic Tissue Characterization and Echographic Imaging of the European Community.*

REFERENCES

Burgess, A.E., Wagner, R.F., Jennings R.J. and Barlow H.B. (1981). Efficiency of human visual signal detection. Science 214, pp.93-94
Burgess, A.E. (1986). On observer internal noise. In: Proceedings of SPIE, Vol. 626, Schneider, R.H. and Dwyer S.J., eds. SPIE, Washington, pp.208-213.
Dainty, J.C. and Shaw, R. (1974). Image Science. Academic Press, London.
Dorfman, D.D. and Alf, E. (1969). Maximum-likelihood estimation of parameters of signal-detection theory and determination of confidence intervalsrating methods data. J. Math. Psychol. 6, pp.487-496.
Egan, J.P. and Clark, F.R. (1966). Psychophysics and signal detection. In: Experimental methods and instrumentation in Psychology. Sidowski J.B., ed. McGraw Hill, New York.
Eijkman, E.G.J. (1979). Psychophysics. In: Handbook of Psychonomics, Vol. 1. Michon, J.A., Eijkman, E.G.J. and De Klerk, L.F.W. eds. North Holland Publ., Amsterdam, pp.303-363.
Green, D.M. and Swets, J.A. (1966). Signal Detection Theory and Psychophysics. Krieger, Huntington.
Hershman, R.L. and Lichtenstein, M. (1967). Detection and localization: an extension of the theory of signal detectability. J. Acoust. Soc. Am. 42, pp.446-452.
Judy, P.F. and Swensson, R.G. (1981). Lesion detection and signal-to-noise ratio in CT images. Med. Phys. 8, pp.13-23.
Metz, C.E. (1978). Basic principles of ROC analysis. Sem. Nucl. Med. 8, pp.283-298.
Patton, D.D., ed. (1978). Seminars in Nuclear Medicine VIII.

Peterson, W.W., Birdsall, T.G., and Fox, W.C. (1954). The theory of signal detectability. Trans. IRE PGIT 4, pp.171-212.

Schulman, A.L. and Mitchell, R.R. (1966). Operating characteristics from yes-no and forced choice procedures. J. Acoust. Soc. Am. 40, pp.473-477.

Simpson, A.J. and Fitter, M.J. (1973). What is the best index of detectability? Psychol. Bull. 80, pp.481-488.

Smith, S.W., Wagner, R.F., Sandrik, J.M. and Lopez, H. (1983). Low contrast detectability and contrast/detail analysis in medical ultrasound. IEEE Trans. SU. 30, pp.164-173.

Starr, S.J., Metz, C.E., Lusted, L.B. and Goodenough, D.J.C. (1975). Visual detection and localization of radiographic images. Radiology 116, pp.533-538.

Swets, J.A. ed. (1964). Signal Detection and Recognition by Human Observers: Contemporary Readings. Wiley, New York.

Swets, J.A., Pickett, R.M. et al. (1979). Assessment of diagnostic technologies. Science 205, pp.753-759.

Swets, J.A. and Pickett, R.M. (1982). Evaluation if Diagnostic Systems: Methods from Signal Detection Theory. Academic Press, New York.

Swets, J.A. (1986). Indices of discrimination or diagnostic accuracy: their ROC's and implied models. Psychol. Bull. 99, pp.100-117.

Thomas, J.B. (1969). An Introduction to Statistical Communication Theory. Wiley, New York.

Thijssen, J.M. and Vendrik, A.J.H. (1968). Internal noise and transducer function in sensory detection experiments: evaluation of psychometric curves and of ROC curves. Perc. Psychophys. 3, pp.387-400.

Thijssen, J.M. and Vendrik, A.J.H. (1971). Differential luminance sensitivity of the human visual system. Perc. Psychophys. 10, pp.58-64.

Van Trees, H.L. (1971). Detection, Estimation and Modulation Theory Part III. Wiley, New York.

Wagner, R.F. and Brown, D.G. (1985). Unified SNR analysis of medical imaging systems. Phys. Med. Biol. 30, pp.489-516.

## Section 2.4
# Applications

# TIME-DOMAIN PHASE: A NEW TOOL IN MEDICAL ULTRASOUND IMAGING

David A. Seggie[1], Sidney Leeman[2], and G. Mark Doherty[3]

*1: University College London, United Kingdom*
*2: King's College School of Medicine and Dentistry, United Kingdom*
*3: ESA-Esrin, Italy*

ABSTRACT

*This paper demonstrates that fluctuations in the temporal phase of medical ultrasound pulse-echo signals can be used as accurate markers of interference-derived artefacts. It is shown that the incorporation of this insight into signal processing techniques can provide an efficient method for noise suppression, and hence maximize the extraction of information concerning the ultrasonic properties of the insonified object. This point is exemplified by applications in both ultrasound pulse-echo imaging and quantitative tissue parameter mapping, where it is shown that a priori knowledge of signal time-domain phase fluctuations leads to a novel speckle reduction technique, and a method for reducing the data necessary for effective pulse-echo attenuation estimation.*

## 1. INTRODUCTION

Conventional medical ultrasound B-mode imaging methods utilize only the envelope of the received pulse-echo sequence, thereby disregarding completely the time-domain phase characteristics of the acquired signals. This paper examines the latent information encoded in the temporal phase or, equivalently, instantaneous frequency (time derivative of phase) of medical ultrasound pulse-echo signals.

The use of signal instantaneous frequency in ultrasound pulse-echo imaging was first suggested by Ferrari et al. (1982). In order to access the possibly useful object-specific information encoded in this signal time-domain attribute, Ferrari et al. (1982; 1984) proposed frequency demodulating the received echo waveforms to form a mapping of signal instantaneous frequency termed an "FM" image. It has since become clear that such mappings do not provide a clear representation of scanned object cross-section (Seggie et al., 1985). As a result, techniques which combine both envelope and time-domain phase information in a single display have been investigated (Seggie, 1986; Aufrichtig et al., 1986). These techniques may be viewed as forms of real zero conversion of the complex zeros associated with the analytic continuation of the pulse-echo signal into the complex time domain (Requicha, 1980). Images formed by these methods are here dubbed real zero conversion (RZC) images. Although preliminary results suggested that RZC images exhibit some advantages over conventional envelope images, most recent experiments (Seggie, 1986) indicate that these advantages are essentially attributable to the non-linear processing of signal envelope inherent in the RZC image formation process, rather than the inclusion of time-domain phase information.

The inability (so far) to utilize pulse-echo signal instantaneous frequency directly for imaging purposes stems from the fact that interference effects are the dominant source of instantaneous frequency modulation. Images formed solely from instantaneous frequency information (FM images) are in effect mappings of strong destructive interference in the received echo waveforms, and therefore do not provide a visually meaningful image of the scanned object cross-section (Seggie, 1986). Given that the contribution of signal time-domain phase to RZC images is of a similar nature, this would indicate that RZC images impart little or no more useful information about the insonified object scattering structure than conventional, albeit non-linearly processed, amplitude images.

The close correspondence between fluctuations in signal instantaneous frequency and destructive interference of echo waveforms, which would appear to severely restrict the utility of FM and RZC images, does nonetheless suggest novel methods for noise suppression in both ultrasound pulse-echo imaging and low resolution quantitative tissue parameter mapping. However, before pursuing these methods it is necessary to elucidate the relationship between signal temporal frequency fluctuations and destructive interference.

## 2. TEMPORAL FREQUENCY FLUCTUATIONS AND INTERFERENCE

Signal envelope, phase, and instantaneous frequency may be described by an analytic signal approach (Gabor, 1946). Given a real, time-varying, band-limited function, $f(t)$, (such as a medical ultrasound pulse-echo signal), the associated analytic signal, $a(t)$, is defined as

$$a(t) = f(t) + jH[f(t)] \quad (1)$$

where $j = \sqrt{-1}$, and $H[f(t)]$ denotes the Hilbert transform of $f(t)$. Signal envelope, $e(t)$, is given by

$$e(t) = |a(t)| \quad (2)$$

and instantaneous phase, $p(t)$, by

$$p(t) = \arg[a(t)]. \quad (3)$$

Signal instantaneous frequency, $p(t)$, is then given by

$$\dot{p}(t) = (1/2\pi)dp/dt = (1/2\pi)d/dt(\arctan(H[f(t)]/f(t))). \quad (4)$$

Note that signal phase as defined above is modulo $2\pi$ and must therefore be "unwrapped" prior to differentiation to avoid discontinuites in $\dot{p}(t)$ whenever $p(t)$ extends beyond $2\pi$. Here, all derived signal instantaneous frequency functions were obtained via the analytic signal approach, using a standard unwrapping algorithm after Schafer (Oppenheim and Schafer, 1975).

The quantities defined in Eq. (2) and Eq. (4) are illustrated in Figs. 1 - 3. Figure 1 shows a segment of a typical ultrasound pulse-echo signal obtained from a tissue-equivalent phantom (Nuclear Associates, model 84-317), using a commercial (KB-Aerotech) 3.5 MHz centre frequency, 13mm diameter, focused (4-10 cm) transducer. The phantom material is a graphite loaded hydrogel claimed by the manufacturers to simulate the ultrasonic properties of human liver tissue with respect to attenuation,

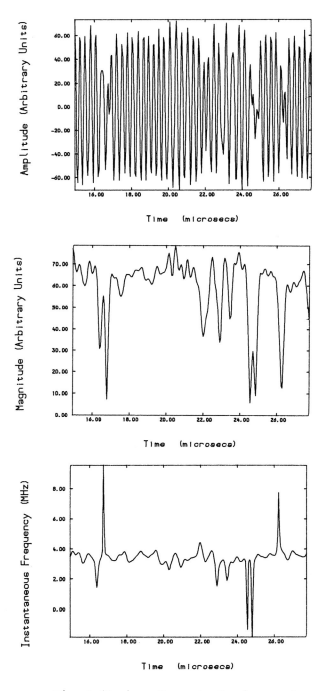

Fig. 1 (Top)     Segment of echo waveform from tissue phantom.
Fig. 2 (Middle)  Amplitude envelope of signal shown in Fig. 1.
Fig. 3 (Bottom)  Instantaneous frequency of signal shown in Fig. 1.

scattering, and propagation velocity. The signal was digitized at a sampling rate of 20 MHz, to a maximum amplitude resolution of 8 bits. The envelope and instantaneous frequency occur when the envelope exhibits a local minimum. This correspondence between fluctuations in signal envelope and instantaneous frequency can be understood in terms of the zero locations of the function obtained by continuation of the analytic signal associated with the echo waveform into the complex time domain (Voelcker, 1966).

The analytic signal, $a(t)$, associated with a real bandlimited echo waveform, may be represented by inverse Fourier transformation as

$$a(t) = \int A(f) \exp(2\pi jft) df \qquad (5)$$

where $A(f)$ is the Fourier transform of $a(t)$. $a(t)$ may be continued into the complex time domain by replacing the real variable $t$ in the previous expression by the complex variable, $z = u + jv$, giving,

$$a(z) = \int [A(f)\exp(-2\pi fv)]\exp(2\pi jfu) df. \qquad (6)$$

The above integral exists for all positive, and all finite negative values of $v$. $a(z)$ is therefore free of singularities in the finite z-plane and is said to be an entire function. Using Schwarz's inequality and the fact that $f(t)$ is bandlimited, it is straightforward to prove that

$$a(z) = O(\exp(k|z|)) \qquad (7)$$

where $O$ denotes "order of", and $k$ is a finite constant related to the bandwidth of $f(t)$. The above expression indicates that $a(z)$ belongs to a class of functions known as entire functions of exponential type. Hence, in a manner analogous to the factorization of algebraic or trigonometric polynomials, $a(z)$ may be expressed as an expansion product and represented unambiguously by its roots or zeros (plus one other parameter) (Boas, 1954). This in turn means that $f(t)$ is specified completely (to within a multiplicative factor) by a knowledge of the real and complex zeros of $a(z)$. If these zeros are distinct, then, in the region of one such zero of order $n_0$, located at $z = z_0$, $a(z)$ may be written as $a(z) = (z - z_0)^{n_0} s(z)$

here s(z) is some zero-free function. It follows that

$$d/dz(\ln a(z)) = n_o/(z - z_o) + d/dz(\ln s(z)). \tag{8}$$

Thus each isolated zero of a(z) will contribute a first order pole to d/dz(ln a(z)) (with a corresponding residue of $n_o$). Now consider the integral

$$\int_c dz [d/dz(\ln a(z))]/z(z - t) \tag{9}$$

where c is a closed contour containing the points z=0, t and may be thought of as an infinite (in the limit) circle centred on the origin. Evaluation of this integral shows that

$$\dot{p}(t) = k_1 + \Sigma_o [n_o v_o/((t - u_o)^2 + v_o^2)] \tag{10}$$

$$d/dt(\ln e(t)) = k_2 + \Sigma_o [n_o (t - u_o)/((t - u_o)^2 + v_o^2)] \tag{11}$$

where $k_1$ and $k_2$ are real constants.

Consider the contribution to $\dot{p}(t)$ and d/dt(ln e(t)) from an isolated zero, order $n_o$, located at $z=u_o + jv_o$. Equation (10) shows that as time goes from $t<u_o$ to $t>u_o$, the contribution to $\dot{p}(t)$ goes through a maximum of $n_o/v_o$ at $t=u_o$. The zero therefore encodes a local excursion in $\dot{p}(t)$ at $t=u_o$; the excursion is positive if the zero is in the upper half of the z-plane, negative if it lies in the lower half plane. Equation (11) shows that through the same time interval, the contribution of the zero to d/dt(ln e(t)) goes from negative to positive, regardless of whether the zero lies in the upper or lower half plane. The zero thus encodes an envelope minimum at $t=u_o$. Hence, from the viewpoint of the complex zero locations of a(z), it is clear that local minima in the pulse-echo signal envelope due to the destructive interference of overlapping echoes from closely-situated scatterers, are accompanied by fluctuations (both negative and positive) in signal instantaneous frequency. (Observe however, that this is not always so. Given a complex conjugate zero pair, located at (u,v), (u,-v): at time t=u the net contribution to $\dot{p}(t)$ is zero. The zero pair nevertheless encode an envelope minimum at t=u).

## 3. NOVEL SPECKLE REDUCTION TECHNIQUE

The relationship between instantaneous frequency fluctuations and image speckle is strikingly illustrated in Figs. 4 and 5. Figure 4 shows the envelope (B-mode) image generated by linearly scanning a 2 x 3 cm² region of the tissue phantom containing a 4 mm diameter cylindrical low scatter zone. A linear magnitude to grey-scale mapping was employed. Different axial and lateral sampling rates result in "stretching" of the image in

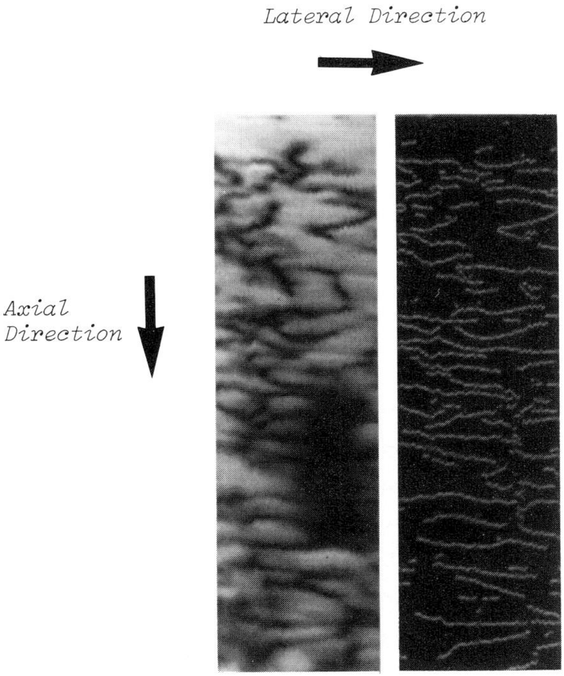

*Fig. 4 (Left)*    *B-mode image of tissue phantom region containing low scatter zone.*

*Fig. 5 (Right)*    *Destructive interference map for Fig. 4.*

the axial direction. Bearing in mind the known scattering structure of the scanned object cross-section, it is clear that the fine image structure exhibited by Fig. 4 is an example of a well developed speckle pattern. Image speckle is the result of intereference of echoes from unresolved scatterers, and it is determined primarily by features of the imaging system rather that of the insonified object. For this reason, speckle is a serious image artefact and its removal is a major objective in the improvement of image quality. The basis of our new speckle reduction technique is the realization that speckle arises from destructive interference and that the presence of speckle can therefore be marked locally by fluctuations in signal temporal frequency. Consequently, this speckle reduction technique is, in contradistinction to all currently utilized methods, deterministic in nature. Figure 5 shows a map of destructive interference effects generated by displaying regions of large instantaneous frequency excursions, as calculated from each echo waveform. Here, only excursions greater than 0.65MHz from the centre frequency of the probing pulse (3.5 MHz) are marked. Note that Fig. 5 provides an almost perfect outline of the dark boundaries of individual speckle elements in Fig. 4.

Having precisely identified the location of speckle in the envelope image, all that remains is to re-assign the image values coincident with large instantaneous frequency excursions, as specified for example by Fig. 5. A number of re-assignment procedures are currently under investigation; to date the best results have been obtained by the following approach. On a line-by-line basis, an image point indicated by the constructed interference map is replaced by the average of the two nearest maxima on either side of the point. This new image then undergoes local one-dimensional averaging only at those points specified by the interference map. Further speckle reduction is achieved by effectively broadening the interference map and repeating the above process. Figure 6b shows the result of employing this speckle reduction technique on the envelope image depicted in Fig. 4. In this case the length of the averaging window was 0.6 mm, (approximately $1.5\lambda$). The speckle reduced image has been axially compressed to avoid image distortion; an axially compressed version of the original envelope image is also provided for comparison, (Fig. 6a).

  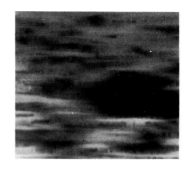

Fig. 6a Axially compressed version of image in Fig. 4.

Fig. 6b Speckle reduced version of image in Fig. 6a.

Figures 6a and 6b show that the effect of the speckle reduction procedure is to de-emphasize, or "fill in", image minima arising from destructive interference, while simultaneously maintaining the delineation of the low scatter zone boundaries. Although the technique in its present form does not completely remove the effect of speckle on the image, there is no doubt that the impression of artefactual fine structure is diminished. It is encouraging to note that even at this preliminary stage, the technique appears to perform at least as favourably as other established speckle reduction methods (Seggie and Leeman, 1986).

4. LOW RESOLUTION QUANTITATIVE TISSUE PARAMETER MAPPING

The frequency dependence of ultrasound attenuation of soft human tissue is thought to have great potential as an indicator of tissue state. Therefore, pulse-echo attenuation estimation figures prominently in most schemes for constructing low resolution quantitative tissue parameter mappings. Over the years, the estimation methods which have received most attention are the spectral difference (Kuc and Schwartz, 1979), spectral smoothing (Robinson et al., 1979), and spectral moments (Fink and Hottier, 1982) approaches. These methods proceed by using simplified models of both wave propagation and the received echo waveforms, to relate tissue attenuation to the measured non-stationarity of backscattered echoes. Specifically, the following assuptions are made. Firstly, that short time duration data segments can be reasonably modeled as the convolution of a propagator function and a tissue reflector sequence. That is,

$$E(i;f) = P(i;f)T(i;f) \tag{12}$$

where E is the signal segment magnitude spectrum, T is an estimate of the true tissue magnitude spectrum, and P is the pulse magnitude spectrum. The second assuption is that the pulse magnitude spectra appropriate for two segments i and j (where j is deeper than i) are related as follows

$$P(j;f) = P(i;f)\exp(-2dQ(f)) \tag{13}$$

where d is the segment separation distance, and $Q(f)$ is the frequency dependent tissue attenuation coefficient. From Eqs. (12) and (13) it follows that

$$Q(f) = 1/2d[\ln(E(i;f)/E(j;f)) + \ln(T(j;f)/T(i;f))] \tag{14}$$

Inspection of Eq. (14) shows that the only term on the right-hand-side which cannot be estimated is the second log ratio term. This term represents the contribution of interference effects to pulse-echo signal non-stationarity. That is, if the echo waveform is considered to be a series of overlapping echoes, the second log ratio term describes the scalloping of the signal spectrum due to the random interference of these overlapping echoes. Experience with real data has shown that this term cannot be ignored, and that it can lead to large statistical fluctuations in the estimation of $Q(f)$. To date, the stratagem which has been adopted in order to minimize these statistical fluctuations is to average over vast quantities of data. For example, in his most recent analysis of the spectral difference approach, Kuc (1984) found it necessary to average $Q(f)$

estimates over more than 3000 echo waveforms. The need to acquire and analyse such large data sets is a major obstacle to the implementation of effective pulse-echo attenuation estimation.

However, the relationship between interference and temporal frequency outlined earlier, suggests a method for reducing the amount of data necessary to produce a stable, reliable $Q(f)$ estimate. Given that large instantaneous frequency excursions pin-point the location of destructive interference in the echo waveform, interference-corrupted data segments can be readily identified and thus excluded from the averaging process.

The feasibility of this approach has been demonstrated by again using tissue phantom data. The attenuation coefficient of the phantom material is claimed by the manufacturers to be linearly proportional to frequency, with a slope value, B, of 0.4 dB/cm.MHz. Figure 7 shows a 2 cm portion of an echo waveform obtained from the phantom; vertical dashed lines have been drawn at intervals of approximately 0.25 cm. The instantaneous frequency for this signal is shown in Fig. 8. Again, vertical lines have been drawn to divide the function into 8 equal segments. Using all 8 data segments, the B estimate obtained via the spectral moments approach is an unacceptably negative -0.03 dB/cm.MHz. After rejecting data segments which exhibit strong destructive interference effects, as indicated by the presence of large instantaneous excursions, the new estimate obtained is 0.17 dB/cm.MHz. Note that this new positive value is now consistent with the manufacturers' claim that the attenuation of the phantom material increases with frequency. Here, data segments containing excursions from the mean instantaneous frequency value of more than 4 MHz (twice the transducer bandwidth) were considered to be corrupted by interference effects and therefore excluded from the analysis. The same data rejection procedure was repeated on a second echo waveform from the tissue phantom, and the results are given in Table 1. Note that the rejection of interference-corrupted data segments significantly improves the consistency of the B estimates from the two echo waveforms. Assuming the initial promise of this error minimization scheme is fulfilled, a drastic reduction in the amount of data necessary for effective pulse-echo attenuation estimation should follow.

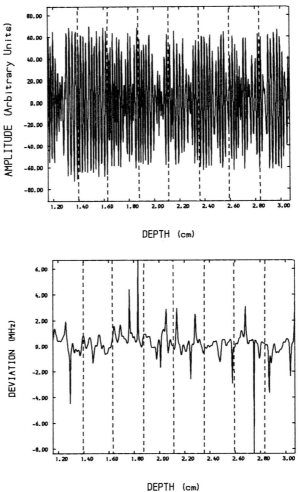

Fig. 7 (Top)  2 cm portion of echo waveform from tissue phantom.
Fig. 8 (Bottom) Instantaneous frequency of signal shown in Fig. 7.

Table 1  β estimates before and after rejection of interference-corrupted data segments

| ECHO WAVEFORM | BEFORE (dB/cm.MHz) | AFTER (dB/cm.MHz) |
|---|---|---|
| 1 | − 0.03 | 0.17 |
| 2 | 0.21 | 0.42 |

## 5. CONCLUSIONS

This paper has shown that in both ultrasound pulse-echo imaging and low resolution quantitative tissue parameter mapping, consideration of the behaviour of temporal frequency provides a means by which the effect of interference-derived artefacts may be minimized. Consequently, both a novel speckle reduction technique and a refinement to current spectral domain pulse-echo attenuation estimation techniques were demonstrated.

## REFERENCES

Aufrichtig, D., Lottenberg, S., Hoefs, J., Ferrari, L.A., Friedenberg, R.M., Kanel, G., Cole-Beuglet, C., and Leeman, S. (1986). Frequency-demodulated US: evaluation in the liver, *Radiology* 160, pp. 59-64.

Boas, R.P. (1954). *Entire functions*, Academic Press, New York.

Ferrari, L.A., Jones, J.P., Gonzales, V., and Behrens, M. (1982). Acoustical imaging using the phase of echo waveforms. In: *Acoustical Imaging*, vol. 12, E.A. Ash and C.R. Hill (eds.), Plenum Press, New York, pp. 635-641.

Ferrari, L.A., Sankar, P.V., Fink, M., Shin, S.B., and Chandler, P. (1984). Use of signal phase in medical ultrasound, *Acta Electronica* 26, pp. 111-120.

Fink, M. and Hottier, F. (1982). Short-time Fourier analysis and diffraction effect in biological tissue characterization. In: *Acoustical Imaging*, vol. 11, J.P. Powers (ed.), Plenum Press, New York, pp. 493-503.

Gabor, D. (1946). Theory of communication, *J. Inst. Elect. Eng.* 93, pp. 429-441.

Kuc, R. (1984). Estimating acoustic attenuation from reflected ultrasound signals: comparison of spectral shift and spectral-difference approaches, *IEEE Trans. Acoust., Speech, Signal Processing* ASSP-32, pp. 1-6.

Kuc, R. and Schwartz, M. (1979). Estimating the acoustic attenuation coefficient slope for liver from reflected ultrasound echoes, *IEEE Trans. Sonics and Ultrason.* SU-26, pp. 353-362.

Oppenheim, A.V. and Schafer, R.W. (1975). *Digital Signal Processing*, Prentice-Hall International, London.

Requicha, A.A.G. (1980). The zeros of entire functions: theory and engineering applications, *Proc. IEEE* 68, pp. 308-328.

Robinson, D.E. (1979). Computer spectral analysis of ultrasonic A-mode echoes. In: *Ultrasonic Tissue Characterization II*, M. Linzer (ed.), National Bureau of Standards, Spec. Pub. 525, pp. 281-286.

Seggie, D.A. (1986). *Digital processing of acoustic pulse-echo data*, PhD Thesis, University of London.

Seggie, D.A., Doherty, G.M., Leeman, S., and Ferrari, L.A. (1985). Ultrasonic imaging using the instantaneous frequency of pulse-echo signals. In: *Acoustical Imaging*, vol. 14, A.J. Berkhout, J. Ridder, and L.F. Van der Wal (eds.), Plenum Press, New York, pp. 487-496.

Seggie, D.A. and Leeman, S. (1986). A deterministic approach towards ultrasonic speckle reduction, to appear in *IEE Proc. - A*

Voelcker, H.B. (1986). Toward a unified theory of modulation, *Proc. IEEE* 54, pp. 340-353.

# PERFORMANCE OF ECHOGRAPHIC EQUIPMENT AND POTENTIALS FOR TISSUE CHARACTERIZATION

J.M. THIJSSEN AND B.J. OOSTERVELD

*University of Nijmegen*
*The Netherlands*

ABSTRACT

*The imaging performance of echographic equipment is greatly dependent on the characteristics of the ultrasound transducer. The paper is confined to single element focussed transducers which are still widely employed in modern equipment. Extrapolation of the results to array transducers may be made to a certain extent. The performance is specified by the 2 dimensional point spread function (PSF) in the focal zone of the transmitted sound field. This PSF is fixed by the bandwidth of the transducer when the axial (depth) direction is considered, and by the central frequency and the relative aperture in the lateral direction. The PSF concept applies to the imaging of specular reflections and the PSF is estimated by scanning a single reflector. In case of scattering by small inhomogeneities within parenchymal tissues the concept of "speckle" formation has to be introduced. The speckle is due to interference phenomena at reception. It can be shown that the average speckle size in the focus is proportional to the PSF above defined for specular reflections. The dependencies of the speckle size on the distance of the tissue to the transducer (beam diffraction effects) and on the density of the scatterers were explored. It is concluded that with the necessary corrections tissue characterization by statistical analysis of the image texture can be meaningful.*

## 1. INTRODUCTION

The advent of "gray scale" display of the echographic information opened new horizons for this still expanding modality of medical imaging. Until approximately ten years ago the equipment contained a compound scanning gantry and a bistable storage oscilloscope for recording the images. This technique enabled to display mainly the outlines of anatomical structures: organs, large bloodvessels, bone, etc., whereas the parenchymal tissues were depicted marginally, if at all. The introduction by industry of the so-called scan-converters (analogue and digital) brought about the display of 16 to 64 gray

levels (4 to 6 bits) and thereby of the weak tissue echo signals. It
should be kept in mind, however, that the range of echo levels from
the highest (bone) to the lowest (nervous tissue) reflectivity
comprises approximately 80 dB ($10^4$ on amplitude scale) according to
Kossoff et al. (1976). So even with 64 gray levels either a selection
(preprocessing), or a so-called "level and window" technique, after
large range logarithmic amplification and digital storage of the echo
signals (postprocessing) is needed . The main reason for the large
difference in reflectivity level of the gross anatomical structures,
as compared to tissue itself, is to be found in the physical nature of
the reflections. The gross structures are acting as a mirror and the
specular reflections are obeying Snell's law. Specular reflections in
general occur if the reflecting structure is much larger than the
wavelength of the employed radiation, i.e. of the order of
millimeters, or more, in case of diagnostic ultrasound . A different
kind of physical phenomenon is present within the tissue, i.e.
diffractive reflections, or better named scattering. Although it is
probable that at high frequencies ($\geq$ 5 MHz) also a contribution at
cellular level is present, it is often assumed that the main cause of
scattering is to be found in the microvasculature and the collagen
matrix of the tissues (Nicholas, 1982). In any case the structures
contributing to the scattering are in the sub-millimeter range of
size.

The interesting question now arises to which extent the echographic
images contain information about the condition of the tissues. Several
authors successfully applied the concepts of speckle caused by
laserlight to the generation of echographic images (e.g. Burckhardt,
1978; Abbott and Thurstone, 1979; Bamber and Dickinson, 1980; Flax et
al., 1981). The analogy of narrowband (pulsed) ultrasound to coherent
light is quite obvious if one realizes that the ultrasonic transducer
produces a coherent radiation as well and, therefore, the far field
can be analyzed according to the Fraunhofer' theory, while at
reception the transducer acts as a phase sensitive receiver. The
before mentioned authors derived the first order gray level statistics
of echographic images in the limit case of a high density of
scatterers in a tissue. Wagner et al. (1983) derived the second order
statistical properties of the echograms, by using the two-dimensional

autocorrelation function of the speckle. Also these authors confined themselves to high scatterer densities and the farfield (or the focal zone in case of a focused transducer). Thijssen and Oosterveld (1985) and Oosterveld et al. (1985) extended this work and investigated both the influence of the characteristics of the sound beam (the so-called beam diffraction effects) and the dependence of the texture and compare the commonly used performance characteristics of echographic equipment to statistical texture parameters of the images, thus presenting the opportunity to answer the important question: can echographic images depict the condition of tissues?

## 2. EQUIPMENT PERFORMANCE CHARACTERISTICS: TWO-DIMENSIONAL POINT SPREAD FUNCTION

Although the echographic display is dependent on many equipment characteristics and control settings (non-linear amplifier gain, time gain compensation, post-processing facilities) the following discussion will be confined to a linear, time-independent transfer of derived echo amplitudes to image gray level. In this case the image quality for specular reflections is exclusively governed by the 2D point spread function (PSF) of the employed transducer. The 2D PSF in the focus is furthermore considered to be separable in an axial direction (i.e. depth) and a lateral direction (i.e. in a plane perpendicular to the beam axis); cf. Wagner et al.(1983). It can be shown that the echo from a point-like reflector placed on axis in the focus is an exact replica of the transmitted sound pulse (Verhoef et al., 1984). This pulse is generally approximated by a Gaussian envelope over a high frequency carrier:

$$v(t) = 1/\sqrt{(2\pi\sigma_t^2)} \exp\left[-(t-t_0)^2/2\sigma_t^2\right] \cos(2\pi f_c t) \qquad (1)$$

where  $v(t)$ = waveform of echo amplitude
$t$ = time
$\sigma_t$ = "standard deviation" of Gaussian envelope
$t_0$ = delay time.

By Fourier transformation the corresponding amplitude spectrum can be obtained:

$$V(f) = 1/\sqrt{(2\pi\sigma_f^2)} \exp[-(f-f_c)^2/2\sigma_f^2] \qquad (2)$$

where  V(f) = amplitude spectrum corresponding to time waveform v(t)
   f    = frequency.
   $\sigma_f$   = "standard deviation" of Gaussian spectrum
   $f_c$   = "centre" frequency.

It can easily be shown, that

$$\sigma_t \sigma_f = (2\pi)^{-1} \qquad (3)$$

so the axial PSF can either be specified in the time domain or in the frequency domain. The transfer from the time domain to the spatial domain (depth range) is made by multiplying $\sigma_t$ by the velocity of sound in biological media (approximately 1500 m.s$^{-1}$), so

$$\sigma_z = 1.5 \; \sigma_t/2 \text{ (mm)}. \qquad (4)$$

Division by 2 is needed, because in echography the ultrasound has to travel a distance twice.

The -6 dB width of the time waveform follows from:

$$\Delta z \; (-6 \text{ dB}) = 2.355 \; \sigma_z \text{ (mm)} \qquad (5)$$
$$= 1.77 \; \sigma_t \; (\sigma_t \text{ in } \mu s)$$
$$= 0.28/\sigma_f \; (\sigma_f \text{ in MHz})$$

which is the commonly employed full-width at half maximum (FWHM) of the PSF in the axial direction and at the same time it can be used to specify the axial resolution.

The width of the sound field in the focus can be analytically derived if continuous wave (i.e. "monochromatic") ultrasound is used (Kossoff, 1979; Wagner et al., 1983). The amplitude is described by a Bessinc-function, the first zeroes of which are found at lateral displacements $x_o$ with respect to the beam axis at

$$x_o = \pm 1.22 \; \lambda_c F/D \qquad (6)$$

where $\lambda_c$ = wavelength at centre frequency, F = focal length of transducer and D = diameter of transducer.

Hence, the FWHM of the lateral PSF will be proportional to the above given displacement of the 1st order zeroes:

$$\Delta x (-6dB) = 1.02 \lambda_c F/D. \tag{7}$$

The question remains whether this expression might be employed when pulsed transducers are considered. The simulations by Verhoef et al. (1984) indicate that this can be done to a fair approximation.
In conclusion: the expressions (5) and (7) approximate the axial and lateral FWHM of the PSF at the focus and the question to be answered now is, how these measures relate to the texture of B-mode echograms. The nearfield behaviour of both the axial and the lateral PSF cannot be analytically obtained, and has to be estimated either by simulations while applying the impulse response method (e.g. Verhoef et al., 1984), or experimentally.

## 3. TEXTURE IN ECHOGRAPHIC IMAGES

### a. *Acoustic Tissue Model*

The tissues are modelled as a homogeneous attenuating medium containing randomly distributed point-like scatterers (i.e. isotropic scattering) of a density n per cubic centimeter. The attenuation is exponential and the attenuation coefficient $\mu(f)$ is proportional to the frequency of the insonating ultrasound:

$$v(z) = v(o) \exp[-\mu f 2z)] \tag{8}$$

where $\mu$ = attenuation coefficient $(cm.MHz)^{-1}$
$z$ = depth (= $ct_0/2$).

Since images are obtained from backscattered energy, the acoustic path length equals two times the depth, i.e. $2z$.
The density of the scatterers is assumed to be relatively high, since the following theory applies to the limit case of an infinitely high density.

### b. *Generation of B-mode images and 1st order statistics*

The ultrasound beam is moved in a single plane, the scan plane, and at every lateral position all the scatterers within the resolution cell (volume comprised by pulse length times effective beam area) at every depth contribute to the echographic signal. The ultrasound transducer is a phase sensitive receiver so, according to the Huygens' principle, a coherent accumulation of backscatterings occurs. This interference process can be described as a summation of a large number of phasors, with a uniformly distributed phase, between 0 and $2\pi$, and of which the real and imaginary parts have a joint, circular, Gaussian probability distribution (Goodman, 1975; Wagner et al., 1983).

The phasor magnitude v, i.e. the demodulated echogram, then has a Rayleigh probability distribution:

$$p(v) = v/\sigma^2 \exp(-v^2/2\sigma^2) \qquad v > 0 \qquad (9)$$

where $\sigma^2$ = variance of phasor = signal power.
The Rayleigh pdf has the following properties:

$$\mu_v = \langle v \rangle = \sqrt{(\pi\sigma^2)} \qquad (10)$$

$$\sigma_v^2 = \langle v^2 \rangle - \langle v \rangle^2 = (4-\pi)\sigma^2$$

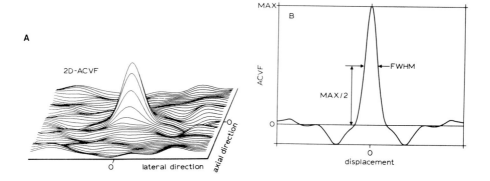

Fig. 1A: *Two-dimensional autocovariance function of B-mode echogram (Oosterveld et al., 1985)*
1B: *Definition of Full Width at Half Maximum (FWHM).*

hence
$$\mu_v/\sigma_v = \sqrt{(\pi/(4-\pi))} = 1.91. \tag{11}$$

This so-called Signal-to-noise Ratio (SNR) thus is a constant. The average signal level $\mu_v$ contains information about the scatterer density.

### c. Speckle, texture and 2nd order statistics

The statistical process underlying echographic image formation produces a "speckle" pattern (2-dimensional) like in optical imaging with coherent individual scatterers cannot be resolved and the speckle pattern will contain only information about the physical properties of the employed radiation source. In other words the texture, i.e. the spatial distribution of light and dark gray levels in the echogram, in the limit case of a high scatterer density, is just a speckle pattern. Before going into the question whether, or not, it is possible at all to obtain tissue information, it is necessary to consider the way to characterize the spatial properties of the texture in echograms.

The average speckle size can be characterized by the autocovariance function (Papoulis, 1965)

$$ACVF(x,z) = \iint_{area} dx'dz' \; (v(x',z')-\langle v\rangle)(v(x'+x,z'+z)-\langle v\rangle) \tag{12}$$

where $\langle v \rangle$ = echoamplitude (gray level) averaged over the image area (see Fig. 1A).
The ACVF can be quantified by its full width at half maximum (FWHM) in the axial and in the lateral direction, respectively (Fig. 1B). It was shown (Wagner et al. 1983) that in the focal zone of the transducer

$$FWHM_{lat} = 0.80 \; \lambda_c F/D' = 0.86 \; \lambda_c F/D \tag{13}$$

with $\lambda_c$ and F defined like in Formulas (6) and (7) and $D' = D/1.08$; also (Wagner et al., 1983)

$$FWHM_{ax} = 2.17 \; \sigma_z = 0.26/\sigma_f \tag{14}$$

Equations (13) and (14) clearly indicate that in the focal zone:
- the average speckle size is exclusively dependent on the transducer characteristics
- the average speckle size is proportional to the point spread function of the transducer (cf. formulas (5) and (7)):

$$FWHM_{lat} = 0.84 \, \Delta x \; (-6 \text{ dB}) \tag{15}$$
$$FWHM_{ax} = 0.93 \, \Delta z \; (-6 \text{ dB}).$$

*d. Dependencies of B-mode image texture*

*(i) Diffraction effect*

In general focussed transducers are employed in medical ultrasound applications, because of the improved resolution in the focal zone. This argument applies to specular reflections and as follows from the previous chapter also to the lateral "speckle" dimension if the density of scattering structures is above a certain limit. The focussing causes a depth dependent intensity distribution of the sound beam. This beam effect induced by focussing is additional

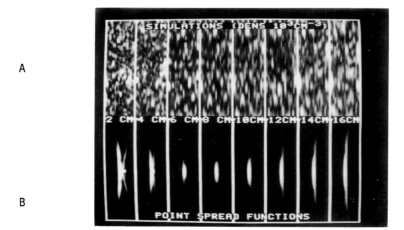

*Fig. 2A: B-mode scans (after gray level normalization) of a scattering medium, with increasing distance to the transducer (from left to right), results of simulations with scatterer density $10^3 cm^3$.*
*2B: Point spread function obtained by simulations with same transducer as in 2A (Thijssen and Oosterveld, 1986).*

to the diffraction pattern produced by an ultrasound transducer. The overall effect of the ultrasound beam on the image texture has been termed the "diffraction effect" (Thijssen and Oosterveld, 1985; Oosterveld et al., 1985).

A clear illustration of the diffraction effect of a focussed transducer is shown in Fig. 2. The bottom row (Fig. 2B) shows the point spread function of the simulated transducer against depth. The gradual decrease of the lateral width (vertical) towards the focus (8 cm) and the increase beyond is obvious. The diffraction effect on the texture can be seen in Fig. 2A. Contrary to the lateral width of the PSF the speckle size is very small near to the transducer (left) and increases gradually going through the focal zone into the far field. Due to the focussing the mean amplitude level of the backscattered ultrasound increases, reaches a (Oosterveld et al., 1985). This effect cannot be observed in Fig. 2A, because the gray levels have been normalized.

*(ii) Attenuation*

The diffraction effect is enhanced by the attenuation of ultrasound. This can be understood from the observation that the attenuation coefficient in most biological tissues is proportional to the frequency (Eq. 7). The attenuation therefore causes effectively a downward shift of the centre frequency, which shift is proportional to the depth. Because the lateral size of the speckle (in the focus) is given by Eq. (13) it follows that

$$\Delta x(-6dB) = 0.86 \ cF/f_c D \qquad (16)$$

where $c$ = sound velocity, $f_c$ = centre frequency.

Therefore, the lateral speckle size will display an additional increase proportional to the depth, due to the attenuation. This effect appeared to be of the order of a factor of 2 when the attenuation coefficient is 0.9 dB/cm.MHz (Oosterveld et al., 1985).

*(iii) Scatterer density*

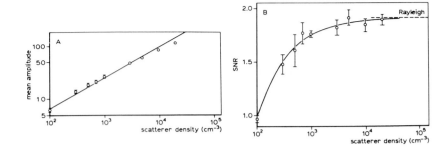

Fig. 3A: Average backscatterer amplitude (gray level) vs. scatterer density. Drawn line: square root dependence.
3B: SNR vs. scatterer density, limit SNR = 1.91 Rayleigh statistic (Oosterveld et al., 1985).

It can theoretically be shown (Flax et al., 1981) and it has been verified by simulations (Thijssen and Oosterveld, 1985; Oosterveld et al., 1985) that the backscatter echoamplitude level is proportional to the square root of the density of the scatterers. In other words the backscattered intensity is proportional to this density (Fig. 3A). Furthermore, the simulations showed a monotonous increase of the texture signal to noise ratio (SNR) with scatterer density until at high levels the theoretical limit of the Rayleigh probability density was reached (i.e. SNR = 1.91). As can be seen in Fig. 3B this limit is approached at densities over $5.10^3$ cm$^{-3}$. It may be remarked that the characteristics of a commercial transducer for abdominal echography were used in the simulations ($f_c$= 3.5 MHz, D = 19 mm, F = 9 cm). The second order statistics, i.e. the average speckle size in axial and lateral directions decreased with increasing density until the limits were reached given by equations (13) and (14), respectively. The limits were like before approached at density levels of the order of $5.10^3$ cm$^{-3}$. The characteristic distance of collagen rich structures in the liver is of the order of 1 mm (cf. Nicholas, 1982). Therefore, if pathological changes bring about an increase of this distance the density will drop below $10^3$ cm$^{-3}$ and changes in the texture might be expected. This effect is convincingly shown in Fig. 4A. This echogram was obtained from an alcoholic liver and it shows a quite homogeneous and fine texture. The bottom image (Fig. 4B) of a cirrhotic liver obtained with the same equipment displays a much coarser texture. So,

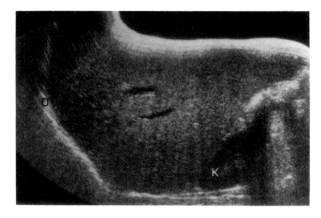

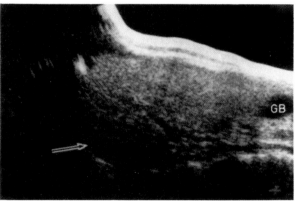

*Fig. 4A: Echogram of alcoholic liver disease, fine texture. K= kidney.*
*4B: Echogram of liver cirrhosis, coarse texture, enhanced attenuation (see arrow), GB= gall bladder (Cosgrove, 1980).*

it might be relevant to perform a quantitative, statistical analysis of B-mode echograms to discover more subtle texture changes than those displayed in these extreme cases. In addition, it should be remarked that the before discussed effects of beam diffraction and attenuation on the texture are obvious in Fig. 4A (top is skin, lateral direction is horizontal, depth direction is vertical).

*e. Conclusion*

-the texture of B-mode images may bear information about the condition of tissues (dependent on transducer characteristics relative to tissue structure)

—the beam characteristics and the attenuation (within the tissues) greatly influence the image texture of echograms and should be corrected for prior to image analysis.

## 4. DISCUSSION

The diffraction and the focussing of the ultrasound beam have been shown to be responsible for the gradual coarsening of the texture with increasing depth. It would be greatly helpful for the diagnostician and absolutely necessary for computer analysis to remove these effects from the echographic information prior to its employment. This can be achieved satisfactory by the so-called multi-focus technique, where an in order to obtain a sequential set of focussed beams at e.g. 5 or 7 depth zones. A second possibility is to measure the diffraction effect vs. depth with a scattering phantom and subsequently correct the radiofrequency data before any imaging processing is carried out. The latter technique has the advantage that also the correction for the frequency dependent attenuation can easily be incorporated. The images thus obtained can be used for one of the two clinical questions: detection and differentiation of focal lesions, or differentiation of diffuse diseases.

This discussion will be confined to the latter question. The question can be translated into clinical terms: to which extent become pathological changes (on a histological level) apparent in echographic images? As an example: in the liver three different classes of reflectors are present: liver cells (< 20 µm), collagen-rich matrix of the triads of Kiernan and thin vessels (characteristic distance of 1 mm) and finally larger vessels (> 2 mm diameter). The wavelength of the 3.5 MHz ultrasound is of the order of 0.4 mm, so clearly the cells give rise to diffractive (Rayleigh) scattering (intensity proportional to $f^4$), the triads to specular scattering ($f^1$ to $f^2$) and the larger vessels to specular reflections. The in vitro backscattering measurements by Nicholas (1982) could be fitted by a three term polynomial, which yields the following contributions: 38% diffractive scattering, 40% specular scattering and 22% specular reflections (frequency range 0.7 - 5 MHz). It is quite conceivable that the

pathological changes in tissues shift this balance thereby letting other characteristic distances become dominant, causing a change in the second order properties i.e. the average size of the speckle (cf. Fig. 4). Also changes in the first order statistics are conceivable from the physics, e.g. due to fatty infiltration, fibrosis, etc. which primarily inflict the reflectivity level. Another important conclusion from the results obtained by Nicholas (1982) is that the tissue model employed in the present paper is probably too simple and a mixed diffractive-specular scattering model should be devised. This was recently done by Wagner and his colleagues (Wagner et al. 1985) who based their approach on the theory developed for laser speckle and an additional specular reflector (Goodman, 1975). They also presented preliminary clinical results (Insana et al., 1985) which demonstrated the potentials of the tissue characterizing parameters extracted.
The data obtained in our simulations can also be employed to indicate the density of diffractive scatters necessary to yield fully developed speckle (i.e. the limits of $1^{st}$ and $2^{nd}$-order statistical parameters). For that purpose the -20 dB amplitude width of the sound beam and of the acoustic pulse are taken. The 99% limit of the parameters is reached when the density is of the order of $10^4$ cm$^{-3}$, i.e. when 100 scatterers are present in the -20 dB "sampling volume" of the sound pulse. This figure is an order of magnitude higher than taken by Wagner et al. (1985), but these authors employed the -6 dB "resolution cell" volume instead. The figure of 100 scatterers can be used to generalize the presented results to the employment of other transducers.

*Acknowledgements.*

*This work is supported by grants from the Netherlands Technology Foundation and is carried out within the framework of the Concerted Action Programme on Ultrasonic Tissue Characterization and Echographic Imaging of the European Community.*


REFERENCES.

Abbott, J.G. and Thurstone, F.L. (1979). Acoustic speckle: theory and experimental analysis, Ultrasonic Imag. 1, pp. 303-324.

Bamber, J.C. and Dickinson R.J. (1980). Ultrasonic B-scanning: a computer simulation: Phys. Med. Biol., 25, pp. 463-479.

Burckhardt, C.B. (1978). Speckle in ultrasound B-mode scans, IEEE Trans. SU-25, pp. 1-6.

Cosgrove, D.O. (1980). Ultrasonic tissue characterization in the liver. In: Ultrasonic Tissue Characterization, J.M. Thijssen (ed.), Stafleu, Alphen a/d Rijn, pp. 84-90.

Flax, S.W., Glover, G.H. and Pelc, N.J. (1981). Textural variations in B-mode ultrasonography: a stochastic model, Ultrasonic Imag. 3, pp. 235-257.

Goodman, J.W. (1975). Statistical properties of laser speckle patterns. In: Laser Speckle and Related Phenomena, J.C. Dainty (ed.), Springer, Berlin, pp. 9-75.

Insana, M.F., Wagner R.F., Garra, B.S., Brown, D.G. and Shawker, T.H. (1985). Analysis of ultrasound image texture via generalized rician statistics. In: Proc. Soc. Photo-Opt. Instr. Engrs (SPIE) 56, pp. 153-159.

Kossoff, G., Garrett, W.J., Carpenter, D.A., Jellins, J. and Dadd, M.J. (1976). Principles and classification of soft tissues by gray scale echography, Ultrasound Med. 2, pp. 89-105.

Kossoff, G. (1979). Analysis of the focussing action of spherically curved transducers: Ultrasound Med. 5, pp. 359-365.

Nicholas, D. (1982). Evaluation of backscattering coefficients for excised human tissues: results, interpretation and associated results: Ultrasound Med. 8, pp. 17-28.

Oosterveld, B.J., Thijssen, J.M. and Verhoef, W.A. (1985). Texture of B-mode echograms: 3-D simulations and experiments of the effects of diffraction and scatterer density, Ultrasonic Imag. 7, pp. 142-160.

Papoulis, A. (1965). Probability, Random Variables and Stochastic Processes, McGraw Hill, New York.

Thijssen, J.M. and Oosterveld, B.J. (1985). Texture in B-mode echograms: a simulation study of the effects of diffraction and of scatterer density on gray scale statistics. In: Acoustical Imaging, Vol. 14. A.J. Berkhout, J. Ridder, and L.F. van der Wal (eds.), Plenum, New York, pp. 481-485.

Thijssen, J.M. and Oosterveld, B.J. (1986). Texture of echographic B-mode images. In: Proceedings SIDUO XI Congress, Capri, in press.

Verhoef, W.A., Cloostermans, M.J. and Thijssen, J.M. (1984). The impulse response of a focussed source with an arbitrary axisymmetric surface velocity distribution, J. Acoust. Soc. Amer. 75, pp. 1716-1721.

Wagner, R.F., Smith, S.W., Sandrik, J.M. and Lopez, H. (1983). Statistics of speckle in ultrasound B-scans, IEEE Trans. SU-30, pp. 156-163.

Wagner, R.F., Insana, M.F. and Brown, D.G. (1985). Unified approach to the detection and classification of speckle texture in diagnostic ultrasound. In: Proc. Soc. Photo-opt. Instr. Engrs (SPIE) 556, pp. 146-152.

DEVELOPMENT OF A MODEL TO PREDICT THE POTENTIAL ACCURACY OF VESSEL
BLOOD FLOW MEASUREMENTS FROM DYNAMIC ANGIOGRAPHIC RECORDINGS

D.J. Hawkes[1], A.C.F. Colchester[1], J.N.H. Brunt[2],
D.A.G. Wicks[2], G.H. Du Boulay[3], and A. Wallis[3]

[1]St. George's Hospital, London, United Kingdom
[2]University of Manchester Medical School, U.K.
[3]Institute of Neurology, London, U.K.

ABSTRACT

*The development of a computer model is described which will predict the concentration of contrast media (CM) in a blood vessel after the injection of CM and hence predict the angiographic appearances of the vessel. This model is used to calculate parametric images of CM concentration versus time and distance along the vessel for a wide range of experimental parameters. Blood velocity estimates are derived from these images at many points, both along the vessel and over the cardiac cycle. The model will be used to investigate the correct strategy for the injection of CM and to predict the precision of the resulting flow estimates.*

1. INTRODUCTION

Blood flow measurements in individual vessels would be of value in a variety of clinical circumstances, including the assessment of the effect of atherosclerosis and other sources of vessel narrowing on flow in individual vessels, the investigation of bypass graft patency, the measurement of total blood flow to an organ and the partition of flow between different vessels.

Although several competing modalities exist for the measurement of blood flow, all previous techniques are valid only in severely restricted circumstances. Many radiographic techniques previously proposed are inadequate in the presence of the high pulsatile flow rates frequency encountered physiologically and pathologically. For a review of different modalities, see, for example, Colchester et al. (1986). However the recent increased interest in X-ray angiography, its decreasing morbidity and, in particular, the advent of high resolution digital fluorography at high framing rates, has given renewed impetus to the search for a reliable angiographic technique for the measurement of intra-vascular flow.

We have developed a method, originally proposed by Colchester and Brunt (1983), which leads to flow measurements in individual vessels from digitised X-ray recordings taken at high framing rates after the injection of contrast media.

We construct a parametric image with distance along the vessel as the y-axis, time as the x-axis and the concentration of injected contrast media within the vessel lumen as the image grey-level. The instantaneous blood velocity at any point along the vessel is equal to the slope (dy/dx) of iso-concentration contours in the parametric image. These velocity estimates can be converted to flow estimates by multiplying by the cross sectional area of the vessel. Averaging along non-branching segments of vessel results in flow measurements at many points over the cardiac cycle.

This technique is currently being validated using phantoms to simulate flow in blood vessels. Work is in progress to extract maximum information on flow from the parametric images and also to combine the flow technique with our method for reconstructing the course of the vessel centreline in three dimensions (Mol et al., 1987; Hawkes et al, 1987) in order to allow measurements in tortuous vessels.

It has become apparent that the correct strategy for the injection of the contrast media is critical for the success of this technique in clinical practice. Our technique demands a significant rate of change of contrast media concentration along the vessel segment, over the whole cardiac cycle, while minimising any artefactual effect due to the injection. The choice of suitable injection strategies has proved difficult due to the infinite variety of possible injection flow profiles, the wide choice of possible sites of injection, and the wide range of expected blood flow profiles, vessel segment lengths and vessel calibres.

Laboratory experiments to investigate injection tactics are time-consuming to perform, expensive in capital equipment and materials, film digitisation is slow and analysis is tedious. The design and construction of an injection pump which would provide the more exotic contrast flow profiles which have been proposed would be expensive. In addition, it is of interest to predict the X-ray quantum limited precision of our technique, for a wide range of blood flows, with data that is free from systematic errors and artefacts.

Errors due to variations in X-ray generator output, X-ray exposure timing and beam hardening effects are equipment dependent and potentially correctable. Although they do not affect the fundamental precision of our technique they are difficult to remove from our experimental data.

For the above reasons we have decided to construct a computational model to enable us to generate realistic parametric flow images and hence to investigate the quantum limited precision of our technique for a range of injection tactics. Having proposed a suitable injection strategy for a given situation, a small number of experiments on phantoms, and 'in vivo', should be sufficient to validate our technique.

An analytical solution for laminar flow velocities under conditions of constant laminar flow (Hagen-Poiseuille flow) in an infinitely long straight vessel with a circular lumen was first described in about 1890 (Lamb, 1932). Womersley (1955) solved the Navier-Stokes equations for a Newtonian fluid to predict laminar flow velocities in pulsatile flow. More recently, Wille (1981) and Wille and Walloe (1982) used the method of finite element analysis, to predict flow in simulated arterial stenoses and aneurysms. We are unaware of any models which have been developed to predict contrast media concentration in an artery, in the presence of pulsatile blood flow.

## 2. METHOD

The computational model was developed on a CVAS-3000 image analysis workstation (Taylor et al., 1986). The core level or kernel of the model computed the concentration of contrast media at successive time intervals at all points along a cylinder of circular cross-section whose radius could vary along its axial length. Assuming that the cylinder is rigid and non-branching, that there is perfect mixing of contrast media at the injection site and neglecting mixing or diffusion thereafter, the concentration of contrast material $C(X,t)$ at a point distance X along the vessel distal to the injection site at time t is given by

$$C(X,t) = Q_I(t') C_I / Q_T(t') \qquad (1)$$

where t' is the time at which the portion of the bolus under consideration was injected and (t-t') is thus the time taken for the fluid at the injection site I to reach point X, $Q_I(t')$ is the injection flow rate of contrast material at time t', $Q_T(t')$ is the total flow in the cylinder distal to the injection site at time t' and $C_I$ is the concentration of contrast material in the injection.

The concentrations along the vessel at time interval $\delta t$ were calculated recursively as follows. The concentration at X at time $(t+ \delta t)$ is equal to the concentration at point $(X- \delta X)$ at time t and the volume of fluid flowing distal to the site of injection in the time interval $\delta t$, $Q_T(t) \delta t$, is given by

$$Q_T(t) \, \delta t = \int_{X-\delta X}^{X} A(x) \, dx. \qquad (2)$$

The cross sectional area of the cylinder $A(x)$ is integrated from X back towards the site of injection until the distance $\delta X$ is found which satisfies Eq. (2).

In the simplest implementation we assume that axial velocity is constant across the vessel lumen (so called 'plug flow'). The injection will have some effect on blood flow proximal to the injection site. Accurate calculation of this effect would require knowledge of the blood pressure and injection pressure at the injection site, the vascular resistance and blood volume proximal and distal to the injection site, the power output of the left ventricle and the elasticity of the arterial system. We have approximated this effect by assuming that the blood flow proximal to the injection site is reduced by a constant proportion of the contrast injection flow rate, this constant of proportionality being one of the input parameters.

Constant laminar flow is modelled by assuming that the blood vessel consists of a series of non-communicating concentric hollow cylinders, or laminae. The blood velocity $v(r)$ for a lamina of radius r is given by

$$v(r) = 2 Q_T(t) \, [1-(r/R)^2]/\pi R^2 \qquad (3)$$

where R is the radius of the blood vessel lumen.

Finally, diffusion of contrast media between adjacent lamina is modelled from the cross-sectional area of the boundary between two laminae, the time interval δt between X-ray exposures, the difference in concentration of contrast material and a factor which is proportional to the coefficient of diffusion.

The computer model permits the interactive setting of 25 different experimental input parameters using the screen text editor. The parameters are grouped into three categories as follows.

(i) Patient related parameters: including blood vessel cross sectional area, length of vessel segment, angle of blood vessel to the X-ray beam, and the volume blood flow rate profile over the cardiac cycle.

(ii) Contrast-medium injection parameters: including concentration of iodine in the contrast injection, injection flow rate, the immediate physical effect of the injection on the blood flow, and the distance from the injection site to the vessel in which flow is to be measured.

(iii) Radiographic parameters: including X-ray beam geometry, X-ray spectrum, X-ray exposure rate and the quantum detection efficiency (QDE) of the X-ray imaging system.

The vessel diameter may vary as a function of vessel length and may be entered interactively or may be predefined and read from disc. The blood flow rates and contrast flow rates over the time course of the experiment may be entered interactively, defined as analytic functions such as a semi-sinusoid or may be predefined and read from disc.

The model generates the parametric image of iodine versus time and distance along the vessel axis. X-ray quantum noise is added according to the radiographic input parameters. For the parametric images presented here we have assumed a monoenergetic X-ray beam of energy 50 keV giving $3.344 \times 10^2$ photons /mm$^2$/μR/frame at the imaging plane, an exposure of 20μR/frame at the imaging plane, a QDE of 50% and the mass attenuation coefficient of iodine of 1.225 m$^2$/kg.

Blood flow velocity is determined from the parametric images either by using a gradient operator or by comparing adjacent columns of the parametric image.

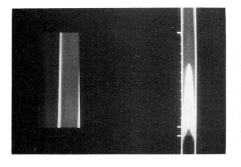 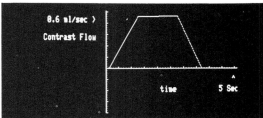

Fig. 1. Generation of a parametric flow image

Fig. 2. Injection flow rate used to generate Fig. 4.

## 3. RESULTS

Figure 1 illustrates the generation of a parametric flow image. The partially formed parametric image is shown on the left, while on the right there is a graphical representation of the contrast concentration across the diameter of the blood vessel lumen. The diameter of the vessel was 5 mm. The blood flow is from bottom to top and the injection site is shown by the lower mark on the vessel outline while the other two larger marks indicate the length of the vessel used to generate the parametric image. The length of vessel analysed was 100 mm. Figure 2 shows a plot of the injection flow rate of contrast media and Fig. 3 the blood flow rate used to generate the parametric image in Fig. 1. Figure 4 shows the completed parametric image after the addition of X-ray quantum noise. Figure 5 shows a parametric image generated using a rapidly pulsed injection whose flow rates are plotted on Fig. 6. The same blood flow rates were used as in the generation of Fig. 4.

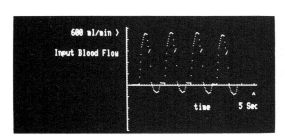 

Fig. 3. Blood flow rate used to generate Fig. 4.

Fig. 4. Parametric flow image including X-ray quantum noise

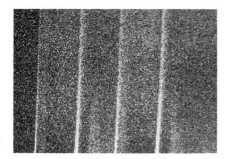

Fig. 5. Parametric image from pulsed injection

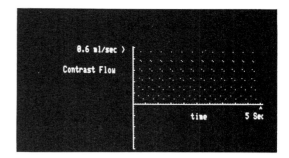

Fig. 6. Pulsed Injection flow rates

In these two cases it can be seen that the most significant features of the parametric image are generated by the pulsatile nature of the blood flow rather than any imposed flow pattern from the contrast injection. It is these features in the parametric image which permit the accurate calculation of velocities and hence flow. Our task is to devise injection tactics which will generate such features over the whole cardiac cycle, or, preferably, several cardiac cycles.

Figure 7 provides a plot of blood flow calculated from a parametric image generated from plug flow, with the same blood flow rates and contrast flow rates as used to generate Fig. 4, but without the addition of noise. The results plotted on Fig. 7 are in close agreement with the bloodflow plotted on Fig. 3, which was used to generate the parametric flow images.

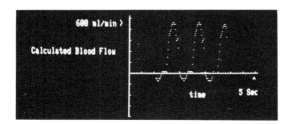

Fig. 7. Calculated blood flow from noise free parametric image.

## 4. DISCUSSION AND CONCLUSION

In all modelling tasks of this nature great care must be exercised in determining the precision of calculations. If calculations are performed at inappropriately high precision, computational time will be unnecessarily high while calculations performed at too low a precision may lead to artefactual results due to rounding and truncation errors.

The model generates parametric images similar in appearance to those generated experimentally. The major limitation at present is the assumption of constant laminar flow. If the Womersley constant $\alpha$ (Womersley, 1955) is less than 0.5, the blood vessel, by analogy with an electrical conductor, acts as a pure resistance to flow and flow would be described by Lamb's equations (Lamb, 1932). If $\alpha$ is greater than about 10, the flow is approximated by 'plug flow'. For most arteries the value of $\alpha$ is intermediate. For the 5 mm vessel modelled in the examples present here, $\alpha$ is about 3.2 for the first harmonic (1Hz) of the vessel blood flow waveform. Accurate computation of the pulsatile flow across the lumen would require the computation of zero order Bessel functions of the first kind with complex arguments. A look-up table of the Bessel functions could be generated but flows would have to be calculated for each harmonic of the blood flow waveform. This would increase current computation time at least sixfold. On the present system each parametric flow image of 256 x 256 pixels takes approximately one hour to generate using 16 laminae to represent the laminar blood flow.

Womersley's theory would simulate flow velocities in an infinitely long tube. Unfortunately, for most vessels in the body, the 'inlet length', (McDonald, 1974) is significant. In this region the variation in blood velocity across the vessel lumen is rather far from parabolic and is more closely approximated by plug flow. Accurate calculation would be prohibitively expensive in computer time on our present system. We may, however, generate vector images for both plug flow and steady state laminar flow, using our model and hence investigate the effect and significance of either extreme of flow pattern in the artery.

At the injection site, we assume uniform distribution of contrast media across the vessel lumen and yet we assume that the catheter and injection do not perturb the shape of the blood flow pattern across the vessel lumen. In practice there will be pure contrast media at the catheter tip and the catheter tip itself will probably produce turbulence. These two effects will tend to cancel as both turbulence and laminar flow lead to dispersion of the bolus. In addition, the parametric image is generated from data some distance distal to the injection site, to permit both uniform mixing and resumption of laminar flow.

Other effects not calculated included the obstruction of blood flow by the catheter, the effect of differences in viscosity and density between blood and contrast material and pharmacological effects of the contrast injection. These may be significant but are considered in more detail elsewhere (Colchester, 1985; Colchester et al., 1986).

In addition to devising injection strategies and predicting the precision of our flow measurements, we hope to use the flow patterns generated by this computer model to generate part of a knowledge base for the analysis of dynamic angiograms. The knowledge base would be used to test and refine hypotheses of blood flow patterns derived from real data.

In conclusion, therefore, we have constructed a computer model of the flow of contrast material and blood in arteries which may be used to generate realistic parametric flow images. This model will require some refinement and the results require experimental verification, yet we believe that the model will enable us to investigate a very wide range of injection strategies and other experimental parameters; thus reducing costs and saving considerable time in development.


*Acknowledgements*

*We gratefully acknowledge the support of the British Heart Foundation for part of this work and S.E.R.C. for a studentship for one of us (DAGW).*

*We are grateful to Barbara Gagen for typing the script.*

## REFERENCES

Colchester, A.C.F. and Brunt, J.N.H. (1983). Measurement of vessel calibre and volume blood flow by dynamic quantitative digital angiography: an initial application showing variation of cerebral artery diameter with Pa $Co_2$, *Cereb. B. Flow Metabol.* 3, pp. S640-641.

Colchester, A.C.F., (1985). The effect of changing Pa $Co_2$ on cerebral artery calibre estimated by a new technique of dynamic quantitative digital angiography. Ph.D. Thesis, University of London.

Colchester, A.C.F., Hawkes, D.J., Brunt, J.N.H., Du Boulay, G.H., and Wallis, A. (1986). Pulsatile blood flow measurements with the aid of 3-D reconstructions from dynamic angiographic records. In: *Information processing in medical imaging,* S.L. Bacharach (ed), Martinus Nijhoff Dordrecht, pp. 247-265.

Hawkes, D.J., Colchester, A.C.F., and Mol, C.R. (1987). The accurate 3-D reconstruction of the geometric configuration of vascular trees from X-ray recordings. In: *Physics and engineering of medical imaging,* R. Guzzardi (ed), Martinus Nijhoff, Dordrecht, pp. 250-256.

Lamb, H. (1932). *Hydrodynamics,* 5th edition (reprinted 1953), University Press, Cambridge.

Mol, C.R., Colchester, A.C.F., and Hawkes, D.J. (1987). Three-dimensional reconstruction of vascular configurations from bi-plane X-ray angiography. Submitted for publication.

McDonald, D.A. (1974). *Blood flow in arteries,* 2nd edition, Edward Arnold, London.

Taylor, C.J., Dixon, R.N., Gregory, P.J., and Graham, J. (1986). An architecture for integrating symbolic and numerical image processing. In: *Intermediate level imaging processing,* M.J.R. Duff (ed), Academic Press, pp. 19-34.

Wille, S.Ø. (1981). Pulsatile pressure and flow in an arterial aneurysm simulated in a mathematical model, *J. Biomed. Engin.* 3, pp. 153-158.

Wille, S.Ø., and Walløe, L. (1981). Pulsatile pressure and flow in arterial stenoses simulated in a mathematical model, *J. Biomed. Engin.* 3, pp. 17-24.

Womersley, J.R. (1955). Method for the calculation of velocity, rate of flow and viscous drag in arteries when the pressure gradient is known, *J. Physiol.* 127, pp. 553-563.

THE QUANTITATIVE IMAGING POTENTIAL
OF THE HIDAC POSITRON CAMERA


D.W.Townsend[1], P.E.Frey[1], G.Reich[2], A.Christin[1],
H.J.Tochon-Danguy[1], G.Schaller[1], A.Donath[1],
A.Jeavons[3]

1: University Hospital of Geneva, Switzerland
2: Union College, Schenectady, USA
3: CERN, Geneva, Switzerland and University of Oxford, U.K.


ABSTRACT


*An important goal of positron emission tomography (PET) is to measure, in vivo and in absolute units, the tissue concentration of a positron-emitting radiopharmaceutical. Over the past few years, such measurements have been made in a wide variety of PET studies using positron cameras constructed of rings of detectors of bismuth germanate or cesium fluoride crystals. Large area detectors, such as those based on the multiwire chamber, have yet to demonstrate such a capability in a clinical research environment. This paper highlights some of the problems of image quantitation that are encountered with the High Density Avalanche Chamber (HIDAC) positron camera.*


1. INTRODUCTION


The major aim of positron emission tomography (PET) is to measure absolute tissue concentrations of positron-emitting radiopharmaceuticals *in vivo*, measurements which are obtained from the count densities in the PET image. However, such measurements generally require positron cameras that have both high sensitivity and good spatial resolution. High sensitivity is important to obtain good statistical accuracy in cases of low tracer concentrations, while good spatial resolution is essential to minimise effects such as partial volume (Mazziotta et al., 1981) which significantly distort local count densities, thereby leading to incorrect estimates of radioisotope tissue concentration. Since these data are the basic input to tracer kinetic models of transport mechanisms in metabolic pathways, erroneous conclusions will result from inaccurate measurements.

Quantitative PET, such as the measurement of the glucose consumption of the brain *in vivo* (Huang et al., 1980), has been performed by crystal-based

ring cameras operating with high sensitivity. However, the volume
resolution (i.e. the smallest resolvable volume element) of such cameras
is, at best, about 0.25 cc. Although considerable improvements in the
spatial resolution of ring cameras have been made in recent years, the
axial resolution, or slice thickness, is still a limiting factor. Thus
partial volume may distort concentration measurements in small-volume
structures such as the caudates (~5.2 cc) and the thalamus (~5.1 cc) in the
brain (Mazziotta et al., 1981).

An alternative technology to the crystal-based positron camera that has
been under investigation for the past few years is the high density
avalanche chamber (HIDAC) detector (Jeavons et al., 1983). Although
currently having a comparatively low efficiency (~12%) for the detection of
511 keV photons, these detectors have both high spatial resolution and full
volume imaging capability. With an axial and in-slice spatial resolution of
2 to 3 mm, the volume resolution is thus around 0.03 cc. However, full
image quantitation will only be possible if the data are properly
normalised and carefully corrected both for photon attenuation and for the
effects of accidental coincidences and scatter. In these respects, the
HIDAC camera performs rather differently compared with most of the
commercially available, crystal-based systems.

This paper highlights some of the problems of image quantitation with the
rotating HIDAC camera. To date, only the intermediate objective to provide
accurate measurements of geometric parameters such as the dimensions and
volume of the organ under study, has been achieved. The delineation of the
surface of the organ has, nevertheless, led to effective three-dimensional
shaded displays. The prospects for full image quantitation with the HIDAC
camera will be summarised.


2. IMAGE RECONSTRUCTION PROCEDURE

During a study, the HIDAC camera rotates around the subject acquiring data
from up to $10^5$, two-dimensional, parallel-ray projections $p(u,v,\theta,\phi)$, where
$(u,v)$ are the projection coordinates, and $(\theta,\phi)$ are the projection angles.
As at most one to two million counts are accumulated during the scan, each

2-D projection, which comprises some $2.10^4$ ray-sums (i.e. (u,v) pairs for given $\theta,\phi$), is statistically poorly determined. Certain projections have unmeasured ray-sums due to the limited angular acceptance of the HIDAC detectors. Since the photons are acquired in a random order, sorting of the resulting data set into a large number of 2-D projections which contain many zeros is thus excluded, if only for reasons of computer storage. Instead, back-projection is performed event by event to give the distribution $f_b(x,y,z)$, which is related to the (unknown) distribution of radioactivity $f(x,y,z)$ by the usual integral equation:

$$f_b(x,y,z) = \int f(x',y',z') \cdot h(x,y,z,x',y',z')\, dx'dy'dz'. \tag{1}$$

The camera point response function $h(x,y,z,x',y',z')$ is not, in general, stationary unless projections with unmeasured ray-sums are eliminated from the back-projection. Using the subset of complete projections, Eq.(1) reduces to the three-dimensional convolution

$$f_b(x,y,z) = \int f(x',y',z') \cdot h(x-x',y-y',z-z')\, dx'dy'dz' \tag{2}$$

where the ideal, normalized camera point response function is given by

$$h(x,y,z) = \Delta / [2\pi \sin\Psi\, (x^2+y^2+z^2)] \tag{3}$$

with $\Psi$ the maximum acceptance angle of the detectors, and $\Delta$ the area of a pixel. The function $f(x,y,z)$ is obtained from Eq.(2) by three-dimensional Fourier deconvolution; closed-form expressions are available for the Fourier transform of $h(x,y,z)$ (Schorr, Townsend, and Clack, 1983), which must include a high-frequency roll-off function to control noise amplification.

a. *Point Source Sensitivity*

The sensitivity for a point source decreases with displacements away from the axis of rotation of the camera. Sources near the edge of the field of view will appear weaker than those near the centre. This effect can be corrected for during back-projection by weighting rays far from the axis more heavily than those close to it (Clack, Townsend, and Jeavons, 1984). The actual weight is therefore related to the perpendicular distance, h, between the ray and the rotation axis, where

$$h = D/2 \cos\phi - S/2 \sin\phi. \qquad (4)$$

The detectors are of side D, separated by a distance S, and $\phi$ is the maximum acceptance angle for a ray passing through a point at a distance h from the axis. The use of back-projection weights affects the normalisation of the reconstructed image. Assuming the weights are normalised to unity at the edge of the useful field of view (e.g. a distance H from the rotation axis), the image should be normalised, not to the actual number of events that are back-projected, but to the number of events falling within the angle $\phi_H$. This angle is the maximum acceptance angle for a ray tangential to the edge of the field of view. Since the acceptance angle at the centre is larger than $\phi_H$, the image is normalised to fewer events than are actually back-projected. The extra events serve to reduce the statistical noise near the centre of the field of view.

b. *Attenuation Correction*

The correction for photon attenuation requires knowledge of the integral of the linear attenuation coefficient $\mu$ for all channels used in the reconstruction of the data. This information is usually obtained from a transmission scan with a 511 keV photon source such as $^{68}$Ge-$^{68}$Ga. If $n_0(k)$ is the measured rate of coincidences for a particular coincidence channel k without the patient in place, and $n_1(k)$ is the corresponding rate with the patient in place, then the correction factor $c(k,\mu)$ for channel k is given by:

$$\begin{aligned} c(k,\mu) &= n_0(k)/n_1(k) \\ &= \exp\{\int \mu(t)\,dt\} \end{aligned} \qquad (5)$$

where the integration variable t is along the particular channel k. In view of the large number of possible channels, some approximations become necessary. A simple assumption is that $\mu$ is constant and independent of t, so that Eq.(5) becomes:

$$c(k,\mu) = \exp\{\mu \int dt\}$$
$$= \exp\{\mu T_k\} \qquad (6)$$

where $T_k$ is the intersection length of the channel k with the attenuating medium. c(k) is then used to weight the backprojection rays for channel k, multiplied by the corresponding point source sensitivity weight (from Eq.(4)).

c. *Accidental Coincidences*

The count rate due to accidental coincidences ($C_a$) may be estimated from

$$C_a = S_1 \cdot S_2 \cdot 2\tau \qquad (7)$$

where $S_1$ and $S_2$ are the singles rates on the two HIDAC detectors and $2\tau$ is the coincidence resolving time. The singles rate (S(i)) for a source of activity $A_i$ located at a point $(x_i, y_i, z_i)$ within the field of view is given by

$$S(i) = \Omega_i \cdot e(E) \cdot A_i \exp\{-\mu_i T_i\} \qquad (8)$$

where $\Omega_i$ is the solid angle subtended by the detector at the point i, e(E) is the efficiency for detection of photons with energy E, and $\mu_i$ and $T_i$ are respectively the mean attenuation coefficient and mean thickness of the attenuating medium. The total singles rate S for a detector is obtained by summing S(i) for all points of activity, i.e. $\Sigma_i S(i)$.

Various methods have been proposed to correct for accidental coincidences (Hoffman et al., 1981). The simplest is to subtract a uniform background at a level estimated from Eq.(7) using measurements of $S_1$ and $S_2$ and the known coincidence time. This procedure does not take into account any potential non-uniformities in the accidental background. An alternative method is based on the measurement of an accidental map, which is obtained by delaying the coincidence signal from one detector so as to exclude almost all in-time signal events. This map is then subtracted from the final image, although since the measurement includes statistical fluctuations, the technique tends to increase the noise in the final image.

*d. Scattered Coincidences*

Good scatter correction is essential for accurate image quantitation. The image background due to scattered events depends in general on both the activity distribution $f(x,y,z)$ and the non-active tissue distribution, on the detector sensitivity to photons with energies below 511 keV, and on the shielding. Area detectors such as the HIDAC can only be shielded from scatters originating outside their field of view; photons which scatter and remain within the field of view may, depending on their energy, be detected. The HIDAC detection efficiency decreases with decreasing energy (Jeavons et al., 1983), although the detectors are unable to measure the incident photon energy because of the conversion process. Small-angle scattering that only slightly reduces the initial photon energy will contribute significantly to the image background. The scattered events produce a structured background distribution that is difficult to correct for.

3. RESULTS

Measurements have been made with a prototype HIDAC positron camera in order to investigate the importance of the above factors for quantitation. Some simple phantoms have been used to study spatial resolution, attenuation and scatter, while liver imaging with $^{68}$Ga-colloids demonstrates the problems arising in a clinical situation.

*a. Line Source*

A 0.5 mm diameter line source filled with $^{68}$Ga was imaged both in air and in a 20 cm diameter, plastic cylinder. In the latter case, the line source

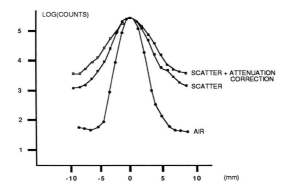

Fig. 1. Line spread functions for a 0.5 mm
diameter source in air and scatter

was reconstructed both with and without attenuation correction. Profiles across the line source are shown in Fig. 1 for the three situations of air, scatter medium and scatter medium with attenuation correction. The (base e) logarithm of the counts has been plotted, normalised to the same maximum value, in order to accentuate the differences at low count levels. The FWHM and FWTM (full-width at tenth-maximum) are respectively 3.5 mm and 5.5 mm in air, 4.5 mm and 8.5 mm in a scatter medium, and 5 mm and 11 mm with scatter and attenuation correction. A low data rate ensures that the contribution from accidental coincidences is negligible. The background level is 9% of the peak source activity with the scattering cylinder in place, increasing to 11% with the additional correction for attenuation.

b. Uniform Cylinder Phantom

A uniform cylinder of diameter 20 cm, filled with a solution of $^{68}$Ga, was imaged and reconstructed with a geometrical attenuation correction (given by Eq.(6)). A profile through the mid-point of a 4mm thick, transverse section of the reconstructed image is shown in Fig. 2. The image has been corrected for accidentals by the subtraction of a smoothed accidentals map. One million counts were collected during the data acquisition.

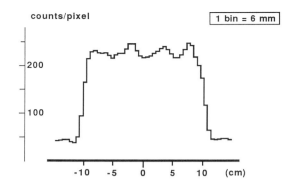

Fig. 2. Profile across a 20 cm diameter uniform cylinder containing $^{68}$Ga

c. Cylinder Phantom with cold spots

A thin cylinder 20 cm in diameter and 2 cm in length, containing three cold spots of diameter 1 cm, 3 cm and 5 cm was imaged under different conditions of scatter. The cylinder was filled with $^{68}$Ga solution. The results are shown in Fig. 3 for the cylinder alone (Fig. 3a), with a 10 cm long cylinder of water (diameter 20 cm) on either side (Fig. 3b) and with the water replaced by a $^{68}$Ga solution having the same radioactivity

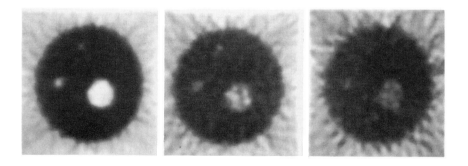

Fig. 3. Section through a cold-spot phantom for different amounts of scatter.

concentration as in the cold-spot cylinder (Fig. 3c). When compared with a region-of-interest outside the cylinder, the background in the 5 cm cold spot was 2%, 15% and 31% of the cylinder activity in the three imaging situations, respectively.

*d. Liver Imaging*

To illustrate the clinical imaging situation, Fig. 4 shows an 8 mm thick transverse section through the liver of a patient injected intravenously with $^{68}$Ga-colloids. The section is viewed from below and is one of a sequence of 20 tomographic sections obtained from a single scan. The four images are: no attenuation correction (top, left); with attenuation correction based on the CT scan dimensions and with a constant attenuation coefficient (top, right); with the same correction as the previous image but with a 2 cm displacement of the centre of the ellipse (bottom, left); as for the previous image, but with less reconstruction smoothing (bottom, right).

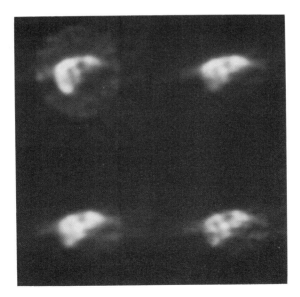

*Fig. 4. A transverse section through a human liver reconstructed under different conditions. The patient was injected with $^{68}$Ga-colloids. The voxel size is 8 mm × 8 mm × 8 mm.*

## 4. DISCUSSION

The intrinsic spatial resolution of the HIDAC detector is 1.5 mm FWHM. After a multi-view reconstruction, the spatial resolution in the image decreases to about 3 mm, in part due to the high-frequency cut-off of the filter. The addition of scatter around the source has a significant effect on the FWTM, which increases from 5.5 mm to 8.5 mm; the slight increase in the FWHM is probably due to penetration into the plastic of the 1.9 MeV positrons from $^{68}$Ga. When an attenuation correction is applied, a further increase in both FWHM and FWTM is observed. The increasing background from scattered events is consistent with the changes in the line spread function; the scattered events contribute to the tails of the spread function, while having little effect on the FWHM. The increase in background with attenuation correction is a consequence of the weighting procedure which does not distinguish between true and scattered coincidences during back-projection. In practice, therefore, the spatial resolution in an image may be up to a factor of 3 worse than the intrinsic resolution of the HIDAC camera.

The geometrical correction successfully compensates for photon attenuation in simple phantoms, as shown by the cylinder in Fig. 2. Good uniformity is seen throughout the 20 cm field-of-view. Such images, however, are only qualitative since they include scattered and accidental coincidences. The accidentals are distributed rather uniformly through the three-dimensional field-of-view, and even at the comparatively high accidental rates of the HIDAC camera, they make only a small contribution to each voxel which may be corrected for by subtracting the background estimated from Eq.(7).

Scatter, however, is a more serious effect since it depends on the actual imaging situation, as shown in Fig. 3. The loss of contrast with increasing scatter is evident; the 1 cm cold-spot, although clearly resolved on the first image (Fig. 3a), cannot be seen in the presence of a large scatter background (Fig. 3c). Figure 3a contains only a small amount of scatter from the thin cylinder itself; Fig. 3b includes photons which originate in the cylinder, scatter in the plastic but are still detected by the HIDAC. Figure 3c includes the additional contribution from the photons that originate in the surrounding activity, scatter and are then detected and backprojected into the cylinder. After correction for the different accidental rates, the scatter contribution for this phantom is seen to lie between 2% and 31%, depending on the distribution of the surrounding activity. In general, the scatter background is structured and even a good knowledge of the mean level is insufficient for true image quantitation. Thus, scatter compensation must be applied during the reconstruction of the data. For this camera, however, methods based on a direct knowledge of the projections would be difficult to apply. An alternative approach might be to modify the filter obtained from Eq.(3) to accomodate the actual point spread function shown in Fig. 1.

In clinical imaging, the situation is even more complex because of the difficulty of applying the geometrical attenuation correction. This is clearly seen in Fig. 4. In the first image (top, left), the ring artifact due to the circularly-symmetric back-projection weights from Eq.(4) is visible, mapping out the 30 cm diameter field-of-view. The shape and appearance of the activity distribution in the liver changes substantially with the inclusion of the attenuation correction (top, right). The ring artifact has been suppressed by the attenuation factor in the back-projection weights, but other streak-like artifacts can now be seen. Movement of the centre of the body ellipse (bottom, left), which is a difficult parameter to measure in practice, has little visual effect, at least within the resolution of these images, i.e. movements of 2 to 3 pixels. Finally, with less smoothing during reconstruction, there is an increase in image noise, although the general appearance of the uptake in the liver is unchanged; the estimated body ellipse can be more clearly seen.

It is evident from these results that the quantitation of PET images from the HIDAC camera will require further work. It is clearly essential to include a transmission scan to provide data for the attenuation correction, if only to obtain a reliable body contour. The measurement of an accidentals map should be included in the data acquisition and subtracted from the final image, although this source of background is potentially less troublesome than scatter. It would appear that fairly high levels of scatter are to be expected (up to 30% for the brain), although this depends on the particular study. A good scatter compensation procedure is therefore needed that takes into account the non-uniform nature of the background due to scatter.

To date, the clinical applications of the HIDAC camera have been limited to those for which the attenuation correction is of lesser importance (thyroid and ENT tumours), and where the target to background ratio is particularly favourable (skull and liver). In the case of the liver, for example, the geometrical attenuation correction provides images in good qualitative agreement with CT and SPECT, although in the absence of a knowledge of the background due to scatter, no absolute concentration measurements can be made. Thus, full image quantitation will depend upon the development of a satisfactory solution to the problems of scatter and the incorporation of a suitable transmission scan into the data acquisition procedure.

*Acknowledgements*

*This projected is supported by the Fonds national suisse under grant number 3.853-0.85. The authors thank Dr Michel Defrise, VUB, Brussels who provided the three-dimensional filtering program.*

REFERENCES

Clack, R., Townsend, D., and Jeavons, A. (1984). Increased sensitivity and field of view for a rotating positron camera, *Phys. Med. Biol.* 29, pp. 1421-1431.
Hoffman, E.J., Huang, S.C., Phelps, M.E., Kuhl, D.E. (1981). Quantitation in Positron Emission Computed Tomography: 4. Effect of Accidental Coincidences, *J. Comput. Assist. Tomogr.* 5, pp. 391-400.
Huang, S.C., Phelps, M.E., Hoffman, E.J., Sideris, K., Selin, C.J., Kuhl, D.E. (1980). Noninvasive determination of local cerebral metabolic rate of glucose in man, *Am. J. Physiol.* 238, pp. E69-E82.
Mazziotta, J.C., Phelps, M.E., Plummer, D., and Kuhl, D.E. (1981). Quantitation in Positron Emission Computed Tomography: 5. Physical - Anatomical Effects, *J. Comput. Assist. Tomogr.* 5, pp. 734-743.
Jeavons, A.P., Hood, K., Herlin, G., Townsend, D., Magnanini, R., Frey, P., Donath, A. (1983). The High Density Avalanche Chamber for positron emission tomography, *IEEE Trans. Nucl. Sci.* NS-30, pp. 640-645.
Schorr B., Townsend, D., and Clack, R. (1983). A general method for three-dimensional filter computation, *Phys. Med. Biol.* 28, pp. 1009-1019.

# THE USE OF CLUSTER ANALYSIS AND CONSTRAINED OPTIMISATION TECHNIQUES IN FACTOR ANALYSIS OF DYNAMIC STRUCTURES

A.S. Houston

*Royal Naval Hospital Haslar*
*Gosport, England*

## ABSTRACT

*Factor analysis of dynamic structures (FADS) extracts physiological information from a series of sequential images using the criterion that physiological data are non-negative. The technique consists of a factor analysis applied to the data followed by an oblique transformation of feature space. The result obtained is non-unique. Set theory, cluster analysis and constrained optimisation techniques are used here to define the transformation more precisely and to provide more realistic information.*

## 1. FACTOR ANALYSIS OF DYNAMIC STRUCTURES

### a. Homogeneous physiological compartments

Let us suppose that we have collected T sequential digital images or frames each with N pixels, e.g. as in a nuclear medicine dynamic study. In the sense of a compartmental analysis let us suppose that we are aware that M distinct physiological compartments are represented in the image and that these are homogeneous, i.e. space-invariant with each compartmental element having the same properties as every other element with respect to time.

Let $d_{it}$ be the value in frame t for pixel i using vector representation of the image ($d_{it}$ describes a dynamic curve (or dixel) as t varies);
$g_{jt}$ be the dynamic curve or factor for compartment j;
$a_{ij}$ the contribution to pixel i from compartment j.

Let $d_{it}$ and $g_{jt}$ be normalised such that $\sum_{t=1}^{T} d_{it} = \sum_{t=1}^{T} g_{jt}$ for all i,j. Then, in the absence of noise, for the i-th pixel

$$d_{it} = \sum_{j=1}^{M} a_{ij} g_{jt}. \tag{1}$$

If $D = [d_{it}]$, $A = [a_{ij}]$, $G = [g_{jt}]$, this is equivalent to

$$D = AG. \qquad (2)$$

Summing (1) over t gives $\sum_{t=1}^{T} d_{it} = \sum_{j=1}^{M} a_{ij} \left( \sum_{t=1}^{T} g_{jt} \right)$ or

$$\sum_{j=1}^{M} a_{ij} = 1. \qquad (3)$$

Since we may suppose physiological factors and their contributions will contain no negative elements, we require

$$\begin{aligned} a_{ij} &\geq 0 \text{ for all } i,j; \\ g_{jt} &\geq 0 \text{ for all } j,t. \end{aligned} \qquad (4)$$

## b. *Factor analysis*

In the absence of noise, it may be demonstrated (Barber, 1980), or regarded as intuitively obvious, that, in order to distinguish among M compartments, we require M-1 discriminators. Thus if we perform a factor analysis (Appledorn, 1987) on the N dixels of the noise-free dynamic study, precisely M-1 orthogonal factors will extract all discriminatory information. These factors are the principal components or eigenvectors of a Karhunen-Loève expansion. Each dixel may then be represented by the sum of a mean curve and M-1 orthogonal factors, i.e.

$$d_{it} = \bar{d}_t + \sum_{k=1}^{M-1} x_{ik} f_{kt} \qquad (5)$$

where $f_{kt}$ is the k-th orthogonal factor;
$x_{ik}$ is the k-th factor loading for the i-th pixel;

and

$$\bar{d}_t = \sum_{i=1}^{N} d_{it}/N. \qquad (6)$$

If $\bar{D} = [\bar{d}_{it} : \bar{d}_{it} = \bar{d}_t]$, $X = [x_{ik}]$, $F = [f_{kt}]$, we have

$$D = \bar{D} + XF. \qquad (7)$$

Due to the orthogonal nature of these factors and factor loadings, the matrices **F** and **X** will contain negative elements. The i-th row of **X** defines a point $(x_{i1}, x_{i2}, \ldots, x_{i(M-1)})$ for pixel i. Such pixel points will plot in an (M-1) dimensional space called feature space.

*c. Oblique transformation*

The relationship between physiological and orthogonal factors may be obtained from Eqs. (1) and (5), i.e.

$$\sum_{j=1}^{M} a_{ij} g_{jt} = \bar{d}_t + \sum_{k=1}^{M-1} x_{ik} f_{kt}. \tag{8}$$

In general, the pixel-independent equation $\sum_{j=1}^{M} a_j g_{jt} = \bar{d}_t + \sum_{k=1}^{M-1} x_k f_{kt}$ represents a transformation in M-1 dimensions from co-ordinate system $(x_1, x_2, \ldots, x_{M-1})$ to $(a_1, a_2, \ldots, a_{M-1})$ with $a_M = 1 - \sum_{j=1}^{M-1} a_j$.

For each j, if $a_j = 1$, let $x_k$ take the value $v_{jk}$, i.e. the value of $x_k$ corresponding to pure compartment j. Hence

$$g_{jt} = \bar{d}_t + \sum_{k=1}^{M-1} v_{jk} f_{kt}. \tag{9}$$

If $V = [v_{jk}]$ and $\hat{D} = [\hat{a}_{jt} : \hat{a}_{jt} = \bar{d}_t]$, we have

$$G = \hat{D} + VF. \tag{10}$$

From Eqs. (1) and (9),

$$d_{it} = \sum_{j=1}^{M} a_{ij} \left( \bar{d}_t + \sum_{k=1}^{M-1} v_{jk} f_{kt} \right)$$

$$= \bar{d}_t + \sum_{k=1}^{M-1} f_{kt} \left( \sum_{j=1}^{M} a_{ij} v_{jk} \right). \tag{11}$$

Since orthogonal factors $f_{kt}$ are independent, by comparing Eqs. (5) and (11), the transformation simplifies to

$$x_{ik} = \sum_{j=1}^{M} a_{ij} v_{jk} \tag{12}$$

or

$$X = AV. \tag{13}$$

In practice the dixels are noisy and Eqs. (2), (7) and (10) are approximations. However, if we can obtain suitable values for the elements of $V$, we may obtain an approximation for the physiological factor matrix $G$ using Eq. (10). Equation (13) will now be adapted to allow us to calculate the pixel contribution matrix $A$ from $V$.

If $\hat{X}$ is the $N * M$ matrix and $\hat{V}$ the $M * M$ matrix defined by

$$\hat{X} = \begin{pmatrix} 1 & x_{11} & \cdots & x_{1(M-1)} \\ 1 & x_{21} & \cdots & x_{2(M-1)} \\ \vdots & & & \\ 1 & x_{N1} & \cdots & x_{N(M-1)} \end{pmatrix} \qquad \hat{V} = \begin{pmatrix} 1 & v_{11} & \cdots & v_{1(M-1)} \\ 1 & v_{21} & \cdots & v_{2(M-1)} \\ \vdots & & & \\ 1 & v_{M1} & \cdots & v_{M(M-1)} \end{pmatrix}$$

then

$$\hat{X} = A\hat{V}. \tag{14}$$

If the rank of $\hat{V}$ is $R < M$, it may be shown that the dimensionality of feature space, in which the points $(v_{i1}, v_{i2}, \ldots, v_{i(M-1)})$ will lie, is $R-1$ and that we have a dependent factor problem, i.e. the physiological factors, or a subset thereof, are linearly dependent. For the time being let us assume that $\hat{V}$ is non-singular. Then

$$A = \hat{X}\hat{V}^{-1}. \tag{15}$$

d. *FADS algorithm*

The principle of FADS is that we find a transformation matrix $V$ which satisfies the non-negativity constraint defined by Eq. (4). The contention is that such a solution for $A$ and $G$ will represent the underlying physiology (Di Paola *et al.*, 1975; Bazin *et al.*, 1979; Barber, 1980).

From Eq. (14) we have

$$\hat{V}^T A^T = \hat{x}^T \qquad (16)$$

and hence

$$\hat{V}^T A^T \hat{x} = \hat{x}^T \hat{x}. \qquad (17)$$

Since the factor loadings are orthogonal, with suitable normalisation of the $x_{ik}$ elements, we have

$$\hat{x}^T \hat{x} = I_M. \qquad (18)$$

Upon introducing

$$W = A^T \hat{x} \qquad (19)$$

we have

$$\hat{V}^T W = I_M. \qquad (20)$$

If $W$ is non-singular, we arrive at

$$\hat{V}^T = W^{-1}. \qquad (21)$$

From Eq. (15), given $V$ we may calculate $A$. From Eqs. (19) and (21), given $A$ we may calculate $V$. This leads to an algorithm to reduce the negativity in physiological factor matrix $G$ and pixel contribution matrix $A$. The steps are as follows:

(i) Assume some initial values for $V$.
(ii) Calculate $A$ using Eq. (15).
(iii) Set negative elements of $A$ to a fraction $s$ of their original value where $0 < s < 1$, e.g. $s = 1/2$.
(iv) Calculate $V$ using Eqs. (19) and (21).
(v) Calculate $G$ using Eq. (10).
(vi) Adjust $V$ to eliminate (or reduce) negative elements of $G$.

We may iterate (ii) through (vi) until the negativity in $A$ and $G$ is sufficiently reduced. This is effectively an apex-seeking routine as the positions of the physiological factors in feature space, defined by the elements of **V**, will define the apices of a linear geometrical configuration.

*e. Dependence on initial estimate of* **V**

It has been shown that, in general, the solution is not uniquely defined by the non-negativity criterion (Houston, 1984), and that the apex-seeking routine will yield merely one possible solution. The solution will depend largely on the initial estimate of **V**.

It is quite possible that this estimate will yield a non-negative, and therefore permitted, solution. In this case the algorithm will accept this as the final solution.

If the initial values of $v_{jk}$ are large, **G** is liable to contain negative elements. Such an estimate of **V** is said to lie outside the domain of non-negative factors. In this case the final solution is liable to lie close to domain limits.

In practice initial values of $v_{jk}$ are kept as low as possible while attempting to ensure that **A** contains few negative elements (Barber, 1980; Di Paola *et al.*, 1982) which are then reduced using the algorithm. If the domain of non-negative factors is tight around points $(x_{i1}, x_{i2}, \ldots, x_{i(M-1)})$, step (vi) of the algorithm will play an essential part. Otherwise, the geometric configuration defined by the apices will be tight around the points and may be some way from the domain limits. It has been demonstrated that this is often the case in practice (Houston, 1984). In such cases, cross-contamination of the contribution images will almost inevitably result where the representation of three-dimensional data in two dimensions causes compartments to overlay.

2. THE USE OF CLUSTER ANALYSIS IN FADS

*a. Existence of a unique solution*

In a recent paper Houston (1985) investigated the validity of the apex-seeking method using set theory. If it is known, a priori, which combinations of overlaying compartments are present in a study, it is possible to predict whether or not the apex-seeking routine with low initial values of $v_{jk}$ will give a good approximation of the "true" physiological solution. It is also possible to predict which further constraints, if any, are necessary to obtain a unique solution. These statements follow from the fact that each set consisting of a combination of overlaying compartments corresponds to an apex, line, etcetera, of the linear geometrical configuration representing the transformation.

In either case, gravitational cluster analysis (Wright, 1977) was used to identify these sets of combinations of compartments in the study. Clusters and combinations of clusters were used to identify the appropriate apices, lines, etcetera, and hence define or constrain the elements of **V**.

The paper also attempted to solve the problem of dependent factors. For example, in the case of a pathological gated cardiac study in nuclear medicine, four compartments, namely ventricle, atrium, pathology and background, may be assumed to be present. Since the background and mean curves are invariably similar in form, the background factor will plot close to the origin in the feature space discriminating among, and therefore containing, the other three factors, i.e. $M = 4$, $R = 3$.

It was shown that a unique solution exists if and only if, for any pixel, only R compartments are present. This solution may be obtained if, in every case, these compartments are known. The geometrical interpretation of these conditions is as follows. For each combination of R apices in feature space (of dimension R-1) a linear geometrical configuration is defined. Each pixel point $(x_{i1}, x_{i2}, \ldots, x_{i(R-1)})$ must be assigned to one and only one of these configurations and no pixel may contain a contribution from any compartment other than those defined by the apices of the configuration. Since this involves increasing the required a priori knowledge to an unacceptable level, some simplification is necessary.

Let us suppose that we can partition the relevant part of feature space using a subset of these configurations in such a way that each pixel point will be contained in one and only one configuration in the subset. The pixel is then assigned to that configuration and is assumed to contain contributions only from compartments defined by its apices. It is not necessary to know all elements of V in order to define such a partitioning as the outer perimeter of the union of all such configurations need not be known.

As before, gravitational cluster analysis may be used to define or constrain the elements of V and thereby, hopefully, define an appropriate partitioning.

b. *Cardiac example*

In the remainder of the paper, the example of a pathological gated cardiac study will be used to demonstrate how the method may be used.

Since M = 4 and R = 3, there will be four compartmental points $(v_{j1}, v_{j2})$ (j = 1,2,3,4) plotting in two-dimensional feature space. Four distinct points will plot in a plane in one of three ways (see Fig. 1). Feature space has been partitioned into small triangles using the points as apices. Note that in case (c) there are two possible partitionings: (i) using triangles ABD

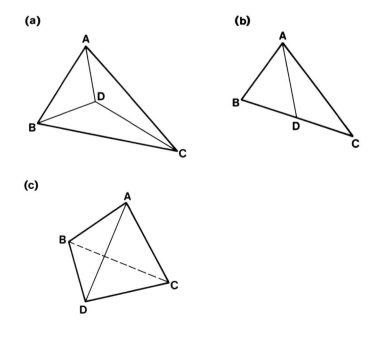

Fig. 1. *Possible configurations for four dependent factors plotting in two dimensions.*

and ACD where the partitioning is shown using a full line; and (ii) using triangles ABC and BCD where a broken line is used. If it can be assumed that each pixel contains only contributions from compartments at the apices of the small triangle containing its pixel point, then a unique solution exists for that pixel. Provided that the pixels have been assigned to the correct triangles, this is equivalent to the assumption that compartments A, B and C do not overlay in case (a); that compartments B and C do not overlay in case (b); and that compartments B and C or compartments A and D do not overlay in case (c). If A corresponds to pathology, B to atrium, C to ventricle and D to background, it appears that the pathological cardiac situation bears closest resemblance to case (b).

Using gravitational cluster analysis it was possible to define the lines BC and DA intersecting at D. Although we have been able to partition feature space, we must now make the unfortunate assumption that atrium and ventricle do not overlay. The price we pay is that, in the final analysis, no pixel will contain contributions from both and hence, in areas of such an overlay, the cross-contamination problem will remain.

## 3. THE USE OF CONSTRAINED OPTIMISATION IN FADS

*a. Unconstrained optimisation*

Let $n_i = \sum_{t=1}^{T} d_{it}$. The contribution of compartment $j$ to $n_i$ is given by $n_i a_{ij}$ or $n_i \left(1 - \sum_{\substack{l=1 \\ l \neq j}}^{M} a_{il}\right)$.

From this relationship, it was shown (Houston, 1985) that, in the case of the pathological gated cardiac study, the contribution of the background compartment to $n_i$ is given by

$$b_i = n_i + p_i z_1 + q_i z_2 + r_i z_3 \qquad (22)$$

where $p_i$, $q_i$, $r_i$ are known functions of the position in feature space of the i-th pixel, the gradients of the partitioning lines (BC and AD in this case) and $n_i$;

$z_j = 1/v'_{j1}$ ($j = 1,2,3$) where $v'_{j1}$ is the abscissa of the j-th apex in feature space transformed by defining D as origin.

If an estimate of the background $[\hat{b}_i]$ may be made, the abscissae of the apices may be chosen to minimise the least squares difference between the background images $[b_i]$ and $[\hat{b}_i]$. Apex ordinates may then be calculated from the line equations.

This was implemented previously using an adaptation of the interpolative background subtraction algorithm (Goris *et al.*, 1976) to estimate background and ignoring the requirement that the apices must lie within the domain of non-negative factors. An alternative method of obtaining an estimate of background using complex contour integration of the count density over the perimeter of an area around the pixel of interest (Nichols *et al.*, 1986) is currently under investigation.

If we make the reasonable assumption that the sign of $z_j$ is known, then $z_j$ will lie within the domain of non-negative factors if and only if

$$z_j \geq z_{dj} \text{ (for } z_j > 0\text{)} \quad \text{or} \quad z_j \leq z_{dj} \text{ (for } z_j < 0\text{)}. \tag{23}$$

Here $z_{dj}$ is the limiting value defined by the domain and may be calculated in each case. This fact has been stated previously (Houston, 1985) but until now has not been used.

### b. *Use of domain constraints*

Unconstrained optimisation, as previously used, will not ensure that the apices lie within the domain of non-negative factors. Clearly this is unsatisfactory since any solution must satisfy the non-negativity criterion. This is remedied by reducing the two inequalities of Eq. (23) to a single inequality.

Since the sign of $z_j$ is known, we may define $z'_j$ such that

$$z'_j = z_j - z_{dj} \; (z_j > 0); \quad z'_j = z_{dj} - z_j \; (z_j < 0) \tag{24}$$

for $j = 1,2,3$. Equation (22) becomes

$$p'_i z'_1 + q'_i z'_2 + r'_i z'_3 = b_i - n_i - e_i \qquad (25)$$

where $z'_j \geq 0$ for $j = 1,2,3$, and $p'_i$, $q'_i$, $r'_i$ (the prime being used to indicate a possible change of sign) and $e_i$ (equal to $p_i z_{d1} + q_i z_{d2} + r_i z_{d3}$) are known.

We therefore require a least squares solution of the equations

$$p'_i z'_1 + q'_i z'_2 + r'_i z'_3 \cong \hat{b}_i - n_i - e_i \qquad (26)$$

over all pixels, or a subset of pixels, subject to the condition that $z'_j \geq 0$ ($j = 1,2,3$). In practice pixels belonging to the background cluster are omitted.

The problem has been reduced to the classical non-negative least squares problem which will now be considered.

*c. Non-negative least squares*

Let **E** be an m * n matrix and **f** an m-vector. In the non-negative least squares (NNLS) problem we require to minimise $\|\mathbf{Ex} - \mathbf{f}\|$ subject to $\mathbf{x} \geq 0$, where **x** is an n-vector. The Kuhn-Tucker conditions for this problem are readily formulated.

An n-vector **x** is a solution for NNLS if and only if there exists an n-vector **y** and a partitioning of the integers 1 to n into subsets A and B such that

$$\begin{aligned}
\mathbf{y} &= \mathbf{E}^T(\mathbf{f} - \mathbf{Ex}); \\
x_i &> 0, \; i \text{ in } A; \quad x_i = 0, \; i \text{ in } B; \\
y_i &= 0, \; i \text{ in } A; \quad y_i \leq 0, \; i \text{ in } B.
\end{aligned} \qquad (27)$$

An algorithm for solving the NNLS problem using the Kuhn-Tucker conditions is given by Lawson and Hanson (1974) and has been tested successfully.

## 5. DISCUSSION AND CONCLUSION

The inclusion of constrained optimisation means that the FADS technique may now be applied generally to pathological gated cardiac studies. In the case of the normal heart, clearly there is no pathological factor. It is important, however, to keep constant the number of orthogonal factors, and hence the number of physiological factors, in order to avoid pre-supposing the clinical situation. The role of this extra factor in the case of a normal heart has been discussed elsewhere (Pavel *et al.*, 1985) and is currently being investigated.

A project to determine the relationship between results obtained by this method and results obtained manually using background subtraction, regions-of-interest, etc., is now under way. The application of the technique to dynamic studies of other organs is also under investigation.

*Acknowledgements*

*The author is grateful to the Gosport Computer Centre for their help in producing this manuscript.*

## REFERENCES

Appledorn, C.R. (1987). Statistical methods in pattern recognition. *These proceedings*.

Barber, D.C. (1980). The use of principal components in the quantitative analysis of gamma camera dynamic studies, *Phys. Med. Biol.* 25, pp. 283-292.

Bazin, J.P., Di Paola, R., Gibaud, B., Rougier, P., and Tubiana, M. (1979). Factor analysis of dynamic scintigraphic data as a modelling method. An application to the detection of metastases. In: *Information Processing in Medical Imaging*, R. Di Paola and W. Kahn (eds.), Inserm, Paris, pp. 345-366.

Di Paola, R., Bazin, J.P., Aubry, F., Aurengo, A., Cavailloles, F., Herry, J.Y., and Kahn, C. (1982). Handling of dynamic sequences in nuclear medicine, *IEEE Trans.* NS-29, pp. 1310-1321.

Di Paola, R., Penel, C., Bazin, J.P., and Berche, C. (1975). Factor analysis and scintigraphy. In: *Information Processing in Scintigraphy*, C. Raynaud and A. Todd-Pokropek (eds.), Orsay, pp. 91-123.

Goris, M.L., Daspit, S.G., McLaughlin, P., and Kriss, J.P. (1976). Interpolated background subtraction, *J. Nucl. Med.* 17, pp. 744-747.

Houston, A.S. (1984). The effect of apex-finding errors on factor images obtained from factor analysis and oblique transformation, *Phys. Med. Biol.* 29, pp. 1109-1116.

Houston, A.S. (1985). The use of set theory and cluster analysis to investigate the constraint problem in factor analysis in dynamic structures (FADS). In: *Information Processing in Medical Imaging*, S.L. Bacharach (ed.), Martinus Nijhoff, Dordrecht, pp. 177-192.

Lawson, C.L. and Hanson, R.J. (1974). *Solving least squares problems*, Prentice-Hall, Englewood Cliffs.

Nichols, K., Shrivastava, P.N., Powell, O.M., Adatepe, M.H., and Isaacs, G.H. (1986). Noninterventional background corrections for scintigrams, *Phys. Med. Biol.*, in press.

Pavel, D.G., Sychra, J., Olea, E., Kahn, C., Virupannavar, S., Zolnierczyk, K., and Shanes, J. (1985). Factor analysis: Its place in the evaluation of ventricular regional wall motion abnormalities. In: *Information Processing in Medical Imaging*, S.L. Bacharach (ed.), Martinus Nijhoff, Dordrecht, pp. 193-206.

Wright, W.E. (1977). Gravitational clustering, *Pattern Recogn.* 9, pp. 151-166.

DETECTION OF ELLIPTICAL CONTOURS

J.A.K. Blokland[1], A.M. Vossepoel[2], A.R. Bakker[2], E.K.J. Pauwels[1]

1: Leiden University Hospital
The Netherlands
2: Rijksuniversiteit Leiden
The Netherlands

ABSTRACT

*In this paper a method is presented to delineate elliptically shaped objects in a scene. The detection algorithm is based on the Hough transformation. The transformation maps feature points onto the parameter space of ellipses. By applying cluster algorithms the best set of parameters of the ellipse can be estimated. The algorithm is able to detect elliptical contours even if they are only partly visualized, like the contour of the left ventricle in Thallium-201 scintigrams of patients with ischemic heart disease. Some results of the application of this transformation to synthetic and real scintigrams of elliptical objects are presented. Such an algorithm is also suitable for application to differently curved contours, as long as the number of parameters describing the contour is relatively low.*

1. INTRODUCTION

The perfused and viable myocardium can be imaged by the use of a radionuclide like Thallium-201 (Ritchie et al., 1977). Ischemic lesions and infarctions then appear as defects in the distribution of thallium within the myocardium. To distinguish between reversible ischemic lesions and infarctions, two images are required: one immediately after exercise, the radionuclide being administered at peak exercise and a second after several hours of rest. Defects in the exercise scintigram which disappear at rest regard regions with mild to moderate ischemic lesions, while remaining defects are related to severe ischemic lesions and irreversible necrosis. This scintigraphic imaging technique is especially valuable in those cases where the results of other noninvasive methods, e.g. ECG-analysis, are ambiguous or suspect (Melin et al., 1984), or when quantitative measurements are required. In spite of the low signal-to-noise ratio of the scintigram, a skilled human observer can adequately interpretate these images visually. Better results can be obtained, however, by the use of computer programs (Maddahi et al., 1981). These programs allow for a quantitative measurement of the distribution of the radiopharmaceutical within the myocardium. In general, the computer aided analysis of scintigrams of the left ventricle is more specific for and more sensitive to multi-vessel ischemic lesions than mere visual interpretation.

To quantify the distribution of the radionuclide it is necessary to delineate the left ventricle. Most of the methods published deal with the delineation of the object as it is visualized. But, especially when the size of the infarct or ischemic lesions is to be determined, it is the contour of the left ventricle that has to be found and not the contour of the visualized part. Inclusion of some preliminary knowledge can guide a contour detection algorithm, but the final result can still be a rather irregularly shaped, nonphysiological contour. The more a priori information about the contour to be found is added, the better the results to be obtained, but in most cases the algorithm looses also its generality. The most important piece of a priori information is the shape of the contour. When the shape is described by a model, only the position and orientation of the contour and some scaling factors have to be determined. The use of parametric models to describe the contour will change the detection algorithm into a parameter estimator, resulting in a contour with the shape of the model and not necessarily of the object as it was visualized in the scintigram.

Elliptical models to describe the myocardium are commonly used (Aarts et al., 1979; Davila and Sanmarco, 1966; Feit, 1979; Pretschner et al., 1979; Vos et al., 1982, 1986). We also adopted this model and developed an algorithm to determine the parameters of the ellipse - the length of both axes, the

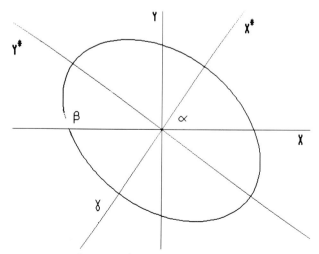

Fig. 1. The rotated ellipse.

co-ordinates of the center and the rotation (Fig. 1) - based on the Hough-transformation. This algorithm is described in the next section. It has been applied to several computer generated images to assess its performance in noise free and noisy environments. The results are presented and discussed in section 3. Also an example of results obtained in real scintigrams of the LV is given. Section 4 presents our conclusions.

## 2. EDGE DETECTION BY FEATURE SPACE TRANSFORM

Many methods to assess the parameters of a model using feature measurements have been proposed. Most of them are based on minimizing cost functions. A widely used method is the minimum squared error estimate, in which the mean squared difference between the measurement and the model is minimized. This can be considered as an example of a global method, using all available feature measurements in one single estimate.

Other methods make use of only one feature measurement to assess the parameters of the model and combine all the single feature measurement based estimates into a final estimate of the parameters. These can be considered as local methods. The Hough transformation (Hough, 1962) which maps feature vectors of the object space into a subspace of the parameter space, is an example of such a local method. This transformation and a Hough-like transformation to detect elliptical contours are described in the following subsections.

*a. The Hough transformation*

When a curve can be described by a parametric form, but with unknown parameter values, a feature space transform can be applied to obtain these values. The transformation can be defined by inverting the parametric form of the curve; then the parameters are a function of feature points. In order to produce feature points usually preprocessing of the data set is necessary. The transformation will map feature points corresponding to the same curve into the same point or region in the parameter space. Then clustering techniques can be applied to the parameter space. The resulting clusters correspond to curves in the original data set. Hough (1962) first gave the following example of such a transformation used for detecting straight lines in an image.

All points on a line satisfy $Y = m.X + b$, for a certain value of $(m,b)$. Suppose an image is preprocessed by a local edge-detection operator to produce a set of feature points $(X,Y,Y')$, where $Y'$ reflects the direction of the edge at the point $(X,Y)$. The next transformation can be defined:

$$m = Y' \qquad (1)$$

$$b = -Y'.X + Y \qquad (2)$$

This transformation pair maps each feature point into one point in the parameter space. All points from the same line will map into the same point $(m,b)$ in the parameter space. After transformation of all feature points clustering techniques must be applied to find the cluster(s) in the parameter space. Each of them corresponds to a straight line in the original image.

*b. A general feature point transformation development scheme*

The Hough transformation and Hough-like transformations are mostly used as a means to detect lines or circles (Duda and Hart, 1972). Shapiro (1978) has published the next general scheme to develop such a Hough-like transformation for two-dimensional curves.

In general, consider a two-dimensional (2-D) curve of arbitrary form:

$$Y = F(\underline{a}, X) \qquad (3)$$

where $\underline{a}$ is the n-dimensional parameter vector.

When the feature vectors have the form $(X,Y,Y')$, two equations result:

$$Y = F(\underline{a}, X) \qquad (4)$$

$$Y' = \frac{d}{dx} F(\underline{a}, X) \qquad (5)$$

Each feature vector leads to a (n-2)-dimensional subspace of the parameter space. From the intersection of all subspaces the best parameter vector(s) can be estimated.

The 2-D contour of an object in an image can be seen as a closed curve. Because these curves are a subset of the set of all possible 2-D curves, this scheme can also be used to develop a transformation to find the contour of 2-D-elliptical objects.

*c. A feature point transformation for an elliptical contour*

If the contour of an object can be modelled by an ellipse, this contour can be described as

$$\frac{X^2}{a^2} + \frac{Y^2}{b^2} = 1 \qquad (6)$$

Differentiation of Eq. (6) gives

$$Y' = -\frac{b^2}{a^2} \cdot \frac{X}{Y} \qquad (7)$$

From Eqs. (6) and (7) the length of both axes of the ellipse can be computed:

$$a^2 = X^2 - \frac{X \cdot Y}{Y'} \qquad (8)$$

$$b^2 = Y^2 - X \cdot Y \cdot Y' \qquad (9)$$

To compute both axes on the basis of the available feature vectors the next transformations can be defined:

$$A^2(X,Y,Y') = X^2 - \frac{X \cdot Y}{Y'} \qquad (10)$$

$$B^2(X,Y,Y') = Y^2 - X \cdot Y \cdot Y' \qquad (11)$$

So, if in each contour point $(X,Y)$ the direction $Y'$ of the tangent is determinable, the transformation described by Eqs. (10) and (11) will map almost any feature vector $\underline{f}=(X,Y,Y')$ into a single point of the parameter space. Because $Y'$ becomes indefinite (plus or minus infinity) or will be equal to zero when a contour point is located on one of the axes, this transformation can only be partially applied to those points; if the contour point is located on one of the axes, the length of the other axis cannot be determined. Because of noise due to discretization and other sources, not all feature points will map into one single point in the parameter space. Instead of an exact computation of the length of both axes by means of the Eqs. (10) and (11), the length must be calculated from the estimated location of an edge point and the estimated direction of its tangent.

All single point based estimates of the axes can be combined in an estimate of the probability density function (pdf) of the lengths of the axes. With this pdf the maximum likelihood estimate can be determined.

*d. A Hough-like transformation for non-centered, rotated ellipses*

The equation of the ellipse (6) used to derive the transformation holds only for ellipses centered at the point (0,0) of the object space with their axes parallel to those of the co-ordinate system. The first constraint however can be easily removed by replacing X and Y in Eqs. (10) and (11) by (X-Xc) and (Y-Yc), where (Xc,Yc) are the co-ordinates of the center of the ellipse. The direction of the tangent will not be changed by this translation.

The second constraint can also be removed by rotating the co-ordinate system until it is parallel again to the axes of the ellipse (Fig. 1). In addition to the co-ordinates of the pixels, also the tangent in an edge point is rotated. The new direction of the rotated tangent can be computed as:

$$Y^{*'} = \tan(\beta - \alpha) = \frac{\tan \beta - \tan \alpha}{1 + \tan \beta \cdot \tan \alpha} = \frac{Y' - \tan \alpha}{1 + Y' \cdot \tan \alpha} \qquad (12)$$

So, both constraints implicitly contained in Eq. (6) can be easily removed by applying co-ordinate transformations, at least when the co-ordinates of the center (Xc,Yc) and the orientation $\alpha$ of the ellipse are known. If these parameters are unknown, the transformation maps each feature vector onto a 3-D subspace of the 5-D-parameter space. And thus a 5-D pdf has to be (partly) estimated.

*e. The direction of the tangent*

The derived transformation maps feature points $\underline{f} = (X,Y,Y')$ onto the parameter space. The required feature points consist of the co-ordinates of a potential contour point and the direction of the tangent in this point. To produce these feature points the image must be preprocessed.

On the assumption of an object with a constant density against a background with also a constant, but lower density the tangent will be orthogonal to the gradient vector. The latter can be estimated by applying two orthogonal gradient operators to the image, resulting in two derivative images DX(X,Y) and DY(X,Y), together the gradient vector image G(X,Y). The gradient size will be proportional to the difference in grey level. Each separate vector will point to the pixels with a higher grey level, i.e. towards the inside of the object, taking into account the assumptions on the density made above. Then the direction of the tangent can be computed as

$$Y' = \tan \beta = - \frac{DX}{DY} \tag{13}$$

Substitution of Eq. (13) into Eqs. (10) and (11) results in

$$A^2(X,Y,Y') = X^2 + X \cdot Y \cdot \frac{DY}{DX} \tag{14}$$

$$B^2(X,Y,Y') = Y^2 + X \cdot Y \cdot \frac{DX}{DY} \tag{15}$$

## 3. RESULTS AND DISCUSSION

To assess the performance of the transformation, several elliptical objects were generated. The objects had a uniform density, except for pixels crossing the mathematical ellipse, describing the contour of the artificial object. A weighted mean of the object and the background density was chosen for those pixels. The weighting factors were chosen proportional to the ratio of the pixel area located inside and outside the mathematical ellipse. The background density was uniform too, but with half the object value. The images were scaled to a density (100 counts per pixel) comparable with those obtained in normal Tl-201 studies of patients and Poisson-noise was added (Fig. 2).

To produce feature points the images were preprocessed by applying a Sobel operator (Duda and Hart, 1973). Cahill et al. (1980) showed that this 3x3 gradient operator is to be preferred in Nuclear Medicine imaging. Because of the coarse image matrix (usually 64x64, or 128x128) larger gradient masks are prohibited. Figure 3 shows an example of gradient and gradient size images. Feature points with gradient vectors pointing to outside the object were excluded from the transformation. From the transformed feature points the 2-D pdf of the lengths of the axes belonging to the supposed center was

Fig. 2. *Artificial elliptical objects. Left: noise free. Right: Poisson noise added. Mean object density: 100. Mean background density: 50.*

estimated by means of a kernel method (Hand, 1982). Next the maximum likelihood estimate (MLE) was determined. Figure 4 shows two examples of these 2-D pdf's.

The algorithm to delineate elliptical objects presented in this paper requires the direction of the tangent in an edge point to estimate the length of the axes of the surrounding ellipse. As long as the object and the background density are uniform and the object has equally shaped edges, a gradient operator can be used to estimate the direction of this tangent. But none of the above-mentioned conditions is present in normal thallium studies. Nevertheless, even in those images the algorithm was able to find the orientation and the center of the myocard surrounding ellipse. Figure 5

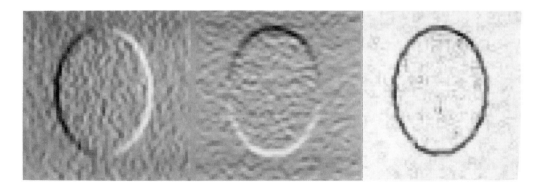

Fig. 3. *Left: Sobel gradient images of the noisy object of Fig. 2. Right: The gradient size image.*

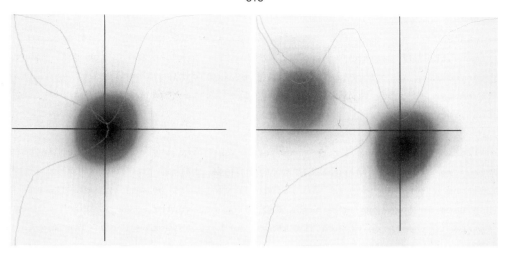

*Fig. 4. Probability density functions. Left: center chosen at the correct position. Right: center is shifted along the a-axis of the ellipse.*

presents an example of a clinical study. To determine this ellipse the following simple search algorithm has been applied.

The user enters the minimum and maximum value for the co-ordinates of the center and for the orientation angle of the ellipse. On this range a coarse grid (5 steps) is defined and the overall MLE is determined. Then the grid is refined on the basis of the found center and orientation, and the overall MLE is determined again. The result shown in Fig. 5 has been found after two refinement steps.

The main drawback of the described algorithm is the time needed to perform the transformation for a given location of the center and orientation of the

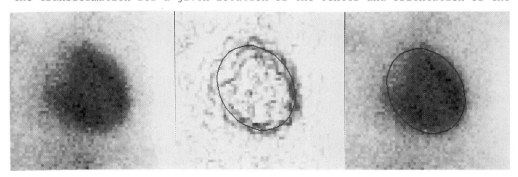

*Fig. 5. A clinical example. Left: the original scintigram. Center: the elliptical contour overlaying the gradient size image. Right: the contour in overlay with the original image.*

ellipse and to estimate the 2-D pdf. Especially the kernel method used to estimate the pdf is time consuming. When more is known about the pdf, other faster methods to estimate the pdf may be applicable. The speed of the algorithm can also be improved by using appropriate hardware, like a floating point processor.

Algorithms based on the Hough-transformation, like the one described in this paper, can also be developed for other object shapes. The main problem regards the dimensionality of the parameter vector to be determined. The ellipse used in our algorithm has a 5-D parameter vector, two of which can be computed if the three others are known, or have assumed values. The search algorithm has to find an optimum in a 3-D space. The use of edge models with many more free parameters is compromised by the time needed to find the best set of parameters.


4. CONCLUSIONS

In this paper a Hough-transformation based method has been presented to delineate elliptically shaped objects.

Even in rather noisy images, like the low density thallium scintigrams of the left ventricle, the algorithm was able to find the (only partly visualized) contour of the object.

Although the method was developed to delineate the myocardium from the background in planar Tl-201 scintigrams of the left ventricle, it can also be used to find the contours of other elliptically shaped objects in 2-D scenes.


REFERENCES

Aarts, T., Reneman, R.S., and Veenstra, P.C. (1979). A model of the mechanics of the left ventricle, *Ann. Biomed. Eng.* 7, pp. 299-318.
Abdou, I.E., and Pratt, W.K. (1979). Quantitative design and evaluation of enhancement/thresholding edge detectors, *Proc. IEEE.* 67, pp. 753-763.
Cahill, P.T., Knowles, R.J.R., Tsen, O., Lowinger, T., and Pouapinya, R. (1980). Evaluation of edge detection algorithms applied to nuclear medicine images. In: *Proc. of 5th Int. Conf. on Patt. Rec.*, Y.T. Chien (ed.), Miami Beach, pp. 1296-1300.

Davila, J.C., and Sanmarco, M.E. (1966). An analysis of the fit of mathematical models applicable to the measurement of the left ventricular volume, *Am. J. Card.* 60, pp. 31-42.

Duda, R.O., and Hart, P.E. (1972). Use of the Hough transformation to detect lines and curves in pictures, *Comm. ACM.* 15, pp. 11-15.

Duda, R.O., and Hart, P.E. (1973). *Pattern classification and scene analysis*, Wiley, New York.

Feit, T.S. (1979). Diastolic pressure-volume relations and distributions of pressure and fiber extension across the wall of a model left ventricle, *Biophys. J.* 28, pp. 143-166.

Hand, D.J. (1982). *Kernel discriminant analysis*, Wiley, New York.

Hough, P.V.C. (1962). Method and means for recognizing complex patterns, *U.S. Patent* 3069654.

Maddahi, J., Garcia, E.V., Berman, D.S., et al. (1981). Improved noninvasive assessment of coronary artery disease by quantitative analysis of regional stress myocardial distribution and washout of thallium-201, *Circulation* 64, pp. 924-935.

Melin, J., Wijns, W., and Detry, J.D. (1984). Probability analysis for noninvasive evaluation of patients with suspected coronary artery disease. In: *Nuclear imaging in clinical cardiology*, M.L. Simoons and J.H.C. Reiber (eds.), Martinus Nijhoff Publishers, Boston, pp. 219-232.

Prewitt, J.M.S. (1970). Object enhancement and extraction. In: *Picture processing and psychopictorics*, B.S. Lipkin and A. Rosenfeld (eds.), Academic Press, New York.

Pretschner, D.P., Freihorst, J., Gleitz, C.D., and Hundeshagen, H. (1979). 201-Tl myocardial scintigraphy: A 3-dimensional model for the improved quantification of zones with decreased uptake, *INSERM* 88, pp. 409-426.

Ritchie, J.L., Trobaugh, G.B., Hamilton, G.W., et al. (1977). Myocardium imaging with thallium-201 at rest and during exercise, *Circulation* 56, pp. 66-71.

Shapiro, S.D. (1978). Feature space transforms for curve detection, *Patt. Rec.* 10, pp. 129-143.

Vos, P.H., Vossepoel, A.M., Hermans, J., and Pauwels, E.K.J. (1982). Detection of lesions in thallium-201 myocardial scintigraphy, *Eur. J. Nucl. Med.* 7, pp. 174-180.

Vos, P.H., Pauwels, E.K.J. and Vossepoel, A.M. (1986). Thallium myocardial perfusion scintigraphy: Influence of perfusion, scatter and photon energy on the detection of lesions, *Cardiovasc. Intervent. Radiol.* 9, pp. 7-12.

# OPTIMAL NON-LINEAR FILTERS FOR IMAGES WITH NON-GAUSSIAN DIFFERENTIAL DISTRIBUTIONS

Miha Fuderer

*Philips Medical Systems Division*
*The Netherlands*

## ABSTRACT

*The statistical properties of the pixels in an image, and the statistical properties of the noise can usually be estimated for a given application. From that knowledge, it is possible to determine an optimal non-linear filter, in the sense of maximal a posteriori likelihood via Bayes' rule. This approach combines several types of non-linear filters (and the linear filter) under the same denominator. The results also reveal that median filtering is optimal for very noisy images with bi-exponential differential distributions.*

## 1. INTRODUCTION

A large number of authors have been optimizing linear noise-reducing image filters (or smoothing procedures), based on the statistical properties of the image and the noise. Wiener filtering is a well-known example of these linear operations.

The disadvantage of a linear filter is that it will blur edges, and, in general, suppress parts of the high-frequency information. To overcome this problem, many types of non-linear filtering are in use, one of them being the median filter (as described by Huang et al, 1978).

An optimization in the mean-squared-error sense is usually not possible in the general case of non-linear filters. However, it is intuitively obvious that, in some way, the best filtering method must depend on the statistical properties of image and noise. If absolutely nothing is known about the image we want to obtain by filtering, then how can a digital filter know where the "useful information" is to be extracted from?

## 2. GENERAL MAXIMUM LIKELIHOOD APPROACH

Suppose we have a hypothetical image $X$, where all pixel values have been corrupted by additive noise to produce the image $Y$. The noise will be assumed to be white and gaussian, as is the case in Magnetic Resonance Imaging (MRI); however, the present approach can be adapted for other noise models.

The basic idea of a noise-reducing filter is to treat the value $y_0$ of pixel $p_0$ of the image $Y$. This value $y_0$ has to be replaced by a value that comes "as close as possible" to the original (noise-free) value $x_0$. To treat pixel $p_0$, the filter will also have to consider the pixel values of some neighbourhood of $p_0$. Let $V = \{p_i \mid i = 1, \ldots, n\}$ be a set of neighbours of $p_0$, chosen in such a way that all its elements have the same statistical properties relative to $p_0$. Since two-dimensional images are usually isotropic, this set will normally contain four pixels (above, below, left, right). In the one-dimensional case it will contain two values, six voxels in the three-dimensional case, etc. One could also think of more fancy data structures, e.g. in a two-dimensional honeycomb image structure, the set would contain the six neighbouring pixels. This means that the distance between the central pixel $p_0$ and its neighbours will be regarded constant.

In this environment, the vector $Y_W$ has been measured, where $Y_W = (y_0, \ldots, y_n)$. $y_i$ stands for the pixel value of pixel $p_i$ and $W$ stands for the set $\{p_0, \ldots, p_n\}$ (this means that $W = V \cup \{p_0\}$). To estimate the noise-free values $X_W = (x_0, \ldots, x_n)$, two more items have to be known. First, it is supposed that it is known how the configuration of the noise-free values originally could have been. In other words, $P(X_W)$ has to be known over the image (which is supposed to be stationary). $P(X_W)$ is the *a priori* probability that the combination of pixel values $X_W$ in the neighbourhood of $p_0$ will occur. Another item that has to be known: in what way can this hypothetical vector $X_W$ possibly be corrupted by noise? In other words, the statistical properties of the noise have to be known, i.e.

$$P(Y_W \mid X_W). \tag{1}$$

This is the probability to obtain the noisy vector $Y_W$, if the noise-free vector is known to be $X_W$.

$P(Y_W \mid X_W)$ can usually be estimated. In this case, the noise is supposed to be white and additive, hence

$$P(Y_W \mid X_W) = \prod_{i=0}^{n} P_n(y_i - x_i) \tag{2}$$

where $P_n$ stands for noise distribution. The value $(y_i - x_i)$ is actually the value of the noise that has corrupted pixel $p_i$. The noise is further supposed to be gaussian:

$$P(Y_W \mid X_W) \propto \prod_{i=0}^{n} \exp\left(-\frac{(y_i - x_i)^2}{2\sigma_n^2}\right). \tag{3}$$

Seemingly, $P(Y_W \mid X_W)$ has no direct use, because it gives the *probability* on $Y_W$ from a *known* $X_W$, while $Y_W$ is *known* (measured) and it would be useful to obtain the *probability* on $X_W$, in order to apply the Maximum Likelihood approach. Fortunately, Bayes' rule solves this problem:

$$P(X_W \mid Y_W) = \frac{P(Y_W \mid X_W) \cdot P(X_W)}{P(Y_W)}, \tag{4}$$

$P(X_W \mid Y_W)$ being the probability on $X_W$ if the (measured) values $Y_W$ are known. $P(Y_W)$ is only a normalizing factor, and does not depend on $X_W$. In the sequel, only the values of $X_W$ that maximize $P(X_W \mid Y_W)$ will be of interest, not the value $P(X_W \mid Y_W)$ itself. Therefore, it will be written as

$$P(X_W \mid Y_W) \propto P(Y_W \mid X_W) \cdot P(X_W). \tag{5}$$

## 3. ASSUMPTIONS ON $P(X_W)$

The general form of $P(X_W)$ is rather difficult to handle. Therefore, an idealisation has to be introduced: $P(X_W)$ will be assumed to depend only on the individual pixel values and on their mutual differences:

$$P(X_W) \propto \prod_{i=0}^{n} P_p(x_i) \cdot \prod_{i=0}^{n} \prod_{j=i+1}^{n} P_d(x_i - x_j) \tag{6}$$

where $P_p$ stands for pixel distribution and $P_d$ for differential distribution. This assumption is a generalisation of the conventional Markov assumption.

In the sequel, however, $P(X_W)$ only has to be determined as a function of $x_0$. So factors that do not contain $x_0$ will be omitted:

$$P(X_W) \propto P_p(x_0) \prod_{i=1}^{n} P_d(x_0 - x_i). \tag{7}$$

This assumption will lead to the privileged situation of the central pixel $p_0$ (and of its true and noisy pixel values $x_0$ and $y_0$ respectively) in the sequel.

In many cases – as in MRI – little can be predicted about an individual pixel value; in other words, the probability distribution $P_p(x_0)$ (or the histogram of a typical image) is rather uniform. For this reason, no use will be made of this probability, so it will be stated that

$$P(X_W) \propto \prod_{i=1}^{n} P_d(x_0 - x_i). \tag{8}$$

The function $P_d(x_0 - x_i)$ (the probability distribution of the adjacent pixel differences) can often be measured and modelled for a specific application. In the following section, the method will be worked out for various model functions of $P_d$.

## 4. MAXIMUM LIKELIHOOD APPROACH FOR SPECIFIC DIFFERENTIAL DISTRIBUTIONS

### a. Gaussian distribution

In this case $P_d$ can be written as

$$P_d(x_0 - x_i) \propto \exp(-\frac{(x_0 - x_i)^2}{2\sigma_d^2}). \tag{9}$$

Substituting this with Eq. (3) into Eq. (5) gives

$$P(X_W \mid Y_W) \propto \prod_{i=0}^{n} \exp(-\frac{(x_0 - x_i)^2}{2\sigma_d^2}) \exp(-\frac{(y_i - x_i)^2}{2\sigma_n^2}). \tag{10}$$

Finding the maximum of expression (10) means finding the maximum of

$$S = \sum_{i=0}^{n} (-\frac{(x_0 - x_i)^2}{2\sigma_d^2} - \frac{(y_i - x_i)^2}{2\sigma_n^2}) \tag{11}$$

since the exponential function is monotonically increasing.

The necessary condition for the maximum of S is given by

$$\frac{dS}{dx_0} = 0 \tag{12}$$

and solving this equation results in

$$x_0 = \frac{y_0/\sigma_n^2 + (1/\sigma_d^2) \cdot \sum_{i=1}^{n} x_i}{1/\sigma_n^2 + (1/\sigma_d^2) \cdot n} \tag{13}$$

which is the value of $x_0$ at the maximum of $P(X_W \mid Y_W)$. However, it is expressed as a function of the (unknown) neighbour values $x_i$. To overcome this problem, it will be chosen to estimate the neighbours' values as

$$\hat{x}_i = y_i \qquad (i \neq 0). \tag{14}$$

This results in an estimate $\hat{x}_0$ of $x_0$:

$$\hat{x}_0 = \frac{y_0/\sigma_n^2 + (1/\sigma_d^2) \cdot \sum_{i=1}^n y_i}{1/\sigma_n^2 + (1/\sigma_d^2) \cdot n}. \tag{15}$$

The important conclusion is here that the maximum likelihood estimate $\hat{x}_0$ can be written as a linear combination of the pixel values of $p_0$ and its neighbours. Furthermore, the lower the signal to noise ratio (SNR) (i.e. the lower $\sigma_d/\sigma_n$), the stronger the filtering action has to be. This is, of course, the same conclusion as can be obtained with the least-squares error approach of linear filters, since Maximum Likelihood and Gauss-Markov estimation coincide for Gaussian signals (see Eykhoff, 1974).

In the extreme case, a very noisy image would have $\sigma_d \ll \sigma_n$, resulting in

$$\hat{x}_0 \approx \frac{1}{n+1} \sum_{i=0}^n y_i \tag{16}$$

i.e. the plain average over itself and all its neighbours. On the other extreme, if the SNR is very high (i.e. $\sigma_d \gg \sigma_n$), then Eq. (15) reduces to

$$\hat{x}_0 \approx y_0. \tag{17}$$

In other words, no filtering is to be done in that case.

*b. Broad uniform distribution*

In that case, $P_d$ is very constant over a wide range:

$$P_d(x_0 - x_i) \propto C. \tag{18}$$

Again, substituting this with Eq. (3) into Eq. (5) gives

$$P(X_W \mid Y_W) \propto \prod_{i=0}^n \exp\left(-\frac{(y_i - x_i)^2}{2\sigma_n^2}\right). \tag{19}$$

Maximizing this expression (with respect to $x_0$) results in

$$\hat{x}_0 = y_0. \tag{20}$$

This is another way of expressing the idea of the introduction: if absolutely nothing can be predicted about the noise-free image, then no filtering can be done.

## c. Laplacian (or bi-exponential) distribution

In many cases a distribution is assumed to be bi-exponential:

$$P_d(x_0 - x_i) \propto \exp(-\frac{\sqrt{2}\,|\,x_0 - x_i\,|}{\sigma_d}). \tag{21}$$

Again, the same procedure of substituting this with Eq. (3) into Eq. (5):

$$P(X_W \mid Y_W) \propto \prod_{i=0}^{n} \exp(-\frac{\sqrt{2}\,|\,x_0 - x_i\,|}{\sigma_d})\exp(-\frac{(y_i - x_i)^2}{2\sigma_n^2}). \tag{22}$$

To find the maximum hereof, only the exponent has to be considered, by virtue of the monotonicity of the exponential function:

$$S = \sum_{i=0}^{n}(-\frac{\sqrt{2}\,|\,x_0 - x_i\,|}{\sigma_d} - \frac{(y_i - x_i)^2}{2\sigma_n^2}). \tag{23}$$

$S$ can be differentiated to $x_0$, so the necessary condition for the extremum becomes:

$$\frac{\hat{x}_0 - y_0}{\sigma_n^2} + \frac{\sqrt{2}}{\sigma_d}\sum_{i=1}^{n} sgn(\hat{x}_0 - x_i) = 0 \tag{24}$$

or

$$\frac{(\hat{x}_0 - y_0)\sigma_d}{\sqrt{2}\sigma_n^2} = -\sum_{i=1}^{n} sgn(\hat{x}_0 - x_i). \tag{25}$$

This derivate is not continuous, so the extrema might be located in the discontinuities $\hat{x}_0 = x_i$. However, it can be shown that S has exactly one maximum, since the left side of Eq. (25) is monotonically increasing while its right side is monotonically decreasing in $\hat{x}_0$.

Again, the intrinsic equation (25) depends on the unknown values $x_i$. So the same approximation will be used as before:

$$\hat{x}_i = y_i \qquad (i \neq 0). \tag{26}$$

With this choice, Eq. (25) can be solved explicitely for extreme values of $\sigma_n$.

*(i)* If $\sigma_n$ is very large (as compared to $\sigma_d$), then the left-hand side of Eq. (25) becomes negligible, resulting in

$$\sum_{i=1}^{n} sgn(\hat{x}_0 - x_i) = 0. \tag{27}$$

This tells, in fact, that $\hat{x}_0$ has to be estimated as the median of its neighbour values.

Since the values $x_i$ are not known, the estimator

$$\sum_{i=1}^{n} sgn(\hat{x}_0 - y_i) = 0 \tag{28}$$

can be used for the image. Strictly speaking, this equation has no exact solution if n is even, but in that case, it can be derived from Eq. (25) that $\hat{x}_0$ tends towards the median value of $\{y_0, \ldots, y_n\}$.

*(ii)* If $\sigma_n$ is very small, then $\hat{x}_0 - y_0$ (i.e. the correction performed by the filter) will be very small. Then, in the non-linear part of Eq. (25), $\hat{x}_0$ can be replaced by $y_0$ (probably there will be no $x_i$ value between $y_0$ and $x_0$). So Eq. (25) becomes

$$\hat{x}_0 \approx y_0 - \frac{\sqrt{2\sigma_n^2}}{\sigma_d} \sum_{i=1}^{n} sgn(y_0 - y_i). \tag{29}$$

This result shows that a pixel value has to be corrected by giving it a (small) shift towards the median of its neighbours.

### d. Lorentzian-shaped differential distribution

In the case of a Lorentzian distribution, $P_d$ can be written as

$$P_d(x_0 - x_j) \propto \frac{1}{a^2 + (x_0 - x_j)^2} \tag{30}$$

where $a$ is a measure for the spread of the distribution (half width at half maximum). However, $a$ is not directly related to the standard deviation $\sigma_d$ of the probability distribution, because — strictly speaking — the standard deviation of a Lorentzian is infinite, while $a$ is a finite value.

Again, the same procedure of substituting this distribution with Eq. (3) into Eq. (5) yields

$$P(X_W \mid Y_W) \propto \prod_{i=0}^{n} \exp(-\frac{(y_i - x_i)^2}{2\sigma_n^2}) \prod_{i=0}^{n} \frac{1}{a^2 + (x_0 - x_i)^2}. \tag{31}$$

Again, the $x_0$ is to be found which maximizes $P(X_W \mid Y_W)$. The extrema can also be found by differentiating the inverse of $P(X_W \mid Y_W)$, i.e.

$$Q = \prod_{i=0}^{n} \exp(+\frac{(y_i - x_i)^2}{2\sigma_n^2}) \prod_{i=0}^{n} (a^2 + (x_0 - x_i)^2) \tag{32}$$

with respect to $x_0$. This results in

$$\left. \frac{dQ}{dx_0} \right|_{x_0=\hat{x}_0} = Q \cdot \left( \frac{2(x_0 - y_0)}{2\sigma_n^2} + \sum_{i=1}^{n} \frac{2(x_0 - x_i)}{a^2 + (x_i - x_0)^2} \right) \bigg|_{x_0=\hat{x}_0} = 0. \tag{33}$$

It can be shown that $d^2Q/dx_0^2$ is always positive under the condition

$$\sigma_n^2 < \frac{n}{4}a^2. \tag{34}$$

In that case $Q$ can have only one single minimum, hence $P(X_W \mid Y_W)$ will have exactly one maximum.

The same problems arise as in the previous subsections: the resulting filtering equation (33) is non-linear (as could have been expected), and it gives $\hat{x}_0$ as a function of $y_0$ and of the (unknown) "perfect" neighbour values $x_i$. Again, the neighbour values are chosen to be estimated as $\hat{x}_i \approx y_i$ before further attempts are made to solve Eq. (33). Furthermore, the weight variables $w_i$ will be introduced,

$$w_i = \frac{2\sigma_n^2}{a^2 + (x_i - x_0)^2} \approx \frac{2\sigma_n^2}{a^2 + (y_i - y_0)^2} \qquad (i \neq 0) \tag{35}$$

whence Eq. (33) can be re-written as

$$\hat{x}_0 \sum_{i=1}^{n} w_i - \sum_{i=1}^{n} x_i w_i + \hat{x}_0 - y_0 = 0 \tag{36}$$

or

$$\hat{x}_0 = \frac{y_0 + \sum_{i=1}^{n} x_i w_i}{1 + \sum_{i=1}^{n} w_i}. \tag{37}$$

So in this case the Maximum Likelihood estimate $\hat{x}_0$ can be expressed as a *non-linearly weighted* average of itself and of its neighbours. There is a remarkable analogy with the Gaussian case (see subsection (a)): there the same expression is used, with $w_i = \sigma_n^2/\sigma_d^2$, constant. For noisy images, this estimate can be approximated as

$$\hat{x}_0 = \frac{\sum_{i=0}^{n} y_i w_i}{\sum_{i=0}^{n} w_i} \qquad (w_0 \hat{=} 1). \tag{38}$$

This is the kernel of the filtering function proposed by Van Otterloo et al. (1986).

## 5. CONCLUSIONS

Via a technique of maximizing the *a posteriori* likelihood via Bayes' rule, several non-linear image smoothing filters have been derived. The results are valid under the assumption that the noise is white, gaussian and purely additive; for the signal, it is assumed that the *a priori* probability density of a neighbourhood only depends on the differential distribution between two pixels. As another constraint (imposed for computational simplicity), the filter takes into account a set of neighbouring pixels, which all have the same statistics relative to the central pixel. This usually means that only the immediate neighbours of any pixel will be considered.

With these assumptions, it has been shown that a Gaussian differential distribution results in linear filtering, while a bi-exponential distribution tends towards the median filter in the case of a very noisy image. Interestingly, the result tends toward the filtering equation that has been proposed by Van Otterloo *et al.* (1986), for the case of a Lorentzian distribution.

The present approach of replacing the unknown neighbour values of a pixel $p_0$ by their measured values is suboptimal with respect to one in which the pixel values in the whole image are estimated simultaneously. The relative suboptimality of the current approach is not known, but the current approach is numerically efficient, and it brings a number of known (but seemingly different) filter methods under the same denominator.

*Acknowledgements*

*The author wishes to express his gratitude to mr. H.J. Woltring for his corrections and advice.*

## REFERENCES

Huang, T.S., Yang, G.J., and Tang G.Y. (1978). A fast two-dimensional median filtering algorithm. In: *1978 IEEE Computer Society Conference on Pattern Recognition and Image Processing,* Chicago, pp. 128-130.

v.Otterloo, P.J., Jansen, A.J.E.M., and Dekker, C.B. (1986). Edge-preserving noise reduction in digital video sequences. In: *Proceedings of the Second International Conference on Image Processing and its Applications,* Institution of Electrical Engineers, London, pp. 248-252

Eykhoff, P. (1974). *System Identification – Parameter and State Estimation,* Wiley, New York.

# PARTICIPANTS

| | |
|---|---|
| Acar, B. S. | Dept. of Engineering Prod., Loughborough Univ. Techn., Loughborough, Leics., LE11 3TU, UK. Tel.: 44-509-263171/ext. 4216 |
| Appledorn, C.R. | Dept. of Radiology, Division of Nuclear Medicine, 926 West Michigan Street, Indiana University Medical Center, Indianapolis, Indiana 46223, USA. Tel.: 1-317-274-1802 |
| Arridge, S.R. | University College Hospital, Dept. of Medical Physics and Bioengineering, 1st Floor, Shropshire House, 11-20 Capper Street, London WC1E 6JA, UK. Tel.: 441-387-9300/ext. 8881 or 441-380-9700/ext. 8881 |
| Barrett, H.H. | Optical Sciences Center, Univ. of Arizona, Tucson, AZ 85721, USA. Tel.: 1-602-621-4425 |
| Barth, N.H. | Joint Center for Radiation Therapy, Harvard Medical School, 50 Binney Street, Boston, Mass. 02115, USA. Tel.: 1-617-732-3726 |
| Belder, M. de | Agfa-Gevaert, R&D Systems Analysis, Septestraat 27, B-2510 Mortsel (Antwerpen), Belgium. Tel.: 32-3-444-3050 |
| Bencivelli, W. | Clinical Physiology Institute, Via Savi 8, 56100 Pisa, Italy. Tel.: 39-50-47231/ext. 14. EM: LFCHP1 AT ICNUCEVM (EARN) |
| Blokland, K. | Academisch Ziekenhuis Leiden, Dept. of Diagnostic Radiology, Div. of Nuclear Medicine, P.O. Box 9600, 2300 RC Leiden, The Netherlands. Tel.: 31-71-263475. EM: BLOKLAND@HLERUL54.BITNET |
| Breckon, W.R. | Dept. of Computing and Mathematical Sciences, Oxford Polytechnic, Headington, Oxford OX3 0BP, UK. Tel.: 44-865-819-677. EM: WRBRECKON@UK.AC.OXPOLY.A |
| Burgess, A.E. | The University of British Columbia, Dept, of Radiology, 10th Avenue & Heather Street, Vancouver, BC Canada V5Z 1M9. Tel.: 1-604-228-7074 |

| | |
|---|---|
| Cappellini, V. | Istituto di Ricerca sulle Onde Elettromagnetische, Via Panciatichi 64, 50127 Firenze, Italy. Tel.: 39-Fir-4378512 |
| Castagmoli, A. | Sec. Medicine Nucleare, Viale Morgagni 85, 50134 Firenze, Italy. |
| Chen, C.T. | The Franklin McLean Memorial Research Institute, Box 433, The University of Chicago, 5841 South Maryland Avenue, Chicago, Illinois 60637, USA. Tel.: 1-312-962-6269 |
| Colchester, A.C.F. | Previously: Dept. of Neurology, Atkinson Morley's and St George's Hospitals, Copse Hill, London Wimbledon SW20 0NE, UK. Currently: Dept. of Neurology, Guy's Hospital, St. Thomas' Street, London SE1 9RT, UK. Tel.: 44-1-407-7600 |
| Defrise, M. | Radio-isotopen Lab., AZ-Vrije Universiteit Brussel, Laarbeeklaan 101, B-1090 Brussels, Belgium. Tel.: 32-24784890/ext. 4404 |
| Edwards, S. | Essex County Hospital, Physics Dept., Lexden Rd., Colchester, Essex CO3 3NB, UK. Tel.: 44-206-69244/ext. 478 or 483 |
| Feldman, M. | Dept. of Mathematics and Statistics, Southern Ill. Univ., Edwardsville, Ill. 62026-1653, USA. Tel.: 1-314-822-3655 |
| Floyd, C.E. | Duke University Medical Center, Box 3949, Durham, North Carolina 27710, USA. Tel.: 1-919-684-6565. EM: DRAECO@TUCC (BITNET) |
| Fox, J. | Imperial Cancer Research Fund, Bldg 20, Equity and Law, Lincoln in Field, London WC1, UK. Tel.: 44-1-2420200 |
| Fuchs, H. | University of North Carolina, Dept. of Computer Science, Chapel Hill, NC 27514, USA. Tel.: 1-919-966-4650. EM: fuchs@CS.UNC.EDU (ARPA) or decvax!mcnc!unc!fuchs |
| Fuderer, M. | Philips MSD, Bldg HOI-5, P.O. Box 10000, 5680 DA Best, The Netherlands. Tel.: 31-40-762858 |

| | |
|---|---|
| Gauch, J. | UNC Chapel Hill, Dept. of Computer Science, Chapel Hill, NC 27514, USA. Tel.: 1-919-942-1033. EM: gauch@UNC.CSNET |
| Gunsoy, S. | Middle East Technical University, Dept. of Electrical and Electronic Eng., Gaziantep, Turkey. Tel.: 90-851-10330/ext. 390 |
| Haar Romeny, B. ter | Academic Hospital Utrecht, Dept. of Radiology, Catharijnesingel 101, 3511 GV Utrecht, The Netherlands. Tel.: 31-30-372164. EM: mcvax!mvaccu!gafheet(USENET) |
| Herman, G.T. | Univ. of Pennsylvania, Dept. of Radiology, 3400 Spruce Street/G1, Philadelphia, PA 19104, USA. Tel.: 1-215-985-4389 |
| Hoehne, K.H. | Institute of Mathematics and Computer Science in Medicine, Universitaets-Krankenhaus Eppendorf, Martinistrasse 52, 2000 Hamburg 20, FRG. Tel.: 49-40-468-3698. EM: F58UKE@DHHDESY3 BITNET |
| Houston, A.S. | Dept. of Nuclear Medicine, Royal Naval Hospital Haslar, Gosport, Hants., UK. Tel.: 44-705-584255/-ext. 2452 |
| Jensen, J.A. | Electronics Institute, Electronics Laboratory, Techn. Univ. of Denmark, Bldg 344, DK-2800 Lyngby, Denmark. Tel.: 45-2-881566/ext. 2841 |
| Kleine Schaars, H. | Philips Medical Systems, P.O. Box 10.000, 5680 DA Best, The Netherlands, Application CT QP 133. Tel.: 31-40-763251 |
| Knesaurek, K. | Clinic of Nuclear Medicine and Oncology, Clinical Hospital 'Dr. M. Stojanovic', Vinogradska 29, 41000 Zagreb, Yugoslavia. Tel.: 38-41-574-666/ext. 541 |
| Koenderink, J.J. | Vakgroep Medische en Fysiologische Fysica, Princetonplein 5, 3584 CC Utrecht, The Netherlands. Tel.: 31-30-533985 |
| Kuijk, S. | Vrije Universiteit Brussel, Dienst Radio-isotopen, Laarbeeklaan 101, 1090 Brussels, Belgium. Tel.: 32-24784890/ext. 4404. EM: sytse@vub.VVCP |

| | |
|---|---|
| Lacana, G. | Sec. Medicine Nucleare, Viale Morgagni 85, 50134 Firenze, Italy |
| Leeman, S. | King's College School of Medicine and Dentistry, Dept. of Medical Engineering & Physics, Dulwich Hospital, East Dulwich Grove, London SE22 8PT, UK. Tel.: 44-1-6933377/ext. 3263 or 3178 |
| Leoncini, G.P. | Istituto di Fisiologia Clinica del C N.R., Via Savi 8, Pisa, Italy |
| Levkowitz, H. | University of Pennsylvania, Dept. of Radiology, International House, 3701 Chestnut Street, Philadelphia, PA 19104, USA. Tel.: 1-215-662-6780. EM: Levkowitz@cis.upenn.edu (ARPA) |
| Licitra, G. | Via Montebello 180, Vittoria 97019, Italy |
| Lobregt, S. | Philips Medical Systems, Scanner Science CT,Gebouw QP-0109, Veenpluis 4-6, 5680 DA Best, The Netherlands Tel.: 31-40-762674 |
| Lokner, V. | Clinic of Nuclear Medicine and Oncology, Clinical Hospital 'Dr. M. Stojanovic', Vinogradska 29, 41000 Zagreb, Yugoslavia. Tel.: 38-41-574-666 |
| Magnin, I.E. | LTSU, INSA 502, 20 Ave. A. Einstein, 69621 Villeurbanne, France. Tel.: 33-78948112/ext. 8607 |
| Natterer, F. | Westfaelische Wilhelms Universitaet, Inst. fuer Numerische und Instrumentelle Mathematik, Einsteinstrasse 62, 4400 Muenster, FRG. Tel.: 49-251-833793. EM: ONM70 AT DMSWWVZA (EARN) |
| Novario, R. | Servizio di Fisica Sanitaria, Ospedale Regionale, Viale Borri 57, 21100 Varese, Italy. Tel.: 39-332-278279 |
| Pizer, S.M. | The University of North Carolina at Chapel Hill, Dept. of Computer Science, New West Hall 035A, Chapel Hill, NC 27514, USA. Tel.: 1-919-929-3641. EM: pizer@unc.csnet |
| Plummer, D. | University College Hospital, Dept. of Medical Physics, 1st Floor, Shropshire House, 11-20 Capper Street, London, WC1E 6JA, UK. Tel.: 44-1-636-5152 |

| | |
|---|---|
| Pupi, A. | Sec. Medicine Nucleare, Viale Morgagni 85, 50134 Firenze, Italy. |
| Rescigno, A. | Previously: Section of Neurological Surgery, 131 FMB, Yale University, School of Medicine, 333 Cedar Street, New Haven, CT 06510, USA. <u>Currently</u>: Facolta di Medicina, Universita di Ancona, Monte d'Ago, 60100 Ancona, Italy |
| Roerdink, J.B.T.M. | Stichting Mathematisch Centrum, Afd. Toegepaste Wiskunde, Kruislaan 413, 1098 SJ Amsterdam, The Netherlands. Tel.: 31-20-5924120 |
| Roney, T. | The University of Arizona, College of Medicine, Dept. of Radiology, Health Sciences Center, Tucson, AZ 85724, USA Tel.: 1-602-721-4056 or 602-626-4916 |
| Swindell, W. | Institute of Cancer Research, Sulton Surrey, SM2 5PT, UK |
| Tan, A.C. | University College London, Dept. of Med. Physics & Bioengineering, 1st Floor Shropshire House, 11-20 Capper Street, London WC1E 6JA, UK. Tel.: 44-1-380-9700/ext. 8875 |
| Thijssen, J.M. | KUN, St. Radboud Ziekenhuis, Inst. voor Oogheelkunde, Philips van Leydenlaan 15, 6500 HB Nijmegen, The Netherlands. Tel.: 31-80-514448 |
| Todd-Pokropek, A. | University College London, Dept. of Medical Physics, Gower Street, London WC1E 6BT, UK. Tel.: 44-1-387-9300/ext. 5319 or 380-9700/ext. 5319. EM: ATODDPOK@UK.AC.UCL.CS |
| Tombeur, D. | Academisch Ziekenhuis Vrije Universiteit Brussel, Faculteit Geneeskunde en Farmacie, Dienst voor Nu-cleaire Geneeskunde, Laarbeeklaan 101, 1090 Brussels, Belgium. Tel.: 322-478-4890/ext. 4404. EM: DIRK@VUB.VVCP |
| Towers, S.J. | MRC Cytogenetics Unit, Western General Hospital, Crewe Road, Edinburgh, EH4 2XU, UK. Tel.: 44-31-322-22471/ext. 124. EM: SIMON@UK.AC.ED.MACVAX |

| | |
|---|---|
| Townsend, D.W. | Division of Nuclear Medicine, University Hospital of Geneva, Geneva, Switzerland. Tel.: 41-22-574532 |
| Trahanias, P. | NRC Democritos, Computers Dept., 153 10 Aghia Paraskevi, Attiki, Greece. Tel.: 30-6513111/ext. 237 |
| Ustuner, K.F. | 803 Verano, Irvine, CA 92715, USA. Tel.: 44-714-856-1628 |
| Vermeulen, F.L. | Rijksuniversiteit Gent, Lab. voor Elektronika en Meettechniek, St. Pietersnieuwstraat 41, B-9000 Gent, Belgium. Tel.: 32-91-233821/ext. 2513 |
| Viergever, M.A. | Delft University of Technology, Dept. of Mathematics and Informatics, P.O. Box 356, 2600 AJ Delft, The Netherlands. Tel.: 31-15-784114. EM: mcvax!dutinfd!dutinfh!maxv (USENET) |
| Voegelin, R. | Institute of Medical Physics, Viale Morgagni 85, 50134 Firenze, Italy |
| Voue, M.P. | Facultes Universitaires N-D de la Paix, Dpt. de Physique, Rue de Bruxelles 61, B-5000 Namur, Belgium. Tel.: 32-82229061/ext. 2798. EM: VOUE at BNANDP10 (EARN) or PVIS at BNANBP11 (EARN) |
| Wankling, P.F. | The Queen Elizabeth Hospital, Dept. of Medical Physics (Electronics), 3rd Floor, South Corridor, Edgbaston, Birmingham B15 2TH, UK. Tel.: 44-21472-1311/ext. 3142 |
| Zuiderveld, K.J. | Academic Hospital Utrecht, Dept. of Radiology, Catharijnesingel 101, 3511 GV Utrecht, The Netherlands. Tel.: 31-30-372164. EM: mcvax!mvaccu!gafheet (USENET) |

# SUBJECT INDEX

0 and 1 objects, 207
3-D, see three-dimensional


Acceleration, of convergence, 326-328
Adjoint operator, 128-129
Algebraic reconstruction technique, 43-65, 56-63, 274, 280, 305-317
Algorithms
    image processing, 401-414
    numerical, for non-linear inverse problems, 354-356
    (various), see name of algorithm
Angiography, 469-478
Apex-seeking routine, in factor analysis, 495-496
Apodization, 113-117
Applications in, see subject area (eg ultrasound)
Approximation by polynomials, 356
Array processors, 102, 221-240
ART, see algebraic reconstruction technique
Artificial intelligence, 241-263
Attenuation
    correction, 482-483
        in positron emission tomography, 324-326
        in single photon emission tomography, 207-210, 271-277, 335
    in ultrasound, 463
Autocorrelation function, 176, 178
Autocovariance function, for speckle size, 461
Average mean squared error, 174


B-mode, ultrasound, 455-468
Back-projection, 112, 251
    operator, 202, 273
Back-to-front, method, 212
Backscatter, echo amplitude, 464
Basis
    functions, 46-47
    of a vector space, 36-37
    pictures, 44-45
Bayes'
    classifier, 148-150
    error, 155
    estimate of maximum posterior density function, 50, 52
    rule, 525
Bifurcations, 79-86
Bilinear basis, 47
Binary array, representation in 3-D, 202
Block-Cimmino method, 311-314

Blocking artefact, 181
Blood flow, 469–478
Blood velocity, 472
Born approximation, 282–286
Born–Neumann expansion, 282
Boundaries, natural object, 95
Boundary
    conditions, 356
    measurement, 353
    smoothing, 376–378
Branch cut, discontinuity across, 286


Cardiology
    applications, 505–515
    use of factor analysis, 498–499
Catastrophe theory, 74
Cauchy
    principal value, 112
    theorem, 272
Central slice theorem, 107–110, 272
    in 3-D, 120–121
Characteristic function, 133
Chi squared, distance, 176
Cimmino algorithm, see also SIRT, 62, 306–311
Classifier, linear, 148–150
Clipping engine, 226–227
Cluster analysis, 491–503
Coded aperture, 339–349
Coding
    block, 182
    DPCM, 176
        predictive, 183
    multi-resolution, 186–188
    non-redundant, 184
    run length, 183
Coincidences, accidental, 483–484
Colour, generalized model, 389–399
Coloured noise, whitening of, 431
Commutative ring, 5–6
Compact support, 138
Compartments, homogeneous physiological, 491–492
Complex domain, 285
Components of a vector, 35
Computer graphics, 197–220
    colour models, 390–395
Computer simulation, of PET reconstruction, 325
Computerized tomography
    applications, 197–220, 309–311, 389
    see also reconstruction
Concatenation theorem, 97
Coincidences, scattered, 484
Condition, Neumann, 354
Condons, 380
Conductivity, of tissue, 352–353

Conjugate gradient, algorithm, 274-275
Consistency conditions, Helgason-Ludwig, 205
Constrained optimization, in factor analysis, 499-501
Containment tree, 203
Contextual shading, 216
Continuity, of functionals, 45
Contours, 249 ,256-260
    detection of elliptical, 505-515
    see also boundaries
Contrast, 336
    mean squared error, 175
Controllers, graphics display, 226
Convergence, 58-59, 63, 323
Convex set, 271
Convolution, 111, 116
    2-D, 273
    algorithm, in reconstruction, 117-120
    integral, 3-5
Cost
    in decision goal, 427-433
    of distortion, 173
Craniofacial surgery, 198-200
Cross-correlation, in perception, 403-404
Cross-correlator, 434
Cytology, applications, 262-263


$D'$, for ROC curves, 419
$D'e$, 419-420
Data compression, 167-195, 205
    ratio, 168-170
Database, image as a, 254-255
Decision strategy, 421
Decomposition, directional, 189-190
Deconvolution, Fourier, 481
Delta function, 20, 118, 209, 344
    2-D Dirac, 345
    derivative of, 113
Derivative, 9-10
    continuous, 15
    of a function, 113
    with an infinite discontinuity, 113
Detectability index, 415-437
Detection, task, 403
Deterministic, estimation procedure, 49
Diagnostic accuracy, 428
Diagonalization
    covariance, 146-147
    simultaneous, 147-148
Differential equations, 27-29
    elliptic partial, 352-353
Differential operator, 8
Differential pulse code modulation, see coding, DPCM
Diffraction, in ultrasound, 462
Diffusion equation, 80-82, 96

Digital radiology, 339-348
Digital subtraction angiography, 167, 469-479, 489
Dimensionality, of Radon transform, 105
Directed contours, 203-204
    representation in 3-D, 203-204
Discontinuities, 13
Discrete cosine transform, 180-181
    in 3-D, 190-191
Discrete scene, definition, 207
Discretization, 45, 181-182
Distance
    and class separability, 155-159
    measures (comments on), 158-159
        (various), see name of distance measure
Distance only shading, 213
Distribution
    broad uniform, 521
    differential, 519
    Gaussian, 520
    Laplacian, 522-523
    Lorentzian, 523-524
    non-Gaussian differential, 517-525
Divergence theorem, 354
Dixel, 491
DOG filter, see difference of Gaussians
Domain, Laplacian, 288
Double hexcone model, 392-395
Dvoretsky conditions, 160-161


Echo waveform, 441-454
Edge
    detection, 507-511
    finder, 258-261
Efficiency, of data compression, 170
Eggermont theorem, 307
Eigensystems, 199
Eigenvalues, 179, 199
Eigenvector, eigenvalue expansion, 145
Electric potential field, 352
Electrical impedance tomography, see tomography, impedance
EM algorithm, see expectation maximization
Encoding
    automatic, 252-254
    by user of image data, 251-252
Entropy, as limit of data compression, 169-172
Equation (various), see name of equation
Error, mean squared, 174
Estimation, of class statistical matrices, 159-161
Ethernet, 167
Euclidean space, 35
Evaluation, of image processing, 401-414
Expectation maximization, EM algorithm, 64, 319-329
Expected cost, optimization, 427-428
Expert systems, 241-266

Factor
    loadings, 495
    analysis, 491-503
FADS (factor analysis of dynamic structures), see factor analysis
False positive rate, 420
Fan
    algorithm, 185
    beam,
        geometry, 204-205
        projection data, 308-309
Feasibility region, 53
Feature extraction, 150-154
Feature point transform, 508-510
FFT, see Fourier transform fast
Fibre optic links, 167
Filling, in graphics, 230
Filter, 257-261
    difference of Gaussians, 258
    function in reconstruction, 117-118
    high pass, 189-190
    matched, 404
    median, 525
    optimal non-linear, 517-525
Filtered back-projection, 116-120, 334
Filtering, 162
    operation, in tomographic reconstruction, 125
        see also reconstruction
    van Otterloo, 524
    Wiener, 517
Finite elements, 356
Fisher discriminant ratio, 153
Flow, laminar, 471-472
Foley-Sammon transform, 153-155
Forward projection, 45
Fourier
    coefficients, 42
    derivative theorem, 112-113
    transform, 29-30, 106-125, 175, 204, 206, 272, 284-286, 423-426, 445
        3-D, 120
        discrete, 30-32
        fast, 32-34
Frame buffer, 233-236
Fraunhofer theory, 455
Frechet derivative, 355
Functional, 44
    correlates, 19-29
Functions
    continuous, of real variable, 5
    Morse, 70-71
Fuzzy
    derivatives, 96
    measures, 101

Gate function, 14
Gaussian
    noise, 405
    probability density function, 417
Generalized function, 112-114
Geometric structure, of local jets, 100
Geometry, engine, 226-227
Gerchberg-Papoulis algorithm, 302
Global properties, 68
Graphics, systems, 221-240
Gravitation cluster analysis, 497
Green's function, 124, 282
Gaussian noise, 423


Hadamard transform, 189
HIDAC, see multiwire chambers
Hidden surface removal, 210-213
Hierarchical
    figure based description, 365-386
    interpolation, 188
Hierarchy
    branch of symmetric axis, 371-375
    "is a", 264
    "part", 262-264
    shape, 262
Hilbert
    space, 38-42, 127 , 295
    transform, 277, 443
Hotelling
    criterion, 155
    trace, 435, 436
Hough transform, 507-509
    for rotated ellipses, 510
Hue, colour, 393
Hyperplane,   107
    geometrical interpretation of ART, 57-59


Ill-posed problems, 43-65 , 127-141, 351-362
Image
    buffers,see frame buffers
    description, 241-263
    interpretation, 249-256
    parametric, 474, 470, 476-477
    real zero conversion, 441-442
    space, 67, 209
        coordinate system, 209-210
    structure, 67-104
    texture, 459-465
    vector, 73-75
    workstation, 221-240, 471
IMOC, see Monte Carlo reconstruction
Impedance tomography, see tomography, impedance
Incomplete
    data, 322
    problems, 43, 127-141

Injection flow profiles, 470
Integral
    equation, 206-207
        of the 1st kind, 138-139
    of an operator, 17-19
    operator, 24, 282
        of the 1st kind, 356
    singular, 286
    transform, 3-42
Interference phenomena, in ultrasound, 455-457
Interpolation, 183, 200-201, 214
    in 3-D, 214
Interpolative background subtraction, algorithm, 500
Interpretation, knowledge based, 241-266
Invariance, 4
Inverse, 8
    imaging in ultrasound, 279-289
    Moore-Penrose, 55, 275, 307
    problems, 43-65, 354-355
    scatter imaging method, 279-289
Inversion
    formula, 272-273
    procedure of Born approximation, 278-289
Irreversible methods, in data compression, 192
Iterations, number of, 59-60
    in reconstruction, 293-304, 326-327
Iterative methods, 43-65
    for reconstruction, 200-201
    see also algebraic reconstruction technique


Jump function, 13-14


Kaczmarz algorithm, 56, 343-346
    see also algebraic reconstruction technique
Kalman, optimal estimate, 162
Karhunen-Loeve
    expansion, 492
    transform, 151-152, 162-163, 176-180, 181
        fast, 178-180
Knowledge based, interpretation, 241-263


Landweber iteration, 130-131, 293-302
Laplace transform, 19, 285-286
Least squares
    minimization, 48-63, 305-306
    minimum norm, 49-52
    non-negative, using Kahn-Tucker conditions, 501
    normal equation, 54, 356
    solution of inverse problems, 48-51

Likelihood
    function, 50, 320–322
    ratio, for ROC curves, 420–424
Limited angle problem, 134–135
Linear equations, solution of, 47–63, 305–317, 355–356
Linearity, 4
    of functionals, 45
Lippman–Schwinger equation, 284–285
Liver imaging, 465, 487
LLF, see likelihood function
Local properties, 68
Loop, see repeat
Lorentzian distribution, 523–524


March theorem, 6–7
Markov assumption, 519
Matched filter, 409–414, 424
Matrices, class statistical estimation, 159–161
Matrix
    block-cyclic, 204
    positive definite, 51
    rank of, 356–358
    system of linear algebraic equations, 47–48
    Toeplitz, 179
Matusita distance, 156–157
Maximum entropy, 51–52
Maximum likelihood, 50, 63
    algorithm, 331–338, 517–525
    estimation, 512–513
    reconstruction, 319–329
Mean-square error, estimate, 162
Measures of fidelity in data compression, 174–176
Medial axis transform, 72
Meta-level knowledge, 246
Minimization, 202–204
    least squares, 50–51
Minimum norm solution, 58
Minimum variance solution, 51–52, 344
    in iterative reconstruction, 345–346
Model
    for blood flow measurements, 469–478
    generalized intensity hue and saturation, 395–397
    hierarchical, 366
Modulation transfer function, 433
Monte Carlo
    simulation, in single photon emission tomography, 331–338
    reconstruction, in single photon emission tomography, 330–337
Moore–Penrose inverse, see inverse, Moore–Penrose
Morozov discrepancy principle, 297–298
Morphology, in interpretation of medical images, 241–263
MRI
    applications, 167, 389
        of detection theory, 432
    use of colour in, 397

Multi-resolution
    coding, see coding, multi-resolution
    of low level visual processing, 261
    processing, 261-264
Multi-parameter
    image display, 396-398
    medical images representation, 389-399
Multiple
    pinhole, see tomosynthesis
    processor architecture, in graphics, 227-228
Multiplicity, 22, 199
Multiwire chambers, 479-490


Neighbourhood, 171
Networks, image, 167-169
Neumann condition, 354
Newton's method, 355
Neyman-Pearson objective, 428-429
NMR, (with apologies) see MRI
Noise
    Gaussian, 50
    image filter for reducing, 517-525
    model, 415
Non-linear
    integral equation, 137
    operators, 354
Non-negativity constraint, 494-496
Norm
    Euclidean, 48
    residual, 311
Normalization, 145-150, 159
Nuclear medicine
    applications, 491-503, 505-515
        see also tomography, positron and single photon emission
Numerical operator, 8


Oblique factors, 493-494
Observer noise, 415
Octree, representation in 3-D, 204-206
Operational calculus, 3-42
Operator, 7-10
    Canny, 258-260
    low level for symbolic encoding, 256-257
    Marr-Hildreth, 257-258
    multidimensional, 23-26
    rational, 20-23
    reflexive, symmetric transitive, 7
    translation, 13-17
Optimal uniform quantizer, (Shannon quantizer), 175
Optimization, criteria, 48-52
Optimization, goal, 150-151, 153

Orthogonal
    functions, 39-42
    vectors, 37
    factors, 492-493
Orthonormal system, 41


PACS, 167-169
Parallel scanning geometry, 203
Parameter
    estimation, 506
    quantization, 264
Partial derivative, 16
Pattern recognition, 241-263
    statistical methods, 143-164
Periodicity theorem, 32
PET, see tomography, positron emission
Phantoms, in positron emission tomography, 485-487
Phase, time-domain, 441-454
Photon transport, 331-333
Physiological factors, 494
Pipeline, in graphics, 226-229
Pixel, basis, 46-47
Pixel planes, 233-237
Point
    critical, 70, 382-383
    saddle, 71, 383
Point spread function
    2-D, in ultrasound, 457-459
    in tomosynthesis, 340-341
    of a linear shift-invariant system, 117
Positron camera, crystal based, 480
Principle of superposition, 4-5
Probability density function, 144-145
    joint, 322
Procedural knowledge, 245
Production rules, 244
Projection, 44, 352
    definition and geometry, 106-107
    onto screen, 209
    parallel ray, 480-481
Psychometric curve, 416-417
Psychophysical, applications, 415-437
Pyramid, 82, 186-188
    Gaussian, 186
    Laplacian, 187
Pythagoras' theorem, 38


Quadratic, optimization, 50-51, 315
Quadtrees, 205
Quantitation, image, in PET, 480-490

Radiology, applications of detection theory, 433
Radon
    space, 106–107
    transform, 43, 105–125, 201–202, 271–277, 293
        attenuated, 271–277
        in 3-D, 107
        inverse, in 2-D, 110–119
        inverse, in 3-D, 120–124
        inverse, in n-D, 124–125
Rank of matrix, 54, 359
Raster, 2-D/3-D, 223–226
Rate-distortion function, 172–174
Rating procedure, 425–426
Rational
    numbers, 7
    operators, 20–23
Rayleigh probability density function, in ultrasound, 460, 464
Receiver operating characteristic curves, 401–414, 415–437
Reciprocity theorem, 31–32, 357–358
Reconstruction
    analytic, 105–125
    block iterative algorithms, 305–317
    discrete, 44–46
    from projections, 279–280
    image, with a multiwire chamber, 480–481
    in impedance tomography, 351–362
    iterative methods, 43–65, 293–304, 343–346
    maximum likelihood
        in positron emission tomography, 319–329
        in single photon emission tomography, 331–338
    Monte Carlo, 331–338
Recursion
    see stop condition,
    see recursion
Regularization, 51, 127–141, 293–304
    for CT, with rotational invariance, 202–204
    for incomplete data problems, 204–207
    optimal, 299–300
    Tikhonov-Phillips, 51, 200, 294
Relaxation
    parameter, 308–310
        in ART, 57–63
    matrices, 307
Rendering
    in 3-D display, 197–220
    of images, 227
    of visible surface, 213–216
Repeat, see recursion
Representation, of knowledge, 241–263
Resolution reduction, 375–379
Reversible methods in data compression, 192
RGB model, see colour, generalized model
Rho filtered layergram, 273
Ridges, 72
    and troughs, 93
Ring of functions, 7

ROC curves, see receiver operating characteristics
Row-action method, 61
Run length coding, 191


S transform, 188-189
Sampling of image and scale space, 89-91
SAT, see symmetric axis transform
Saturation, colour, 393
Scalar product, 41
Scale
    hierarchy, 260-261
    inner and outer, 75-77
    space, 78-80, 89
Scatter correction, in SPECT, 335
Scattering
    amplitude, 283
    in ultrasound, 279-289, 455-468
    theory, 353-354
Scene
    binary, 200
    cubic, 200
    discrete 3-D, 200
Schrodinger equation, 353
Second order statistics, 464
Segment-end-point representation in 3-D, 202-203
Segment separation distance, 450
Segmentation, 201-208, 257-260, 377-379
    of 3-D scenes, 197-202
Self-information, 169
Self-normalization, 323
Sensitivity, point source, 481-482
Shading
    of images, 221-240
    using colour, 391
Shannon's
    noiseless coding theorem, 171
    sampling theorem, 260
Shape
    description, 365-386
    hierarchy, 262
Signal detection, 415-437
Signal to noise ratio, 174-175, 258, 336, 521
    of ROC curves, 419
    in SPECT, 336
SIRT, see Simultaneous Iterative Reconstruction Technique
Simultaneous iterative reconstruction technique, 62-63, 305-317
    see also Cimmino algorithm
    see also Landweber
Singular
    matrix, 47
    vectors, 295-297
    value decomposition, 296-302, 358-360
        truncated, 198-200
    values, 197-209, 275, 360

Singularities, 112-114
Smoothed local symmetries, 379-380
Smoothing, 375, 517-525
Sobel operator, 511
Space, infinite dimensional, 38
Speckle
    in ultrasound, 455-468
    reduction, 447-449
SPECT, see tomography, single photon emission
Spectral, description, 86
Spectrum, of tissue magnitude, 450-451
Split and merge, 72
Stability, infinitesimal,(genericity), 69
Stereo pairs, 217
Stereoscopic presentation, 217-218
Stochastic optimization function, 49-50
Stop condition, see loop
Strategic knowledge, 246
Surface
    definition, 206
    tracking, 208
SVD, see singular value decomposition
Symbolic representation, 241-263
Symmetric axis transform, 367, 368-386
    multi-resolution, 375-386
System performance, 405
Taylor expansion, 100
Text generation, 224
Texture, 249
    in ultrasound, 461-465
Theorems (various), see name of theorem
Thermal noise, 432
Three-dimensional
    data compression, 190-191
    display, 197-220
    graphics, 221-240
    symmetric axis transformation, 371-386
    visualization of grey scale, 381
Thresholding, for segmentation, 201-202
Tikhonov-Phillips, see regularization
Time of flight, reconstruction in PET, 319-329
Tissue characterization, using ultrasound, 455-468
Tissue parameter mapping, 449-451
TOFPET, see time of flight
Tolerance limits, 53
Tomography
    diffraction, 280-284
    impedance, 351-362
    industrial, 206
    longitudinal, 339-350
    positron emission (PET), 207, 271-272, 319-329, 332, 479-490
    reconstruction methods, 43-65
    single photon emission computerized (SPECT), 208-210, 271-277, 331-338
Tomosynthesis, 339-350

Topological structure, 67-105
Topology, of level sets, 72-73
Transducers, in ultrasound, 455-465
Transfer, of bit aligned block, 223
Transformations, in computer graphics, 221-228
Transformations, operational, 10-13
Transforms (various), see name of transform
Translation operator, 13-17
Transversality, 69-70
Tree machine, using processor per pipeline, 229-230
Tretiak-Metz formula, 274
True positive rate, 420
Two-alternative forced choice procedure, 426-427


Ultrasound
    applications, 279-289
    applications of detection theory, 431-432
    Doppler, use of colour, 398
    equipment, 455-468
    imaging, 441-454
Unbounded operator, 128-129, 193
Unconstrained optimization, in factor analysis, 499-500
Unsharp masking, 409-414
User interface, 247


Variance, 175
Vector
    random scan system, 222-224
    space, 34-38
Visual
    model of the human system, 401-414
    perception criteria, 175
    processing, low level, 261
VLSI, in graphics, 221-240


Wishart density, 161


X-ray coded source tomosynthesis, 339-349


Z-buffer, algorithm, 216, 228-229

# NATO ASI Series F

Vol. 1: Issues in Acoustic Signal – Image Processing and Recognition. Edited by C. H. Chen. VIII, 333 pages. 1983.

Vol. 2: Image Sequence Processing and Dynamic Scene Analysis. Edited by T. S. Huang. IX, 749 pages. 1983.

Vol. 3: Electronic Systems Effectiveness and Life Cycle Costing. Edited by J. K. Skwirzynski. XVII, 732 pages. 1983.

Vol. 4: Pictorial Data Analysis. Edited by R. M. Haralick. VIII, 468 pages. 1983.

Vol. 5: International Calibration Study of Traffic Conflict Techniques. Edited by E. Asmussen. VII, 229 pages. 1984.

Vol. 6: Information Technology and the Computer Network. Edited by K. G. Beauchamp. VIII, 271 pages. 1984.

Vol. 7: High-Speed Computation. Edited by J. S. Kowalik. IX, 441 pages. 1984.

Vol. 8: Program Transformation and Programming Environments. Report on an Workshop directed by F. L. Bauer and H. Remus. Edited by P. Pepper. XIV, 378 pages. 1984.

Vol. 9: Computer Aided Analysis and Optimization of Mechanical System Dynamics. Edited by E. J. Haug. XXII, 700 pages. 1984.

Vol. 10: Simulation and Model-Based Methodologies: An Integrative View. Edited by T. I. Ören, B. P. Zeigler, M. S. Elzas. XIII, 651 pages. 1984.

Vol. 11: Robotics and Artificial Intelligence. Edited by M. Brady, L. A. Gerhardt, H. F. Davidson. XVII, 693 pages. 1984.

Vol. 12: Combinatorial Algorithms on Words. Edited by A. Apostolico, Z. Galil. VIII, 361 pages. 1985.

Vol. 13: Logics and Models of Concurrent Systems. Edited by K. R. Apt. VIII, 498 pages. 1985.

Vol. 14: Control Flow and Data Flow: Concepts of Distributed Programming. Edited by M. Broy. VIII, 525 pages. 1985.

Vol. 15: Computational Mathematical Programming. Edited by K. Schittkowski. VIII, 451 pages. 1985.

Vol. 16: New Systems and Architectures for Automatic Speech Recognition and Synthesis. Edited by R. De Mori, C.Y. Suen. XIII, 630 pages. 1985.

Vol. 17: Fundamental Algorithms for Computer Graphics. Edited by R. A. Earnshaw. XVI, 1042 pages. 1985.

Vol. 18: Computer Architectures for Spatially Distributed Data. Edited by H. Freeman and G. G. Pieroni. VIII, 391 pages. 1985.

Vol. 19: Pictorial Information Systems in Medicine. Edited by K. H. Höhne. XII, 525 pages. 1986.

Vol. 20: Disordered Systems and Biological Organization. Edited by E. Bienenstock, F. Fogelman Soulié, G. Weisbuch. XXI, 405 pages.1986.

Vol. 21: Intelligent Decision Support in Process Environments. Edited by E. Hollnagel, G. Mancini, D.D. Woods. XV, 524 pages. 1986.

Vol. 22: Software System Design Methods. The Challenge of Advanced Computing Technology. Edited by J.K. Skwirzynski. XIII, 747 pages. 1986.

# NATO ASI Series F

Vol. 23: Designing Computer-Based Learning Materials. Edited by H. Weinstock and A. Bork. IX, 285 pages. 1986.

Vol. 24: Database Machines. Modern Trends and Applications. Edited by A. K. Sood and A. H. Qureshi. VIII, 570 pages. 1986.

Vol. 25: Pyramidal Systems for Computer Vision. Edited by V. Cantoni and S. Levialdi. VIII, 392 pages. 1986.

Vol. 26: Modelling and Analysis in Arms Control. Edited by R. Avenhaus, R. K. Huber and J. D. Kettelle. VIII, 488 pages. 1986.

Vol. 27: Computer Aided Optimal Design: Structural and Mechanical Systems. Edited by C. A. Mota Soares. XIII, 1029 pages. 1987.

Vol. 28: Distributed Operating Systems. Theory und Practice. Edited by Y. Paker, J.-P. Banatre and M. Bozyiğit. X, 379 pages. 1987.

Vol. 29: Languages for Sensor-Based Control in Robotics. Edited by U. Rembold and K. Hörmann. IX, 625 pages. 1987.

Vol. 30: Pattern Recognition Theory and Applications. Edited by P. A. Devijver and J. Kittler. XI, 543 pages. 1987.

Vol. 31: Decision Support Systems: Theory and Application. Edited by C. W. Holsapple and A. B. Whinston. X, 500 pages. 1987.

Vol. 32: Information Systems: Failure Analysis. Edited by J. A. Wise and A. Debons. XV, 338 pages. 1987.

Vol. 33: Machine Intelligence and Knowledge Engineering for Robotic Applications. Edited by A. K. C. Wong and A. Pugh. XIV, 486 pages. 1987.

Vol. 34: Modelling, Robustness and Sensitivity Reduction in Control Systems. Edited by R. F. Curtain. IX, 492 pages. 1987.

Vol. 35: Expert Judgment and Expert Systems. Edited by J. L. Mumpower, L. D. Phillips, O. Renn and V. R. R. Uppuluri. VIII, 361 pages. 1987.

Vol. 36: Logic of Programming and Calculi of Discrete Design. Edited by M. Broy. VII, 415 pages. 1987.

Vol. 37: Dynamics of Infinite Dimensional Systems. Edited by S.-N. Chow and J. K. Hale. IX, 514 pages. 1987.

Vol. 38: Flow Control of Congested Networks. Edited by A. R. Odoni, L. Bianco and G. Szegö. XII, 355 pages. 1987.

Vol. 39: Mathematics and Computer Science in Medical Imaging. Edited by M. A. Viergever and A. Todd-Pokropek. VIII, 546 pages. 1988.